Construction Contract Preparation and Management

D1336089

Construction Contract Preparation and Management

From Concept to Completion

Geoff Powell

Senior Lecturer in Construction Contracts,
Contract Management and Procurement,
Coventry University, UK

First published 2012 by
PALGRAVE MACMILLAN

Palgrave Macmillan in the UK is an imprint of Macmillan Publishers Limited, registered in England, company number 785998, of Houndmills, Basingstoke, Hampshire RG21 6XS.

Palgrave Macmillan in the US is a division of St Martin's Press LLC, 175 Fifth Avenue, New York, NY 10010.

Palgrave Macmillan is the global academic imprint of the above companies and has companies and representatives throughout the world.

Palgrave® and Macmillan® are registered trademarks in the United States, the United Kingdom, Europe and other countries

ISBN 978–0–230–27379–5   paperback

This book is printed on paper suitable for recycling and made from fully managed and sustained forest sources. Logging, pulping and manufacturing processes are expected to conform to the environmental regulations of the country of origin.

A catalogue record for this book is available from the British Library.

A catalog record for this book is available from the Library of Congress.

10 9 8 7 6 5 4 3 2 1
21 20 19 18 17 16 15 14 13 12

Printed in China

I dedicate this book to my daughters,
Leona, Charlotte and Chloe.

Short Contents

Contents

Acknowledgements

Crown copyright items, particularly in Chapter 8 on safety, the definition of designers in Chapter 10 and extracts from the Utilities Contracts Regulations in Chapter 13 are reproduced by kind permission under the Open Government Licence. The licence can be viewed at: www.nationalarchives.gov.uk/doc/open-government-licence/.

Preface

Why this book?

Construction industry professionals produce terms of contracts all the time – whenever they write a specification clause or put a dimension on a drawing for example. Unfortunately, most graduates leave university without realising this, or the significance of it. Many graduates will also be involved in construction contracts either for contractors or in a contract administrative role. Again it is very important that they understand the day-to-day significance of contract terms and the contractual consequences of their actions.

This book seeks to give civil engineering, building and construction undergraduates and postgraduates, as well as new professionals in the industry, this knowledge and understanding in an easily accessible and concise form.

What's covered?

The book begins with general principles of contracts, concepts such as the importance of time and programmes, payment and pricing mechanisms, and processes such as tendering and the importance of stakeholders. It develops these concepts in detail in an explanation of the two most common forms of contract in use today, NEC3 ECC and JCT SBC05.

The EU Procurement Directive applies to most of the work of the civil engineering side of the industry, that sector being employed largely by public bodies and utilities. It is a subject largely ignored by the student textbooks of today, but is of the greatest significance to those in industry. This directive has probably been the major driver behind the extensive use of framework agreements. The directive and framework agreements are both explained in this book.

Two other important topics are covered, site investigation and the CDM regulations. Not only are these crucial to the success of all industry projects, but they also underpin project work in our universities. This book gives guidance in both areas.

Contract law and negligence are covered in order to provide the background to the rest of the book and because of their importance to the construction professional. However this is not a law book and it is not intended to be.

Told from a hands-on perspective

Although I am now a lecturer at Coventry University, the bulk of my working life to date has been spent in industry. After graduation I worked for a contractor and then as a bridge designer, before becoming chartered. I then worked for Severn Trent Water Company for 33 years, progressing from design and project engineer to a senior manager with 200 staff and a capital programme of hundreds of projects per year. Consequently I can very much take into account the perspective of a senior client representative when discussing the investment process and its contract issues. This is why this book discusses projects from concept to completion, rather than just focusing on contract administration during construction. Now and then I include real-life examples from my experience so that readers can understand how things really work, along with more general examples so that they can relate more immediately to the different issues raised. In this way I hope that readers will develop a more holistic and a more realistic view of contract matters, their interaction and their significance to the client.

During many years as a manager, I have seen at first hand the misunderstandings and contractual mistakes that younger professionals make; frequently very expensive mistakes. This is obviously not deliberate but a consequence of inadequate understanding of this subject. This is perhaps a result of university courses, particularly in civil engineering, which concentrate largely on design and analysis and have little coverage of contract matters, producing an imbalance which was presumably not the intention of UK-Spec as published by the Engineering Council.

I have been a Senior Lecturer for the last four years, teaching at third year and MSc level. I now realise how difficult it can be to teach these concepts. I hope that my extensive industry experience together with my recent exposure to teaching has produced a book that will be of real value and benefit both to university students and young professionals in industry. I hope that they will then truly understand how contracts work and the part they personally play in taking a project from concept to completion.

Geoff Powell CEng, MICE, MCIWEM
Coventry, September 2011

1

The construction industry

Aim

This chapter introduces the concept of the construction professional and explains the nature of construction.

Learning outcomes

On completion of this chapter you will be able to:

>> Relate the basic nature of the construction industry.
>> Describe how this has produced the characteristics of the construction industry.

Introduction

Imagine for a moment a world without 'us'. A world without safe water to drink, sewers, waste and sewage treatment; electricity and gas supplies, telecommunications; roads, railways, airports, canals, docks and harbours; schools, hospitals, offices, shops and factories; houses; art galleries, churches and cathedrals; and you see a world without the construction professional. You see a world without civilisation.

But construction is not a straightforward endeavour. It is very complex, with numerous interconnecting professions and activities requiring skill, knowledge, coordination and expertise. This is the realm of the construction professional both within civil engineering and within the built environment.

This book seeks to describe the activities of the construction professional, the importance of these, and their interrelationship. Much of the academic training of civil engineers concentrates on design modules, often mathematical, largely because of their complexity and difficulty. Although students of the building professions have a broader academic training they still spend a great deal of time learning about design and its application. However, to convert these designs into constructed assets requires a very different skill set based on contracts, management and people – concepts that this book attempts to explain.

Whilst it is possible to study each contractual aspect in isolation, an understanding of their connectivity and mutual dependence is crucial to visualising how construction projects are conceived, designed and constructed. Critically, too, it is the employer (for example the local council, the supermarket chain, the electricity company) who pays for the construction project. So ensuring that the project truly delivers the requirements of that employer is both a crucial role and a major challenge for the construction professional. This will be explored in detail in this book.

The construction industry

The construction industry covers an enormous range of projects, from those of small value, to projects costing billions of pounds; from simple to very complex technology; from projects lasting days, to those lasting years. Despite this diversity, all projects have certain common features. Once a construction professional understands these features, he or she can apply them to projects of great variety.

Some characteristics of construction

It is important to understand why construction and the construction industry are different from manufacturing. This understanding can help us appreciate the need for different processes and skills in our industry. Of course there are similarities and there are many initiatives to bring some manufacturing thinking (such as standardisation and prefabrication) into the construction industry, to improve its efficiency. We all buy manufactured products: cars, clothes, televisions, computers, mobile phones, pots, pans and kettles. At first, it is easy to think that the construction industry is the same as manufacturing in that both processes produce a tangible end product. However construction does in fact have many differences, which construction professionals need to appreciate if they are to be successful in their endeavours:

- The construction industry is characterised by a wide range of contributing professions, each with its own entry standards, ethical standards and professional pride.
- 'Civil engineering' and 'building' are seen as separate professions in the UK, and their practitioners have different academic training at our universities.
- The industry is 'trade-based', with a high number of temporary and mobile workers who actually carry out the construction itself, whereas manufacturing tends to have employees who work together from year to year.
- There is an industry shortage of competent labour, and construction and professional personnel.
- Unlike manufacturing, which sells ranges of products to a multitude of customers, construction projects usually have only one *customer* – the employer – but many users and other stakeholders.
- The EU Procurement Directives, applying to all 'public procurement', have an enormous effect on tendering and selection, materials specification, and employers' selection processes for suitable suppliers, manufacturers, contractors and consultants (see Chapter 13).

- Where these directives do not apply, contractors often bid (tender) for projects on a 'one-off basis'.
- In 'one-off' projects teams often come together for the first time. This can make initial communication difficult, so misunderstandings are frequent; whereas in manufacturing teams the working relationships tend to be more permanent.
- Contracting has a low 'start-up cost', compared with manufacturing. This can result in new entrants who may be less experienced and competent than established companies and have more potential for insolvency.
- Most construction projects are built 'outside', not in a factory environment. Thus ground conditions, weather and access issues produce much of the 'risk' inherent in construction.
- Construction projects often require 'temporary works', to facilitate the construction of the 'permanent works'. These temporary works are usually designed by the contractor and are removed before contract completion.
- Many projects have a high individual impact on the environment and the public, and hence there is much legislation that regulates and controls construction activities. By contrast in manufacturing, the appropriate legislation forms part of the initial design of a product, which is then repeated many times during production.
- Construction projects usually require the lease or purchase of land.
- All projects require adequate and safe access. The impact of projects on local road networks must be considered both during construction and when completed.
- Environmental and ecological factors must be considered, both for ethical reasons, and to satisfy regulatory bodies (see Chapter 7).
- Sustainability and minimisation of the carbon footprint of our projects is becoming increasingly important and the subject of new legislation.
- Construction is a dangerous activity, and there is a constant need to improve health and safety.

The result of these characteristics of construction

These characteristics of construction produce many of the processes and procedures which form an essential part of our industry. It is important for construction professionals to be familiar with them in order to design and build successful projects.

- An appropriate and comprehensive investment process is essential to control projects from concept to completion (see Chapter 2).
- The identification and management of stakeholders is essential (see Chapter 3).
- Parties to contracts must be clear about their rights and obligations (see Chapter 5).
- Thorough site investigation is indispensable (see Chapter 7). This investigation occurs prior to the preparation of the drawings and other documents that form the contract.
- Contract strategies must be selected to match the individual features of projects (see Chapters 9 and 11).

- Acceptable and understood conditions of contract are essential so that the allocation of risk is clear (see Chapters 21 to 26 on NEC3 ECC and Chapter 27 on JCT SBC05, two forms of contract in the construction industry).
- Contract documents must be clear, comprehensive and issued in good time (see Chapter 6).
- Fair and well-structured tendering procedures are necessary, to select appropriate contractors, designers and suppliers in the first place (see Chapter 6).
- Payment mechanisms must be efficient, fair and understood (see Chapter 14).
- Good project management is fundamental if projects are to be run effectively (see Chapter 16).
- Clear, comprehensive and up-to date design and construction programmes are essential. They need to cover all activities and show their durations, criticality and interrelationship (see Chapter 16).
- Team formation and team working across the whole project are essential to avoid misunderstandings, disputes and unnecessary cost. Team working only happens when the people involved in the project make a clear commitment to it, and constantly strive to improve it (see Chapter 11).
- Health and safety and the Construction (Design and Management) Regulations 2007 are now of great importance. These are commonly called the 'CDM regulations' or 'CDM' in the industry. Their proper application is a legal requirement on all but the smallest projects. Chapter 8 deals with these regulations.
- Construction projects usually have an impact on the environment and there is extensive and increasing legislation to improve the activities of construction in this area. Construction professionals should always seek to improve the environment with their projects, and make them as sustainable as possible.

Mechanisms and enablers

Construction professionals are involved in all stages of developing a project from its first formation to hand-over to the employer. In fact they drive the project from concept to completion. This process is often called the **investment process** and will be discussed in Chapter 2.

This book will examine many of the important **mechanisms** in this investment process and indirectly refer to many **enablers**. Mechanisms are the processes or systems by which the constructed asset is realised. Enablers are the 'oil in the wheels' of the mechanisms. They are attributes, qualities and skills such as respect, trust, good communication, leadership, cooperation and teamwork. Most of the work of the construction professional is concerned with mechanisms, and it is all too easy to concentrate on these and neglect the enablers.

In the previous decade or so, the enablers have become increasingly important in construction. Although not labelled enablers, they were identified in government reports on the industry such as those by Sir Michael Latham (*Constructing the Team*, 1994) and by Sir John Egan (*Rethinking Construction*, 1998). The New Engineering Contract, NEC3 ECC, widely used in Civil Engineering projects, begins with clause 10.1: 'The employer, the contractor, the project manager and the supervisor shall act as stated in this contract, and in a *spirit of mutual trust and co-operation*.' So this clause begins with a mechanism and two enablers.

Mechanisms

As outlined above, mechanisms are the processes or systems by which the constructed asset is realised, so they include pricing and payment mechanisms, project plans and programmes, critical path networks, resource scheduling, and all the other tools of project management. Modern procurement has seen the development of many other mechanisms, such as key performance indicators (kpis), business process mapping, stakeholder analysis, collaboration tools and benchmarking. The law is largely a mechanism and sets down rules and procedures by which future cases are judged. Similarly, much of this book is devoted to the forms of contract used in the industry and how to administer and manage them. An understanding of the mechanisms used in the construction industry is fundamental to success.

Enablers

As described above, enablers are the 'oil in the wheels' of the mechanisms – they help the mechanisms work smoothly. Attributes, qualities and skills such as respect, trust, good communication, leadership, cooperation and teamwork are all enablers. By the installation of good mechanisms, and the development of suitable enablers, we produce an effective and efficient investment process. It is possible to argue that many of the problems of the construction industry in previous decades were an over-concentration on mechanisms and the neglect of enablers.

The contract strategy called 'partnering' (described in Chapter 11) in various forms is much used today. Partnering concentrates on enablers such as trust, cooperation and continuous improvement, in addition to having a form of con-

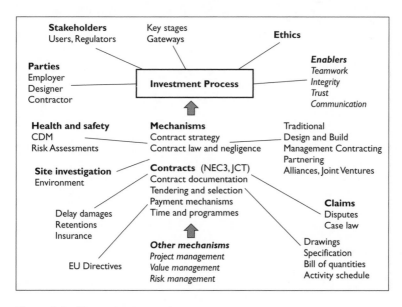

Figure 1.1 The mechanisms in the investment process.

tract based on mechanisms. Enablers need to be developed and nurtured especially if confrontation and diminished relationships are to be avoided if projects start to go wrong. Construction is a complex and difficult process and many projects have problems which need solving. Solutions can be found more easily if good enablers are in place and parties to those contracts can cooperate.

The complexities of the investment process

The investment process, whereby needs and opportunities are converted through a construction project into a constructed asset, and all the mechanisms and enablers that go to make it are very complex. What can make all the facets of the investment process harder to understand is that many of these mechanisms are not serial – they do not build up, one on another. This is unlike most university courses where, for example, second-year mechanics builds on the theories and knowledge of first-year mechanics. The mechanisms and enablers of the investment process interconnect in a complex way. One way of visualising this is by means of the spider diagram shown in Figure 1.1. This diagram shows the mechanisms that are the subjects of this book and their interrelationships.

Chapter summary

- The differences between construction and manufacturing have a major influence on our systems and processes, particularly our forms of contract.
- Effective construction processes depend on effective mechanisms.
- For success these mechanisms must be supported by enablers.
- The investment process is complex because it is so interconnected and not serial in its make-up.

2

The investment process

Aim

This chapter will describe the Investment Process as a whole, by showing the key stages and activities that take a project from *concept to completion*.

Learning outcomes

On completion of this chapter you will be able to:

>> Describe the dependencies and interrelationships in the investment process.
>> Name the key players in the investment process and differentiate between their roles.
>> Describe the key stages in the investment process.
>> Evaluate the contribution that you personally make as a construction professional.

The investment process in brief

This is where it all begins . . . with an idea – an idea on the part of the 'employer', which needs translating through the investment process into concrete, steel, bricks and mortar, into a new asset to serve the employer's needs. The employer is the person, company, local authority or agency who has the need and who will pay for the design and construction of the new asset. The employer is often called 'the client' by design organisations such as consulting engineers or architects. Effective construction professionals explore the employer's need, provide advice, and then develop the employer's requirements into the design for a construction project. The contractor will build or construct this project.

It is very important to remember that construction professionals tend to concentrate on the design and construction of *their projects*. However the employer is not usually interested in the construction project *in itself*. Employers are interested in whether it satisfies their needs and delivers a business benefit in return for their investment. In many ways it is like buying a house. Most people are not particularly interested in the construction materials used, or the design of the roof. What they want is a functional house, that is nice to look at, with a good layout, efficient to heat and easy to sell when they wish to move on. This book will continue to remind construction professionals to focus on the right things. In that way they will deliver much better results for their employer or client and hence be more successful in their careers.

The investment process is a conversion process. It converts the intangible into the tangible. The employer will begin with a **business opportunity**, or a **business need.**

A **business opportunity** is often an idea to generate more growth, sales or profit; or to develop new markets, diversify, reduce risk or beat the competition.

A **business need** is usually the necessary replacement or improvement of aging assets. It could also be a new IT system, management system or work needed to meet new legislation.

Business opportunities and needs that require constructed assets

Of course not all business needs and opportunities require constructed assets. Some might require new machinery or a new IT system for example. Most manufacturing enterprises have relatively low investment in constructed assets, unless they need a new factory or modifications to an existing one. However local authorities and utilities, for example, have a massive range of constructed assets such as roads and bridges, pipes, cables, power stations and treatment works. The maintenance, extension and replacement of these constructed assets provide a high proportion of work for the construction professional, particularly civil engineers. Equally, professionals in the building side of the industry will be concerned with hospitals, schools, office blocks and houses for example. So here we are analysing the investment process for constructed assets.

Business needs and opportunities start as intangible ideas that will be *developed* through the investment process into the constructed asset as shown in Figure 2.1. As a result the investment process is often called the development process, the procurement process or one of many other terms. However calling it the investment process focuses on the organisation with the need, and who will pay for their completed asset – the employer.

Construction professionals convert these intangible ideas into tangible assets such as schools, hospitals, buildings, roads, railways, utility assets and all the manifestations of the constructed world.

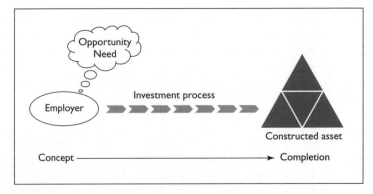

Figure 2.1 From concept to completion.

All design and construction is part of an investment process of one sort or another. The work of engineers, architects, building surveyors, construction managers, builders, contractors and all other construction professionals is to contribute to this process. They mainly do this through design, construction or organisation and management. Other professionals work in associated areas such as geotechnical, surveying, renovation, hydraulic modelling and a variety of specialist advisory and investigation work. A very small 'one-off' project, even a new garden wall for example, will actually be part of an investment process, possibly with only two stages (design and construction), even if the project participants do not realise it at the time.

The investment process goes as a minimum from **concept to completion** of a constructed asset. There are usually seven stages. Within these stages are a number of other very important processes, such as contract strategy and tendering which will be described in detail in later chapters. Occasionally the investment process goes beyond the seventh stage where the asset is contractually complete and is taken over by the employer in at least a further two stages – optimisation and review. These last two stages are usually organised by the employer, and may not always involve the construction professional. Nevertheless, they are very important, and they are shown as the last two of the nine stages in Figure 2.2 below. In this figure the first stage 'identification of need' is abbreviated to IoN.

So the investment process covers the identification of the need for the project itself, the consideration of the different options available, the design process, tendering and forming a contract once the plans are in place, and finally the actual construction of the project. Note just how many steps there are before the construction proper can begin at stage 7.

Programmes of projects

The investment process is becoming increasingly complex, particularly for employers with **programmes** of projects. Programme management must be distinguished from project management. The latter is concerned with a single project; the former, with the control and phasing of many simultaneous projects, and the associated funding requirement. It's the difference between building one new hospital, and building and maintaining several hospitals, care homes and surgeries.

Where programmes of work are concerned, the employer will have a *long-term* requirement, and it is well worth designing an investment process to suit their

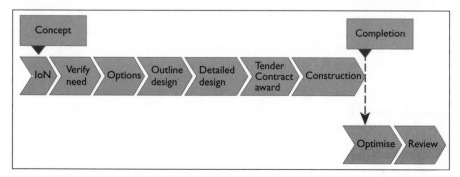

Figure 2.2   The nine stages of the investment process.

particular needs. It costs a lot of money to design and run an efficient investment process, but it is money well spent if it results in the effective procurement of a large number of projects. Many of the larger employers in the UK have annual construction programmes of between £100 million and £500 million pounds to provide new or replacement assets. This is very serious investment, and must be done well. These massive construction programmes provide much of the work of construction professionals in the UK and so are very important.

Purchasing and contract strategies

Most large employers in the UK will also have a **purchasing strategy** which may impact on the work of the construction professional. 'Purchasing' in this context is the process of obtaining goods, equipment and services. In many major employers procurement is divided between purchasing departments and contract (or engineering) departments. Purchasing departments mainly obtain manufactured goods and equipment through the expertise of professionals called 'buyers'. These buyers may buy anything from paperclips to company cars, whereas contract or engineering departments procure and organise design and construction work as part of an investment process for constructed assets. So purchasing departments normally buy *finished* goods, or straightforward services that can be easily defined; whereas contract or engineering departments manage the whole process of procuring a new constructed asset that has to be designed and constructed *from an initial concept*, often on a 'one-off' basis. These are very different processes and require fundamentally different experience and professional training to manage them well. There can be tensions between purchasing and contracts (engineering) departments because their work can overlap. However, a wise employer will place responsibility on the basis of whether items are bought complete, or whether they have to be developed, designed and constructed.

Many of an employer's purchasing strategies have no impact on the investment process described here, for our investment process relates to *capital* expenditure on *construction* projects. So a purchasing strategy for paperclips or company cars is not relevant. However, where an employer has a purchasing strategy which includes manufactured items to be *built into* construction projects such as boilers, pumps, electrical panels and so on, then the investment process for constructed assets must take account of this.

Many employers will purchase 'repeat-items' that are used in many projects in bulk, to achieve economies of scale. They may also use a **framework agreement** lasting several years where the total value of the goods falls within the Public Contracts or Utilities Contracts Regulations. A framework agreement is simply an overall agreement to buy or build specific items whenever a specific order is placed (each specific order will be subject to the overall terms of the framework agreement). Framework agreements are explained in Chaper 12.

Arrangements to include these manufactured goods must be carefully built into the investment process for constructed assets and any relevant **contract strategy**. Contract strategy is the *arrangement and choice of contract types*, between the employer and contractor, and all other contracting parties, such as designers and sub-contractors. Contract strategies are described in detail in Chapters 9 and 11.

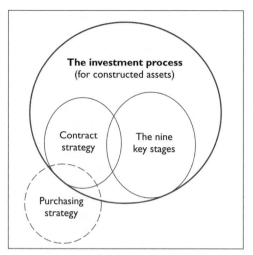

Figure 2.3   The investment process (for constructed assets).

Thus a university with a large building programme of many lecture theatres may have a purchasing strategy to buy overhead projection units under a framework agreement organised by its purchasing department. The university may well have a separate contract strategy for the construction of new buildings. Any projects for building construction would then have to include this purchasing framework agreement under specific contract terms in their contract strategy wherever they build lecture theatres with overhead projection units. This is a simple example. In practice it is a major feat of organisation to coordinate the many inputs to a construction project, especially for a complex building or a shopping centre for example.

Figure 2.3 shows the investment process, and within it the contract strategy. As discussed above, the purchasing strategy may impact on some projects and hence has to be taken account of in the contract strategy. Figure 2.3 also shows the impact of the nine key stages, which are described in detail below.

Key stages in the investment process

Key stages and gateways

The investment process, whereby needs and opportunities are converted through a construction project into the constructed asset, can be split into **key stages**. Each stage normally has an employer approval process, or **gateway**. This gateway allows progression to the next stage. The work done in every stage is an expense to the employer, and the employer therefore needs the opportunity to change or stop the project, before proceeding to the time and expense of the next stage. This decision point for the employer is the **gateway approval**.

Approval at gateways at the end of each stage will usually be dependent on the output from the tools and processes that contribute to the analysis of a construction project such as reports, cost estimates and risk assessment. Whilst the employer may have external technical advisors (such as consulting engineers or architects) the employer, through directors or senior managers, actually has the responsibility of making the gateway approvals. Designers, architects and other construction professionals can advise, but only the employer can decide whether the project is developing in a way that should meet their needs, at a price they are willing to pay and to a timescale that is satisfactory. Whilst the employer may not be that familiar with the construction process, they will certainly understand business drivers, risks and needs. In fact, the employer will understand them better than external advisors, and this can be a problem, which will be discussed below.

Though we talk usually of seven stages in an investment process, at its most simple the investment process has only two stages, *design* and *construction*. Even a project that appears to be *only* construction will actually have design elements. Let us consider again the construction of a garden wall. The *design* element would actually include the brick strength, the wall thickness, the mortar type and strength, and the foundation depth and size. So there are two stages even in this simple project. If we look at UK construction many centuries ago we see that most churches and cathedrals were constructed by master masons and other trades-people, who 'designed' their structures as they built them, on the basis of their acquired experience and training. There were no design calculations as such, so the design element was rather blurred into the construction process. Of course, this is not the case today, where because of complexity and specialisation, design and construction have evolved into two different branches of the construction professions.

In order to really understand an employer's ideas and needs, or to capitalise on opportunities, a lot of effort, time and resource must be applied to the stages *before* construction. Hence without the third stage, *option assessment*, the design may not be the most appropriate for the employer's needs. A thousand pounds spent on option determination and analysis, or effective design, can save many thousands of pounds of construction expenditure. Options need to be identified and explored in full, to ensure the most appropriate project goes forward into design. The option selection and design stages are particularly important times in which to understand and debate the employer's ideas and needs in detail. Challenge is often a good tool to expose real needs and objectives. Once 'concrete is poured', change becomes very difficult and expensive.

The expense of moving a project through the key stages

Moving a project through the key stages is expensive simply in terms of the time spent by designers, the employer's staff and other professionals together with the cost of the various investigations themselves (ground investigation, ecology surveys, locating electricity and gas services and so on). It is worth remembering that even a new graduate will cost about £300 a day, with **overheads.**

It costs a serious amount of money to employ people. The obvious cost is their salary, but overheads can double or even triple this sum of money. In a design organisation overheads will include national insurance and any other statutory payments, training costs, the cost of office accommodation, heat, light and IT systems, administration systems, the staff needed to administer payroll and any pension arrangements, and all the staff who do not directly earn fees. Then of course there is advertising, the costs associated with employing new staff and possibly redundancy, bank holidays, sick leave and staff leave. Finally, fee-earning staff are rarely able to work solidly for fees every hour they are at work, and a utilisation of 80 per cent is not uncommon. Hence a graduate on £25,000 per annum will cost about £60,000 per annum. This is the case whether the graduate is employed in an employer organisation such as a county council, or a private organisation such as a consulting engineer or a firm of architects. Senior people such as directors can easily cost £1,000 to £1,500 per day with overheads. This is one reason why having a suitable number of key stages is important so that these costs can be controlled.

Once a project is on-site, a contractor's overheads tend to be lower because many of the accommodation and other costs are recovered as part of the contract payments. Staff costs too are generally charged to the contract and hence paid by the employer as part of the regular contract payments. A contractor's overheads for headquarters staff will be similar to those for a design organisation but the headquarters overhead is much lower as a total proportion in a contractor because most staff are employed on-site.

The design of the investment process – balancing up costs, risk and complexity

The number and complexity of the investment process stages should allow the employer to commit some money, have some work and investigations done to firm up ideas, and then to proceed (committing more money) *or to stop*. In this way, expenditure and risk can be planned, managed and phased. Of course, the resources involved in controlling each key stage, and the reports, discussions and meetings, all cost money too. Hence there is a balance between the cost and complexity of a project, and the number of stages in the investment process.

A small, straightforward project can only justify a few stages whereas a large, complex project can support many stages. A really well-designed investment process will be *appropriate* for the type, number, variety and complexity of the projects involved. There is no sense in having a nine-stage process for a small project costing a few thousand pounds, such as a domestic garage, but for the Olympic Stadium, or Crossrail (a new railway system tunnelling under London linking west to east) it is a very different matter. The other thing to remember is that putting a number of little projects together will often be justification for a more extensive investment process. So to take a simple example, a housing estate of two hundred houses is a very different investment proposition from a single house, and will warrant a more complicated and thorough investment process.

> All investment processes have much similarity. The ones with a few key stages *just combine* the stages of the more complex investment processes.

It is also worth considering whether *a number* of quite different investment processes are appropriate for an employer with a *range* of dissimilar work. Thus a water company may have different investment processes for infrastructure assets (water mains and sewers), and non-infrastructure assets (treatment works). This is because the **investment cycles** are different (more on this later in the chapter), they have different **asset lives** and they tend to have different specialist contractors.

Asset life is the average length of time before that type of asset is likely to need replacement. Thus a sewer is expected to last 80 years, whereas mechanical plant is only expected to last 20 years before replacement. Finally, water mains and sewers have a much greater interface with landowners, highways and the public, and need contractors who are skilled in these matters as well as in building things.

It is all a matter of balance and appropriateness. Ideally, an investment process would be designed to meet each specific investment need, but realistically companies do not want the expense and complexity of numerous different processes. So somewhere along the line a sensible compromise is necessary.

> The cost and complexity of the investment process must be balanced with the number, nature and complexity of the projects involved.

Importantly, because of the impact of the EU Procurement Directive (described in Chapter 13), most large **public-sector** or **'ex-public sector' employers** operate framework agreements, which usually last for four or five years. Thus a number of one-off projects are joined to form a large programme of work. Any large programme of work will justify a well-thought-out investment process. Public-sector employers are local authorities, the highways agency, government departments and other publicly-owned bodies. 'Ex-public sector' employers are mainly the utilities such as gas, electricity and water companies, who were all privatised in the late twentieth century.

Detail on the nine stages of the investment process

> The nine stages of the investment process
>
> - **Stage 1 – Identification of need (IoN)**
> - **Stage 2 – Verification of need**
> - **Stage 3 – Assessment of options**
> - **Stage 4 – Outline design or design development**
> - **Stage 5 – Detailed design**
> - **Stage 6 – Tender and award of contract**
> - **Stage 7 – Project delivery (construction)**
> - **Stage 8 – Optimisation**
> - **Stage 9 – Post completion review**

As we have described, for the construction professional, stages 1 to 7 represent the real involvement, because they take the project from 'concept to completion'. Completion of the project and handover to the employer come at the end of stage 7 of course. Two more stages after completion may be added by the employer.

> Many of the stages repeat work from a previous stage, but go into more detail. More detail requires additional time and money.

Stage 1 – Identification of need (or IoN)

The employer has a

- **Business need:** this could be to replace old plant or buildings, make health and safety improvements, solve operational problems, meet new legislation, satisfy employees, users or regulators and so on.
- **Business opportunity:** to develop new products, grow the business, diversify, make more profit, beat the competition or take less risk.

At this stage, the employer has to crystallise the 'need or opportunity' and have it recorded in a suitable format. The record document will usually be a report or business needs statement.

Gateway 1 is usually to register the identification of need (IoN) in the company's investment process.

Stage 2 – Verification of need

Frequently this stage is carried out by the employer's own staff, sometimes with the help of external advisors, and sometimes exclusively by advisors. External advisors at this stage are frequently firms of consulting engineers or architects, but may be management consultants or quantity surveyors. The employer is trying to ensure that there is a real and credible 'need, idea, opportunity' in the first place, and that it will stand up to scrutiny. Once the employer embarks on stage 3 onwards, large amounts of money will be committed. If a project needs stopping, it is best to stop it at the end of stage 2.

Typical elements of stage 2 are:

- Justification of need or explanation of opportunity.
- Nomination of a project sponsor in the employer's organisation.
- Preliminary identification of likely stakeholders.
- Preliminary identification of risks/benefits/costs/timescales/constraints.
- Consideration of funding/finance/budgets.

Gateway 2 approval is usually by the employer's directors.

Stage 3 – Assessment of options

This stage is often carried out by the employer's own staff but external advisors may be used. External advisors in stage 3 are frequently firms of consulting engineers or architects, since option identification and assessment are fundamentally an engineering process. The other main professional who is usually involved at this stage is the accountant, who will often be involved in modelling the impact of whole life costs, often using discounted cash flow (DCF) techniques. These basically convert revenue costs into the equivalent capital cost, to allow any options which are a combination of capital and revenue to be compared. Take an example of two options for boilers to heat a house, one being efficient but costing £2,000, and the other being less efficient but costing £1,500. The capital costs differ by £500. The revenue costs are the annual running costs in terms of paying for gas and boiler maintenance. A DCF analysis would convert these revenue costs over a boiler life (its asset life) of say ten years into an equivalent capital value. Thus the options can be properly compared.

Potential options are identified, and assessed for:

- Feasibility
- Value
- Cost
- Cost against benefit
- Risk
- Health and safety implications
- Environmental impact
- Operational impact.

In order to produce a business case, it is necessary to give for each option:

- Description
- Preliminary arrangement drawings
- Likely capital cost (and whole life cost)
- Likely timescales and project completion date
- Benefits (such as income, profit and risk-reduction)
- Risks
- Dependencies
- Health and Safety implications
- Funding/finance/budgets
- Recommendation of a person to be the **project manager**.

The business case should always include a recommended option, with reasons.

Gateway 3 approval is usually by a sub-group comprising *some* of the employer's directors, plus senior managers. This sub-group is often called an investment group and is likely to have senior representatives (including the director where appropriate) from the main employer departments who are affected by capital investment. These departments will probably include engineering, finance, operations, purchasing and possibly human resources (HR) and marketing. It is very important that all departments that will be affected by or involved in a project are represented at a suitable level of seniority. This is to ensure 'buy-in' and to ensure the project takes into account the needs and possible constraints of all departments of the employer.

Stage 4 – Outline design

This stage may be carried out by the employer's own staff but is more usually undertaken by external advisors. External advisors at this stage are frequently firms of consulting engineers or architects since the stage is concerned with design. Stage 4 is sometimes called *design development*.

Typical elements of stage 4 are:

- Preliminary design, layouts, drawings, specifications.
- Preliminary risk assessments of business risk.
- Preliminary health and safety risk assessments

- Some specialist work may be needed at this early stage to identify risks or design requirements such as ground conditions, asbestos, condition statements of retained buildings.
- Identification of specialists needed
  - Tender documents for any specialists
  - Management of the work of any specialists.
- Preliminary enquiries.
- Initial view of the planning authority.
- Identify significant environmental issues.
- *'Stats'* (see below for explanation).
- Land/access requirements.
- Assessment of the potential impact of the Public Contracts Regulations or Utilities Contracts Regulations, which are described in Chapter 13.
- Contract strategy, which is shown as overlapping the nine key stages in Figure 2.3.
- Form of contract and suitable tenderers.
- Preliminary programme showing activities, programme constraints and dependencies.

'Stats' is a word used in the industry. It originates from the name 'statutory undertakers'. These were electricity and gas boards, water authorities, the telephone network and so on. Despite privatisation into private companies, these providers of essential services retain statutory powers and obligations. In recent years some companies providing cable television and also cross-country oil mains have also come under the *stats* definition. So for the construction professional, stats are companies who may have buried mains and sewers, cables and so on, which could represent a hazard and which may have to be diverted. Occasionally the stats also have surface equipment (particularly overhead power cables) which could affect the design and construction of the project. Diverting stats equipment is usually expensive and can take many months. This is why it is important to identify any significant stats issues at this early stage in the investment process.

Gateway 4 approval is usually by the same group of senior people who made the decisions at gateway 3 (see above).

Stages 3 and 4, and gateways 3 and 4 are very important. In these stages the employer's needs are discussed, recorded and examined. Options are identified and assessed. The impact on the employer's staff, assets, finance, operations and markets is determined. Full consultation with affected departments in the employer's organisation is crucial. Finally a formal system of 'sign-off' is advisable, where senior representatives of the relevant departments in the employer's organisation sign off the proposals, to record their agreement and satisfaction. If all this is not done properly, there can be a real problem at the end of stage 7 when the completed project is handed over. This is discussed later in this chapter.

Stage 5 – Detailed design

This stage *may* be carried out by the employer's own staff. However in recent years most major employers have 'outsourced' their own design departments and now rely on external companies to carry out this work. External companies used in stage 5 are almost always firms of consulting engineers or architects. They will usually be the same companies that carried out the work at stage 4, thereby ensuring continuity. This stage is heavily weighted towards civil and building design, and is *the major input from designers.*

Typical elements of stage 5 are:

- Final calculations, design, layouts, drawings, specifications.
- Pricing mechanism in the contract which may be price schedules, activity schedules or bills of quantities, depending on the form of contract used. These are all explained in detail in Chapters 14 and 15.
- Other technical schedules and documents.
- Finalisation of planning requirements, environmental issues, stats diversions, land or access needs.
- Programme, but now in greater detail and including an estimate of a construction programme.
- Health and safety issues; particularly compliance with the CDM regulations and including design risk assessments, updating the Health and Safety file and all necessary requirements of the CDM coordinator (the CDM regulations are covered in detail in Chapter 8).
- Finalisation of any information from specialists that should be included in the final design such as ground parameters, access restrictions and any other constraints on the construction process.
- Final pre-tender estimate of cost.
- Finalisation of a list of tenderers, using one of the tendering options described in Chapter 6.
- **Prepare tender documents.**

Gateway 5 approval is often automatic, provided the pre-tender estimate is within a percentage (often 10 per cent) of the expenditure approved at stage 3. If the pre-tender estimate is outside this range, then a referral back to stage 3 is usually needed, in case items need to be omitted, or the specification changed in order to come back within budget. It is even possible, albeit expensive, to reconsider rejected options again.

In most organisations it is assumed that the project manager ensures that the detailed design meets all the agreed requirements of the outline design. Good designers and project managers continue to meet the employer and keep them informed. They must inform the employer if any significant changes are necessary. Thus most investment processes have a 'mini-approval system', to fast-track the agreement of changes.

Stage 6 – Tender and award of contract

The main elements of stage 6 are:

- Invite **tenders** by sending the contract documents to suitable contractors (normally between three and six) for them to price
  - A suitable tender period must be allowed, depending on the size and complexity of the project. This period is usually between four and twelve weeks.

- Priced tenders are received from contractors.
- Analyse and assess tenders (this is usually done by the design organisation).
- Resolve any tender qualifications.
- Prepare a tender report and send it to the employer.
- The employer will then decide whether to award the contract.
- The employer signs the contract and so does the successful tenderer, who now becomes the 'contractor' for the project.

A tender is the name given to all the contract documents, when they are sent to potential contractors to price up. In legal terminology, the contractor's returned priced tender is the *offer*, the employer's signature on the contract is the *acceptance* and the price given by the contractor in return for constructing the project is the *consideration*. Tendering and tender documents will be described in Chapter 6.

Gateway 6 is when the employer (usually represented by a director or senior manager with written authority to accept tenders) awards the contract by signing the relevant documents personally. Obviously, the director or senior manager should always act within the authority given to them under the company's rules. These rules will often prescribe at what level of management within the company a contract can be signed. So in many companies only the managing director can sign the largest contracts.

Stage 7 – Project delivery (construction)

This is a major project stage, but usually involves little input from the employer. He or she is usually represented by a *contract administrator* (called the project manager, architect or engineer, depending on the form of contract used). The contractor is the major player in this stage, and will usually employ dozens of staff and operatives on-site, both the contractor's own and sub-contractors. However, there are a number of other important roles and obligations, as outlined below.

The **contractor** must produce a 'construction phase health and safety plan' to meet the CDM regulations (more again in Chapter 8), before commencing the construction of the project.

The **contract administrator** administers the project, and updates the employer at intervals on progress, and any issues. The contract administrator's role is to implement the *contract terms* for access to site, payment, changes, additional time and so on depending on the form of contract used. The contract administrator also issues further drawings and instructions as necessary, which are usually produced by the design organisation. The contract administrator notifies the employer as required by the contract (often by sending the employer copies of contract certificates).

The **supervisor** (this role used to be known as the resident engineer), who usually works for the designer organisation, supervises quality and ensures that the contractor is working to the contract documents.

The **designer** who works for the design organisation should still be involved if further or amended drawings or details are needed. They may also need to be consulted if anything may affect the original design, particularly actual ground conditions, tests on piles or foundation material, concrete mix design, the contractor's temporary works design and any changes that could impact on the original design.

In some forms of contract such as NEC3 this contract administrator is called the project manager, which can lead to some confusion. In traditional contracts such as ICE7 and many JCT contracts, there is also a supervisor on-site (usually called the resident engineer in ICE contracts), who was often assisted by 'clerks of works' and 'inspectors'. Such 'supervision' of the contractor typically added three or four per cent to the project cost. Today, many contracts have dispensed with the supervisor role, and rely on the professionalism of the contractor, or the contractor's quality control systems, or sometimes on an external site audit team, who visit sites on a random sample basis. However a good resident engineer can also provide valuable advice, a 'checking function' and a second opinion on the contractor's proposals. A good resident engineer can also assist the contractor in many other ways, such as discussion of working methods, management of the interface with any employer's operational staff, discussion of health and safety considerations, programme and record keeping. A good and experienced resident engineer can be invaluable to the contractor, and can make a large contribution to a successful project. So there is a real argument as to whether the 'saving' of three or per cent for the resident engineer (and their staff) is simply spent elsewhere on inefficiency.

The final stages of construction are:

- Performance testing of plant and equipment.
- Commissioning, training and handover; the employer's staff are usually involved in this very important process.
- Production of 'as-constructed' drawings and details.
- Hand-over of the health and safety file (as required by the CDM 2007 regulations) to the employer for safe-keeping and future use.
- The employer will usually add details of the constructed asset to databases for future operation, maintenance and possibly income.

Gateway 7 is not an approval gateway as such. It is take-over by the employer (and their staff) of the new asset and its inclusion on the company records.

There can be a major problem here, if the completed asset is not really what the employer wanted. This issue is discussed later in this chapter.

Contractually, we have now reached 'COMPLETION'.

However there are usually two more essential stages, which may involve the designer and contractor in a support function. These last two stages are 'employer-led'.

Stage 8 – Optimisation

Optimisation is the tuning, refinement, adjustment, realignment of the constructed asset to optimise revenue cost, operation, maintenance, income generation (letting, rent) and so on. If this stage is carried out, it is usually by the employer's own staff with assistance from external advisors and/or the contractor (usually under a separate contract).

Stage 9 – Post completion review

There may sometimes be a ninth stage. This is carried out after the project is complete. A small team of the project participants hold a meeting to review 'lessons and learning points', to feed into other similar projects in the future. This is designed to prevent the same mistakes being repeated, and good practice to be recorded for future use. In practice, it is difficult to document the 'learning', and make it accessible enough to others, for this stage to work well. Some companies use systems such as 'knowledge management' (KM), to try and capture learning on IT systems. KM can be ponderous and expensive, if not set up well. Having said this, it is always a good thing to seek feedback in order to improve.

An unacceptable outcome

This can be a real problem, and a well-designed investment process minimises the likelihood of it happening. With a poor or non-existent investment process, this problem can occur with great frequency.

For example, suppose you can afford to buy a hand-made suit from a Saville Row tailor. You are therefore the employer, and the tailor is both the designer and contractor. You may not know much about suits (in terms of fabrics, styles and patterns) so you rely on the tailor to assess your needs and wishes, and to advise you. When he discovers that you want the suit for an important formal occasion (like a degree ceremony), he advises a dark colour and a formal, re-strained style. You agree on a dark blue, three button, single-vent suit with a fine pin-stripe. You have now reached *gateway 4*. The suit is designed and constructed. A month later you go to pay and collect it (you are now at the end of *stage 7*). It has been made to your agreed specification, and properly 'constructed'. Contractually the tailor has met his obligations, so you have to pay. You have two problems. Firstly, the suit is darker and more formal than you had imagined, and secondly your new partner likes men in grey suits, and loathes blue ones.

This little example illustrates a number of points relevant to the construction investment process.

- Your needs were not fully explored. Were all the various options investigated, and explained to you (such as different colours, styles and fabrics)?
- You were not given an opportunity to see *an example* of the suit; even if had been shown to you in a different size, made for someone else. In construction, employers can be shown drawings, models, computer simulations, so they can really visualise the project.
- You were not consulted again during detailed design and construction. In this example there was no sign-off at gateway 5.
- Who is the employer? In this example you are, but there has been a change in management of one of your 'departments' (your new partner!), and they were not consulted.

In real life what often happens is that the project is completed to the agreed specification, and one of the 'receiving departments', such as operations, safety or estates, do not like certain features. *Contractually*, the employer must pay for the completed project *if it meets the contract and specification*. The employer may then have to undertake a further contract to make modifications to satisfy these staff. All this can be avoided by a well-designed investment process with proper stages, gateways, consultations and sign-off.

A key point to remember is that construction professionals are used to reading drawings. Most other professions involved in the investment process such as accountants, HR and operations are not used to reading drawings. Effective construction professionals ensure that their ideas are clearly understood by all. This may necessitate more than just normal engineering drawings – possibly 3D views, photo-montages, 3D animated CAD simulations, and models could all be used to help communicate ideas. Of course, the extent of these specialist communications has to be justified by the size, complexity, cost or impact of the project.

The main thing, as always, is to *communicate clearly*, taking full account of the 'audience'. The project manager needs to ensure that everyone clearly understands and accepts the proposals, the issues and how they will be affected by the completed project.

Change

Change always costs money. There is the abortive work, the disruption, the lost time and the cost of the change itself. However, in construction, changes are frequently necessary throughout the investment process. We are usually designing a one-off product from an initial concept. Sometimes these changes will be major; sometimes they will be minor refinements and improvements. It is essential to retain flexibility, but to remember that change costs money. The further we move through the stages, the harder it is to change, and the more expensive it is. Change at stages 1 to 4 is fairly cheap and easy. At stage 5 (detailed design) it is more expensive, but completely revising a finished design may only cost three or four per cent of the total project cost.

Where change becomes most expensive is during the construction stage. It is quite simple to amend a CAD drawing, but taking a jack-hammer to break out concrete is a very different matter. All good investment processes recognise these facts about change, and are designed to make any big changes early, preferably before stage 5, and certainly never as late as stage 7 (construction). In this way the *cost* of change increases and the *scope* for change decreases as we move through the investment process. This is shown in Figure 2.4.

> The further you move along the stages of the investment process, the harder it is to change, and the more expensive change becomes.

Involvement and responsibility

The involvement of the main players in the investment process is not constant. Different people are involved at different stages, and their time input is different. It is very important to estimate and plan for those who will be involved, for how much of their time and when. The project manager needs to map out this in-

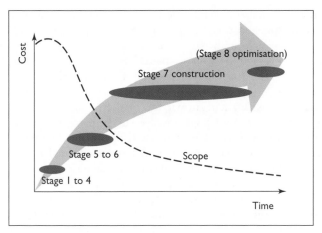

Figure 2.4 The cost of change.

volvement, and should talk to all those who need to be involved to ensure that they are available and can allocate the time required. The project manager needs their 'buy-in' to the investment process, and they need to understand how important their input is.

Figure 2.5 illustrates the typical involvement in the investment process of the various contract parties and staff working for a utility such as gas, electricity or water. Buildings and offices will be similar, only for 'operator', read 'occupier'.

We see that the employer and the asset investment staff are heavily involved in stages 1 to 3, less involved in stage 5, and then more involved at take-over at the end of stage 7. The designer is most involved in stages 4 and 5, and usually involved in stages 3 and 6, together with an input into stage 7. In traditional contracts (which separate design from construction), contractors do not come onto the scene until stage 6, when they receive a tender document to price, but their main input is obviously stage 7, the construction stage. In more modern contract

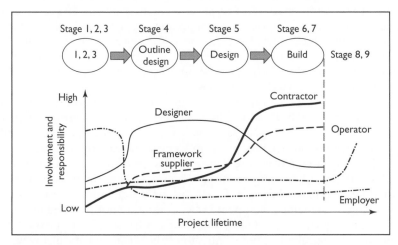

Figure 2.5 Involvement and responsibility from concept to completion.

strategies such as management contracting or framework agreements, the contractor can be involved in the design stage to give advice on 'buildability' and programme issues, which can be very valuable. The same argument is true of suppliers who can be involved in the early stage if they are on framework agreements. Contract strategies and framework agreements are described in detail in later chapters.

Operational staff, with their knowledge of the issues around operating and maintaining capital assets ought to be involved in the early stage of the investment process, not just at hand-over. Unfortunately, operational people are usually so busy 'operating' that they may not find sufficient time to make a proper input to project design and option assessment in the early stages. The result can be lack of operational 'buy-in', and even rejection of the finished project. Operational staff are the key player after the end of stage 7 (take-over), but as we have seen in 'an unacceptable outcome' above, their earlier involvement and 'buy-in' to the project are essential.

If the project involves construction work on an existing operational site, or within an existing building, very careful attention must be given to the health and safety of occupants and employees, and any disruption to their activities must be minimised. This type of project can be particularly difficult to plan and coordinate well.

Investment cycles (or capital programmes)

The investment process is not linear. It usually has to accommodate investment cycles. These investment cycles are often called capital programmes or investment programmes by major employers. All of the large utilities have investment cycles, often five years in duration. Once their prices are agreed for an investment cycle with their regulator such as Ofwat or Ofgen, the utility embark on the design and construction of the constructed assets in the next investment cycle. These investment cycles represent a major issue for funding the capital projects within them, but this is not our concern here. We are concerned with the effect this has on the staff and resources employed in the key stages of the investment process.

It used to be common for most major civil engineering employers like county councils, district councils, government departments and the utilities (such as gas, water or electricity) to employ large numbers of *their own* designers. Consultants were only used for specialist work, or for **'peak-lopping'**. Peak-lopping means smoothing the work flow to your own employees to ensure that they are always fully-utilised (in other words, always busy).

If we take a five-year investment cycle as an example, this normally begins with a lot of design work, which reduces in years four and five. Consequently there is little actual construction work in year one because the earliest projects are still in the design phase. Construction picks up to peak in years three, four and five. Hence these major employers use their own designers throughout the five-year cycle, but often bring in extra resources (consultants) in the peak design years one to three. This makes life easier for the big employer, but not for the consultants working to the same investment cycle, as they are not required in the later stages of the project. The staff utilisation problem has simply been passed down the line to the consultant. Figure 2.6 shows this diagrammatically, where the lower part of each bar is design and contract administration work, usually carried

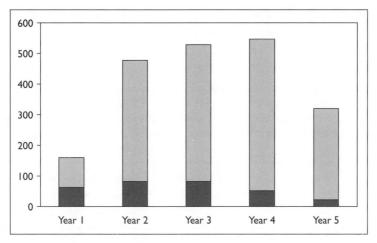

Figure 2.6   A five-year investment cycle.

out by the design organisation. The upper part of the bar is construction work. The figures on the *y*-axis are expenditure in millions.

The staff utilisation issue is why many consultants diversify into different types of work and develop a wide client base. Designing gas mains for example is similar to designing water mains, but the gas and water industries have different investment cycles. Hence the consultant's gas/water engineers can be kept busy. Consultants also diversify into different types of clients such as private/public and national/international, and different contract strategies such as *design and build* where they obtain work via contractors rather than from the large employer organisations.

Despite outsourcing of design departments, most large employers still retain a number of smaller but crucial departments, often employing engineers. An investment appraisal department will often do the work in stages 2 and 3. A finance department will take overall responsibility for approving and releasing expenditure. A contract department will write specific contracts, and possibly award contracts to successful tenderers and get involved in any dispute procedures. An audit department will ensure that company rules are followed. A health and safety department will set overall standards and provide an audit and monitoring service. Many employers still employ their own project managers, particularly for NEC3 contracts. All of these departments have an input to make in the investment process. So the 'employer' involved in the investment process is not one person, but probably the representatives of half a dozen departments or more.

The cyclical nature of the investment programmes of utilities and similar organisations presents a major problem for contractors, and also suppliers. So a contractor working for the water industry will see little work in year one of the cycle, more work in year two, and potential work overload in years three and four, with a tail-off in year five. What makes the situation worse for the contractor is that the whole UK water industry operates on the same cycle, because of regulation. Hence in many ways the situation is even worse for the contractor than the design organisation. Again, contractors diversify and as far as possible try to ensure that their staff is multi-skilled, so that they can move from one type of project to another.

Typical costs

As the project progresses, the employer will accrue *ever greater* costs. The best investment processes balance the costs of each stage (the amount of time and work done), with the benefit of *increased project definition and certainty*. **Greater accrued cost usually increases certainty.**

> The more time and money that are spent on project definition and initial investigations, the greater the certainty of the likely risk and cost of the project.

Typical costs as a percentage of the total project cost might be

- Stages 1–3   2%
- Stage 4   (outline design)   2%
- Stages 5 and 6   (detailed design and tender)   6%
- **Stage 7**   **(construction)**   **75%**
- Stage 8   (optimisation)   1%

- Contract administration/supervision   **4%**
- Legal/land/diversion/stats etc.   10%

Building projects are often more complex than civil engineering projects with more interfaces and contributing professions and may have a higher proportion spent on professional fees.

'Cost certainty'

Certainty of the final *construction* cost is almost impossible to achieve, despite the best efforts of everyone involved. In estimating the likely cost of a project, many factors will come into play. Even the unlikely instance of an *exactly duplicate* project would still produce problems in estimating, largely because of different access, ground and weather conditions. Imagine perhaps a repeat warehouse project on a site with similar access and ground conditions. If the *same contractor* is approached for a price, it may still be different for a number of reasons:

- Inflation may have increased prices generally.
- The labour market (costs and availability of labour) may have changed.
- There may have been changes in legislation which increase costs (such as waste or environmental legislation).
- Levies or charges may have changed.
- Prices of some materials and equipment may have increased.
- The contractor's 'need for work' may now be different; the contractor may be 'too busy' or have underutilised resources (the opposite effect).

• Most importantly, the contractor may now realise that some items were mispriced in the previous project, and wishes to correct mistakes on this project (or at least, not repeat them).

But if the project is re-tendered to *different* contractors, they will all take a fresh look at the project and apply all their own estimating techniques and overheads relating to their own company. In this situation it is quite usual to have a tender price variation on a well-specified project of between 5 and 10 per cent.

The example above, of an exactly duplicate project, rarely occurs, and the nearest we usually get to it is a *similar* project. To make an approximate estimate of a similar project, we can use factors to reflect relative size, assess differences in access, ground conditions and the cost of different specified items.

At stages 1 to 4, estimates are usually produced using previous project prices, factored as necessary; or built up from elements of previous projects (such as piling, roads, structure, finishes and so on). A stage 3 estimate should be within about 20 per cent of the tendered price. At stage 5, especially if traditional contracts with bills of quantities are used, an estimate within about 5 per cent of the tendered price should be possible. Once the contract is awarded at stage 6, the costs are still not certain because the construction phase has the most risks. Any cost variation here largely depends on the comprehensiveness of the site investigation, the quality of the contract documents and the form of contract used.

The cost change during stage 7 (construction) depends on the conditions of contract used. These distribute the *construction risk* between the employer and the contractor. This is a complex subject and will be discussed in detail in later chapters. However, on a well-designed and specified project, with minimum change or unexpected events, a final cost outturn within 5 per cent of the original price at contract award is a good outcome. Badly run, poorly organised projects based on incomplete designs can easily outturn at over 30 per cent above the original price at contract award. This is guaranteed to upset the employer.

Timescales

Projects vary so much that definite advice on timescales is difficult to give. However, where land purchase or planning consents are not required, stages 1 to 6 account for about 50 per cent of the total time, and stage 7 (construction) accounts for the other 50 per cent. A typical, straightforward project of about a million pounds will take about 18 months from concept to completion. If planning consents are needed, at least another two or three months should be added. For large or contentious projects the whole process could take a year or even many years. Where land acquisition is needed, it is usual to add a year or more to the timescale. Thus a larger project of say five to ten million pounds, with land acquisition will take from two to four years from concept to completion.

Many new construction graduates are involved in the detail design stage (stage 5), and may easily forget that this stage is probable in only 20 to 25 per cent of the total project timescale.

The final factor affecting timescales is the difference between '*working time*' and '*waiting time*'. Both types of time need to be programmed in. *Working time* is the time taken to produce a piece of work (such as a design) and can be reduced by using more resources (more designers). Whereas *waiting time* is not influenced by resource input. Waiting time can be such things as the two to three months to obtain planning approval, or the six weeks given to tenderers to price the contract documents. In the early approval stages, waiting time can often be 'waiting' for the next meeting of the employer's investment approval group. This is the group that gives gateway approval. These groups will usually meet at monthly intervals, and may take a further day or so to communicate their decision to the project manager. Allowing for the timing of employer approvals can be a very important factor in the design of a master project programme, and the project manager would be well advised to make full allowance for it.

Employers and external advisors

As discussed above, most employers will use external organisations for advice, and will usually use them to carry out the design of the project. Most of the 'external advisor' work on a construction project is in stages 4 and 5, the outline design and detailed design stages. In these stages firms of consulting engineers or architects are usually used. Other external advisors such as management consultants, quantity surveyors, accountants and lawyers may also be engaged. The real difficulty is that none of these advisors understand the employer's business as well as the employer does. So a consulting engineer, probably a chartered civil engineer, will know about design and producing contract documents, but is unlikely to understand the finer points of retailing, renting business premises or running a large utility. This makes it very difficult to really understand the employer's needs properly, and hence satisfy the employer with the constructed asset. This is a real challenge for all construction professionals. They should always remember that the focus is not just on design or construction, but on exploring, understanding and then satisfying the employer's need.

If the employer's need is not really understood and satisfied then the construction project will fail, however well it is designed and built.

Chapter summary

- The employer begins with a business need or opportunity: a concept.
- Construction professionals develop this concept into a completed asset.
- The process of taking a construction project from concept to completion is the investment process.
- The investment process includes contract strategy, the preparation of contract documents and the use of a form of contract. Construction professionals are involved in or drive these essential elements.
- The investment process should be designed to suit the number, cost and complexity of the projects which will fall within it.
- The investment process has up to nine major stages.
- It is normal to have a gateway approval system between each stage, managed by the employer. At each gateway the project can be amended or stopped.

- The further along the investment process that we travel, the more expensive change becomes.
- Many people and organisations are involved in the investment process and their coordination and management are crucial to success.
- Certainty of final contract cost is very hard to achieve.
- Projects have much longer timescales from concept to completion than people may imagine. A year is often a minimum and two years or more is not unusual. Major projects can take many years.

3

Parties and stakeholders

Aim

This chapter aims to outline the nature of construction contracts and to explore in more detail the roles and responsibilities of the parties and stakeholders engaged in or associated with such contracts.

Learning outcomes

On completion of this chapter you will be able to:

>> Define and differentiate the roles of parties and stakeholders.
>> Describe the range of design and construction services commonly employed in a construction contract.
>> Assess the responsibilities placed upon employers, designers and contractors when entering into construction contracts.
>> Propose appropriate lists of stakeholders for different types of projects.

Parties and stakeholders

The investment process is made up of people and organisations that carry out, or pay for, work as part of a legally binding contract. The people or organisations with these contracts are called **parties** to those contracts. 'Parties' is a legal term indicating someone with rights or obligations under a contract. In the main they will be employers, designers, contractors, sub-contractors and suppliers.

Most construction affects other people or organisations who are not parties to these contracts. Such people and organisations are commonly known as **stakeholders**. Some stakeholders are simply *affected* by a project, but many others will have a *direct influence* on it in some form or another. Many stakeholders will have a statutory function to discharge. Communication with stakeholders and management of the relationship with them are fundamental skills of effective construction professionals.

Parties are the *direct* contributors to the investment process and include
- Employers (also called the purchaser or client)
- Designers
- Contractors (sometimes called builders)
  - Sub-contractors
  - Manufacturers
  - Suppliers
  - Other professionals

Stakeholders contribute to or are affected by the investment process and include
- Regulators
- Interest groups
- Neighbours
- Internal stakeholders who may be managers or users

Parties

There are many organisations and people involved in construction projects. There will always be an employer, a designer and a builder. The builder is usually called the contractor. Just occasionally these may be individuals. Normally they are members of organisations but the *organisation* is called the employer, designer or contractor. So when we use these terms, we are generally referring to an organisation not an individual. All three of these organisations (the employer, the designer and the contractor) will have contract relationships connected with the procurement, design and construction of the construction project. It is the nature of these relationships, and the roles and obligations within them that are the primary subject of this book.

Contract obligations are enforceable in the courts under English contract law, and there are sanctions (usually financial) on people or companies who do not perform these obligations. However, contracts should not be about courts of law, they should be concerned with ensuring that parties to them understand their obligations and have a clear idea of what they are to do, to what standard and when.

The term '**party**' indicates a person or company with a legal, contractual relationship.

There are many different types of contract used in the construction industry. There are contracts between employers and designers, employers and contractors, and between contractors and their suppliers and sub-contractors. This book will concentrate on the main contract – the one between the employer and the contractor.

Figure 3.1 indicates the contract relationships as solid arrows. It shows what is known as the **traditional strategy** in which the employer has a construction

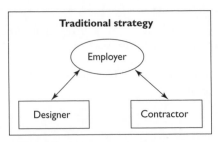

Figure 3.1 The traditional strategy.

contract with the contractor and a separate design contract with the design organisation. 'Traditional' simply means a contract strategy that has been used for many years in the industry; it does not imply that this strategy is outdated. In fact it is still one of a number of major contract strategies used today which are described in Chapters 11 and 13.

The construction contract could be the NEC3 Engineering and Construction Contract (often called NEC3 ECC or simply NEC3 or ECC) as published by the Institution of Civil Engineers or one of the many JCT forms of contract as published by the Joint Contracts Tribunal such as the JCT Standard Building Contract (JCT SBC05) or the JCT Minor Works Contract for smaller and less complex projects. NEC3 contracts are generally used in civil engineering and JCT contracts in the building side of the industry. These organisations publish a large range of contracts for use on construction projects and the construction industry is very familiar with them.

A different sort of contract is used for design services. This is the contract between the employer and the designer. Many of the established professional bodies publish suitable forms of design contract, and NEC publish the NEC3 Professional Services Contract which is written to be compatible with their construction contract NEC3 ECC.

The web of contracts

This is only the start of the contract relationships however. Most contractors will employ sub-contractors, and all will use suppliers of plant and materials both for the construction itself, and possibly for the supply of permanent plant, equipment, fixtures and fittings in the completed project.

Developing this idea further in Figure 3.2, we see at the top of the diagram the same triangle of employer, designer and contractor. The diagram shows some of the contracts that may exist for the particular sub-contract for formwork, reinforcement and concrete (FRC). Many main contractors sub-let this work to people who specialise in it. The FRC sub-contractor is just one of potentially dozens of sub-contractors.

As we see, the FRC sub-contractor has contracts for materials, plant and labour. In turn, many of their suppliers, such as the concrete supplier, will also have contracts and sub-contracts. Multiply this diagram up to cover a tower block for example, and we can easily understand how such a project can have hundreds of contract relationships. So in most construction projects, a complex web of contracts, each with their own 'parties', is set up. Each of these parties

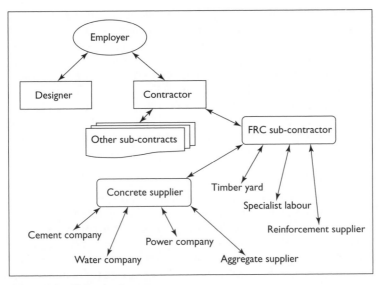

Figure 3.2 The web of contracts.

will have rights and obligations under their own contracts. All of these parties will, if necessary, have recourse to the courts and the application of contract law. However, the role of construction professionals is to manage their contracts properly within the terms of those contracts. In this way, recourse to the courts, which is time-consuming and very costly, is rarely necessary. Construction is a complicated activity and needs competent and capable professionals to manage it well.

Employers (purchasers, clients)

In many ways this is the most important role. The employer is always a party to one or more contracts because the employer is the person or organisation that commissions and pays for the work. Employers are also called 'the purchaser' in some forms of contract and are often called 'the client', particularly by design organisations. Employers are also called the client in the CDM regulations (see Chapter 8). This book will use the word 'employer', which is the term used in the main forms of contract in general use today. The word 'employer' is used to denote the fact that the employer 'employs' the contractor: not directly as a member of the employer's staff, but as a party to a construction contract.

The number of contracts to which the employer is a party depends on who designs the project and whether more than one contractor is selected to construct it. It also depends on whether specialist suppliers are contracted directly to the employer.

> The employer has to decide what is wanted, and pay for it; this is often not as simple as it sounds.

The employer is usually a company or a government organisation such as a water company, the highways agency, a city council or a county council or a body responsible for the provision of hospitals, schools and so on. The employer may also be a developer, a retail chain, a supermarket, a company wanting new offices or a housing association. In the building side of the industry it is common for the employer to be an individual; possibly an individual householder requiring a house extension, garage or renovation work. Whenever someone wants a new or replacement constructed asset they become an employer of construction services. These services are usually design and construction, but they may also include advisory work, site investigation, evaluation of possible options and so on.

Who represents the employer?

The fact that many employers are large organisations produces the additional problem in many cases, of determining *who* exactly represents the employer in this organisation; who makes the decisions, who gives the approvals at the various 'project gateways', who organises the site and access to it and who will manage the interface with the employer's staff and operational functions? Employers with an effective investment process will always take great care to nominate responsible individuals or groups of people within their organisation. They may be known as the 'sponsor' or 'investment committee' or similar terms. It is very important to ensure that the employer's contractors and the design organisation know exactly who these responsible people are, what their powers are, and what information they need to make decisions and how long this may take.

Another frequent complication will be the interrelationship between the construction project, and the employer's organisation, operational functions and other ongoing activities. Once again, defined lines of authority and properly designated individuals or committees will make the management of projects as equally clear *within* the employer's organisation as *outside* it.

The employer's brand or image and cultural fit

Increasingly employers are concerned that contractors should enhance the employer's brand or public image, and do nothing to detract from it. In this context, employers will also seek contractors and consultants with a 'cultural fit'. In other words, the employer will look for organisations which have similar values, methods and processes to the employer's own, largely because this makes mutual understanding easier and improves teamwork. This can put added pressure on a contractor's organisation that, of necessity, works for many employers, especially if these employers have dissimilar organisational cultures from those of the contractor.

Finally many larger employers modify traditional forms of contract, and have developed their own procurement systems, and contract strategies, into which contractors, consultants and suppliers have to fit. This can represent a real and difficult challenge for these organisations.

Designers and design organisations

The 'designer' is generally a shorter way of saying the 'design organisation' which is always a party, either to a contract with the employer, or to a contract with the contractor. The first is often called a 'traditional contract', whereas the latter is where the contractor has responsibility for the design as well as the construction. This is what is known as a 'design and build' contract or 'D and B' contract for short. Designers are often independent organisations of architects, consulting engineers or building surveyors. However they may be employees of the employer. Rarely, designers may be employed as part of the contractor's organisation itself.

Where designers are employees of the employer they must always act in a professional manner and try to avoid conflicts of interest. When administering a contract, they must remain unbiased and independent. Ensuring this independence is a difficult and common issue in the industry, but most contract administrators take this part of their role very seriously.

Design organisations often fulfil a wide range of essential functions in the investment process that are not simply concerned with design. These might include:

- Investigating and advising on different options.
- Producing layouts and architectural features for approval.
- Producing all necessary drawings and specifications.
- Producing a bill of quantities or other pricing mechanism.
- Making estimates of cost.
- Project management.
- Providing contractual and procurement advice.
- Advising on methods of selecting contractors.
- Assessing tenders.
- Administering contracts.
- Supervising work on-site.
- Assisting in the final stages of commissioning and hand-over.

In simple terms, design organisations do everything necessary for the creation of the new asset, except actually build it. This essential and complex activity is carried out by the contractor.

Many designers are also expected to undertake the role of project manager. The roles of designer and project manager have many differences, and many people are not suited to both. In their separate ways, both roles demand good communication skills. However, whilst design can be a solitary and intellectual activity, project management demands skills in leadership, organisation, planning and coordination; it is 'people-orientated'. This separate project management skill base has become one contributing factor in the development of the project manager as a role and profession in its own right.

Contractors (builders)

Contractors, or builders as they may be known, are always a party to a contract with the employer, designer, sub-contractors and suppliers, or a combination of these. The contractor organises all the construction processes in order to produce the finished asset.

Contractors range from small organisations to immense multi-national companies employing thousands of people. It is easy to focus on the obvious work of a contractor; that of constructing the asset. What we see is the 'work on-site'.

This work involves the coordination of numerous trades, sub-contractors and plant and machinery suppliers. It also involves dealing with ground conditions and the vagaries of weather and climate. However, it is important to remember that contractors also have many headquarters functions, and employ construction professionals in that capacity. Such headquarters functions are estimating (producing prices for tenders for new work), planning and programming, contract and commercial advice, purchasing of goods and materials, plant hire and control, health and safety, business development and marketing together with the overall control and management of the site work itself.

Recent developments have led to the increased use of sub-contracting. One of the main reasons is the reduction in the directly employed labour of the general contractor and the desire to offset risk to other organisations. Other reasons are the move of some contractors towards specialisations such as management and coordination, whereby sub-contractors do most of the construction work itself. There has also been some polarisation towards either specialisation into even narrower fields, or the offering of wider integrated packages such as design and build. Successful contractors tend to be flexible and fast-moving with highly developed skills in management and coordination and a deep understanding of risk and commercial issues. They will use these skills to develop their organisations to meet whatever challenges and opportunities the construction industry of the future might offer.

Sub-contractors

Sub-contractors usually have contracts with the main contractor for items of work such as 'formwork, reinforcement and concreting' (FRC), or for installation of plant, or specialist work, or even the provision of labour only. Sub-contractors are generally known as 'subbies' in the construction industry. In building there will be many sub-contractors often based on trades such as plastering, painting, and electrical work and so on. Some companies are organised to do sub-contract work only, and form business relationships with the larger contractors who employ them.

Some main contractors will also do sub-contract work for other contractors – often as specialists. Additionally, many contractors will do sub-contract work for other contractors who have been successful in their tenders with an employer, and who wish to redistribute parts of the workload. Clearly this latter case must be handled carefully, as few employers wish to go through a careful tendering process to select contractor A, and then find that the work itself has largely been sub-let to contractor B. Many forms of contract have restrictions on the main contractor sub-letting work to other contractors, particularly the whole of the work. The contract administrator's permission is usually required for all sub-contracting, but under most forms of contract, this permission should not be unreasonably withheld.

Manufacturers and suppliers

All contracts have suppliers of materials (bricks, concrete and steel for example), products and components (windows and doors for example) and frequently of manufactured plant (such as pumps, panels, lifts and boilers). These suppliers usually have contracts with the contractor, or one of the sub-contractors. A further difficulty can arise with manufactured plant, since it is usually *designed by the manufacturer*, rather than the 'design organisation' or the contractor. The organisation and coordination of these manufacturers and suppliers require great skill and planning if they are to be done effectively. Under most types of procurement, this coordination is the contractor's responsibility. Although many manufacturers and suppliers specialise in construction industry requirements, they are essentially in the manufacturing sector, rather than the construction sector. These sectors have many differences, and this can lead to misunderstandings if not properly managed.

Other professionals

The construction industry has a wide range of contributing professions. In civil engineering, most practitioners have first degrees in civil engineering. Later they may specialise, possibly with an MSc, into fields such as structural design, hydraulics, project management or transportation.

The 'building' side of the industry tends to be more specialised however, and a wide range of different degree qualifications are available. Architects are usually responsible for the overall concept and appearance of a building, whilst architectural design technologists provide all the development and details to produce the design. Quantity surveyors tend to cover the financial side of the industry and take-off quantities, do monthly measurement and financial certification of work, and give advice on contracts and claims for loss and expense. A number of modern contracts use target cost or cost-reimbursable options. Quantity surveyors and cost-consultants are increasingly used in the validation of these costs which can be a more complex activity than measurement by bills of quantity in traditional contracts. Building surveyors specialise in the design and structure of buildings, especially renovation and conservation, and are often involved in contract work which requires a complex array of different trades within existing buildings.

All of the professions discussed above have their own professional institutions. These institutions maintain a professional knowledge base, advise members and develop training and accreditation programmes for the admittance of new members. They also give advice to politicians, represent their section of the profession and provide a focal point for members.

A major focus on all construction projects today is health and safety, and the reduction of accidents to the workforce. Most design organisations and certainly contractors will employ specialist health and safety professionals on their own staff, or utilise one of the many health and safety consultancies for advice and monitoring. One of the key roles in the Construction (Design and Management) Regulations 2007 is that of the CDM coordinator. There are now an increasing number of companies that specialise in carrying out this role.

Many construction projects now include operation and maintenance of the asset as well as design and construction. Much work for the government is carried out under PFI (the private finance initiative). Professionals employed by banks and other financial institutions play a fundamental role in these projects. Many construction projects are checked by an employer's auditor, who will often be an accountant, but may be an engineer who specialises in financial and contractual matters. Legal professionals may advise on specific issues in contracts, but particularly become involved if a problem on a construction contract becomes a 'dispute'.

The project manager

Good project management is essential in all areas of the industry, and increasingly people undertake formal training in this area after graduation and a number of years' experience. Many project managers belong to the Association of Project Managers, or APM, who are one of the construction industry's professional institutions. The project manager may be an individual or organisation with a direct contract link to the employer. However, the project manager will usually be an experienced member of the organisation that designed the project. The role is not just to organise the design, but to coordinate all aspects of the programme and to ensure that proper communication with all affected people is taking place, and that the project is planned and organised effectively from concept to completion. The increasing complexity of projects and rising expectations of employers and the public continue to put more demands on the skill of the project manager.

The role of the project manager is to understand the place of all the above participants in the investment process and also the importance of stakeholders, as described below. Good management of stakeholders and satisfaction of their needs are an essential part of the project manager's role, and a prime feature of a good investment process. Plans for the management of all stakeholders must form a major part of any master project programme.

The term 'project manager' is also used in some forms of contract such as NEC3 ECC to denote a formal contract administration role. This role will be described in detail in the chapters on NEC3 ECC contracts.

Stakeholders

Construction projects always have an impact on other organisations and people. A new building in a city, for example, will affect members of the public who see it, people in neighbouring offices and the transport system itself. It will change the view of the skyline and cast shadows over streets and other buildings. Its construction will create noise, dust and waste, which must all be minimised. It will inevitably affect the environment. If nothing else, it will use power and water and generate waste, heat and discharges to the sewerage system.

We have discussed the parties to our construction contracts. However, many other people and organisations are *affected by our projects*. These are the stakeholders.

> The term **'stakeholder'** indicates those with an interest in the project or contribution to it, or a regulatory role. It also includes those affected or influenced by the project.

A stakeholder is not a party to any of the construction contracts for the project. A stakeholder has no direct contractual link into the project, but is usually very

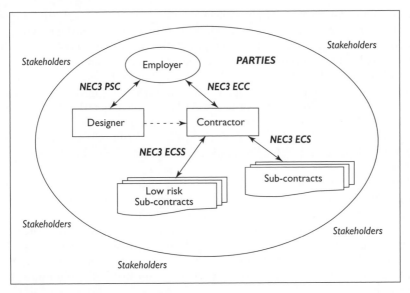

Figure 3.3 Parties and stakeholders.

important nevertheless. It is easy to concentrate on the direct contractual participants in the investment process and ignore the stakeholders. However, effective construction professionals always identify and communicate with stakeholders. As far as they can, effective construction professionals address stakeholder concerns and satisfy stakeholder needs. In this way, stakeholders can become supporters of the project, rather than objectors to it.

Figure 3.3 shows the parties to some of the main contracts in the web of contracts as connected by arrowed lines. The NEC contracts that might be used are shown, and are depicted in italics as NEC3 PSC, NEC3 ECC, NEC3 ECS and NEC3 ECSS. They will be explained in Chapter 20. For building projects JCT contracts would be used, but the relationships would be the same. The stakeholders are outside the web of contracts. The stakeholders have no direct contracts with the members of the investment process, but they are very important nevertheless. As we touched upon earlier in this chapter, there are four main categories of stakeholders:

> Regulators
> Interest groups
> Neighbours
> Users

Regulators

The importance of regulators is a recent phenomenon. Regulators ensure that construction professionals obtain the requisite legal permissions and satisfy all relevant legislation. Regulatory bodies are usually set up and financed by the government. There are many instances where a structure/road/building can affect the freedom, privacy, or rights of an individual, or the environment. Legislation of many types has evolved to regulate the activities of those who wish to build. So planning legislation controls the appearance of buildings, and building control regulates the safety and usability, health and safety legislation controls safety, environmental legislation protects (and seeks to improve) the environment, waste legislation controls the disposal of surplus material, and so the list goes on. Many of these regulators work in departments of the local county council or district council. Others are part of a national body such as the Drinking Water Inspectorate, the Environmental Agency or the Health and Safety Executive.

Obtaining regulatory approvals is usually a major feature in overall project programmes, and its importance must never be underestimated, either by designers or project managers. Since projects cannot proceed without regulatory approval, which often influences the *design* or *appearance* of a project, it is usual for either the employer or the design organisation to obtain these approvals. However on small projects, or projects using the design and build contract strategy, it is the contractor who is responsible for obtaining these approvals.

Finally many former 'public bodies', now private companies, in the power, telephone, gas and water industries are *regulated* to ensure they carry out their functions, and make proper charges to do so. (Ofwat and Ofgem are examples of regulators in the water and power industries.) This is a different type of regulation which only affects such companies. However, utility companies (such as gas, water, telecommunications and electricity) are important to construction professionals for two reasons. Firstly, these 'utilities' are a very significant purchaser of construction and design services in the UK, which they use to improve and extend their infrastructure. They have massive capital programmes with a total construction expenditure of many billions of pounds per year and so they are a major employer of the construction industry's services. Secondly, in addition to having their own capital programmes, they have some statutory powers, and act rather like a regulator when anyone wishes to divert or connect to their systems of cables, pipes, or sewers.

Interest groups

This stakeholder can be very important. There are many voluntary groups in the UK who take a special interest in areas or wildlife that can be affected by construction. So for example many counties have 'badger societies', who look after the interests of local badgers. Badgers are protected by UK environmental legislation, and must not be killed or injured. Badgers live in a network of tunnels called a 'sett' and often have an alternative sett some distance away. Badgers affected by construction activities have to be found and relocated. Great crested newts, orchids, bats and many other species of flora and fauna have their own special interest groups. These interest groups can be very useful to the construction professional, if they are involved and consulted to ensure that the project meets environmental legislation. In recent years, the extent and complexity of environmental legislation have led to the rise of a large number of specialist consultancies in the UK, who employ qualified environmentalists, biologists and other professionals to advise on compliance with the relevant legislation.

Neighbours

Neighbours are people living adjacent to the site of the project, or in the vicinity. Neighbours can be very vocal in their objections to a project, and can influence regulators such as planning authorities. Neighbours can also be a great supporter for a project or even an instigator of new projects. There are many instances of 'neighbour pressure' initiating construction projects such as bypasses, new bridges to improve access, land regeneration and new parks and open spaces.

However, neighbours often have negative views of construction projects. They can be affected by increased noise or traffic, decreased property values, changing amenity, reduction in open spaces or even light and many other factors that cause them concern. Effective construction professionals consider neighbours in their designs and construction plans. They communicate with neighbours, try to address their concerns and wherever possible improve their situation. There are no greater supporters than neighbours who actually want a construction project because they see an improvement in their lives.

Users

In a sense we all use a building when we pass by it or drive around it, for it forms part of our landscape. Here however the term 'users' is meant to represent the *direct users* of the constructed asset. These are tenants of a building, workers in a factory or office, doctors and nurses in a hospital or operations people in a utility such as gas or water. In other words, the *users* are those who will actually use the constructed asset. Users also include the managers of operational functions affected by the project.

For many companies, their new buildings will become offices or factories for their employees. They want positive reactions, and an efficient and motivated workforce. The constructed asset must be safe, operable, habitable, easily maintained and it should provide a pleasant environment. It is therefore essential that the construction professional consults these users during the design process, in the construction process, and particularly in any training and hand-over process. This consultation is frequently done with the people who manage these internal groups of stakeholders. These managers may represent the views of their staff, but they will certainly have a direct input to the project to ensure that it meets the needs of their departments or does not adversely affect their operational functions and responsibilities.

Another feature of many construction projects is that they extend, limit or demolish existing installations where operational or office staff must continue working. The management of safe access, demarcation of construction from staff work areas, shut-downs, connections and so on becomes an essential project activity, requiring the cooperation of users and their managers, and careful project management.

Many smaller building projects will be commissioned by domestic users. The domestic user needs special care and attention by the construction professional since they will rarely have any direct professional knowledge of the construction process. They will therefore need guidance through all stages of the process and their needs and requirements must be carefully identified and addressed.

Dealing with stakeholders

The identification and management of stakeholders are essential to the success of most construction projects. Satisfying stakeholders may mean making adjustments to the design of the project or the phasing and type of construction planned or to the selection of construction plant. 'Managing stakeholders' is a study in its own right. However, in brief, effective project managers do the following, commencing at an early stage of the project:

- Identify stakeholders both within the employer's organisation and outside it.
- Ensure that the legal role of all regulatory stakeholders is understood by the project team.
- Produce a stakeholder management plan which will list stakeholders and rate them on a scale from supporters to objectors.
- Begin initial communication with stakeholders.
- Reassess the supporter/objector scale.
- Have a stakeholder management plan to try to move the objectors into supporters (there may be time and financial constraints here)
  - This may mean making some concessions or adjustments to the project
  - It may also mean restricting working hours, noise or traffic.
- Ensure that all *regulatory* stakeholders are satisfied; this is a legal requirement.
- Continue to communicate with stakeholders, possibly with regular meetings or newsletters.
- Address stakeholder concerns during construction by reducing noise, mud on roads, traffic congestion.
- Have a good training and hand-over process to ensure that users, operational staff and their managers are satisfied.

Chapter summary

- All construction projects are subject to contract law, and have 'parties' to those contracts. A 'party' is a legal term for a person or organisation with rights and obligations under those contracts.
- Almost all construction projects will also have 'stakeholders'. Stakeholders are people or organisations with an interest in the project or contribution to it, or a regulatory role. The word also includes those affected or influenced by the project.
- The main parties to contracts will include employers, designers, contractors, sub-contractors, manufacturers, suppliers and members of other related professions.

- Stakeholders will include regulators, interest groups, neighbours and internal stakeholders who may be managers or users.
- It is very important for construction professionals to understand that projects are not just about 'designing' or 'building' but they involve a vast array of professions, regulators and interested parties who must be considered.
- Effective construction professionals always identify and consult all stakeholders.
- Regulatory stakeholders must be satisfied.
- Other stakeholders should be satisfied, or at least have their concerns identified and addressed as far as is possible.
- Neglecting the importance of stakeholders is one of the causes of construction project failure.

4

An introduction to contracts

Aim

This chapter *introduces* the concepts of contract law, terms of a contract and forms of construction contract.

Learning outcomes

On completion of this chapter you will be able to:

>> Describe in outline the requirements for the formation of a contract.
>> Differentiate between terms and representations.
>> Propose examples of contract terms.
>> Differentiate between express and implied terms.
>> Explain the features of forms of contract.

Making contracts

Have you made a contract today? We make contracts all the time, often without realising it. We make contracts when we buy something or perform a service for money for example. Thus on most days we will make contracts when we buy a newspaper or a book, get on a train or bus, buy petrol, buy a parking ticket, buy a coffee or a sandwich, go to a café for lunch and so on.

Whenever people have a commercial agreement to carry out work for payment, they have a contract. The people or organisations that make the contract are called **parties**.

The **formation of the contract** is based on an *offer* to do something (in our case to carry out design or construction work), an *acceptance* of this offer and *consideration* (payment in return for the design work or the construction work). There are a number of other requirements which will be explained below.

In the construction industry we make a contract during the process of **tendering** and contractor selection as described in Chapter 6. Tendering is the industry word for inviting contractors to submit priced bids based on the employer's tender documents. The tender is the collection of documents, priced up by the tenderer and returned to the employer as an offer to do the construction work. These tender documents are usually prepared by the **design organisation**. It is important that this is done correctly in order to ensure we have a valid contract with all the important terms agreed. Much of the work of the design organisation (firms of architects, consulting engineers or building surveyors for example) is in ensuring that the drawings, specifications and other tender documents are comprehensive and clearly define all the construction work that must be done and to what standards of quality.

Once we have made a valid contract, the parties to that contract have rights and obligations as described in Chapter 5. The parties to the construction contract are the employer and the contractor. The **terms** of a contract seek to define these rights and obligations. Importantly they usually set down procedures for payment and also for the resolution of difficulties or risks that *might* arise. Without contract terms, parties to a contract who could not agree would have to go to court to settle an outcome. This is very time-consuming and costly; it is not unusual for a typical High Court case in the construction industry to take two years, and cost between £50,000 and £100,000.

Formation of a valid contract

A valid contract in English law has a number of essential elements. These essential elements will be described very briefly below, but are covered more comprehensively in Chapter 17. The elements of a valid contract are:

- Offer and acceptance (this is the agreement)
- **Consideration**
- An intention to create a legal relationship
- Genuine consent (no **vitiating factors, discussed below**)
- The parties must have **capacity**
- Legal formalities (where required)

Offer

The person or organisation that invites tenders (the employer) is actually asking contractors to submit an offer based on the tender documents. Hence the tender documents are called an 'invitation to treat' in legal language. When the contractor (here called the 'tenderer') returns the tender documents containing prices for the work, this is the 'offer' in legal terminology.

Acceptance

Since the employer will become a party to a contract with the contractor, it is the employer who decides which tender to accept. However it is usually the design organisation acting for the employer which will assess and evaluate the tenders and recommend which one the employer should accept. The employer's acceptance has to be clear and it must not introduce any new terms.

Consideration

All contracts must have consideration. Consideration is a legal word that usually means money, paid by the employer in return for the construction work. Payments are usually made to the contractor on a monthly basis until the construction work is complete.

Intention to create a legal relationship

The parties must intend to create a legal relationship. Hence in domestic arrangements it is assumed that there is no such intention. However, in commercial agreements this intention is assumed. It would clearly be nonsense for the employer to spend large amounts of money having a design and drawings produced, if the employer did not wish to create a legal relationship to have the work constructed by the contractor.

Vitiating factors

These would include such issues as 'mistake' or 'misrepresentation' of the facts and prevent a proper contract being formed. Vitiating factors are described in detail in Chapter 17.

Capacity

Capacity is concerned with whether parties entering into the contract are legally capable of doing so. This is rarely an issue in construction contracts and often relates to minors (a party being too young) or possibly the insane.

Legal formalities

A contract can be made by a simple verbal agreement. However in construction where we have complex and expensive contracts this is rarely sufficient. So a typical formality is the requirement for both parties to sign a 'form of agreement'. Other formalities might include signature before witnesses for example.

Terms and representations

Once the contract is made, it is important for both parties to be clear as to what they have to do. These rights and obligations are defined by the **terms** of the contract. However, another important concept is that of **representations**, as described below.

Representations are statements that are made *before* a contract is formed. Hence if I were buying a second-hand car, representations might include 'one careful driver', 'never driven over 70 mph', 'fully garage serviced' and so on. Representations usually induce someone to enter into a contract.

In construction, representations could be statements made to a tenderer on a site visit such as 'there is no history of vandalism here' or 'the ground is good and you should not need to use dewatering plant'. Clearly such statements could affect a tenderer's price for the contract and hence they should be correct or, preferably, not made at all.

However, much of the work of construction professionals is administering or complying with the terms of a contract. The terms are not just the words in the contract itself (such as **NEC3 ECC** or **JCT SBC05**) but include statements made on drawings or in specifications or other documents.

> The contractor and the employer should comply with the terms of the contract.

We write terms of construction contracts most frequently when we write specifications and when we produce drawings for a particular contract. A dimension on a drawing is a term of the contract, with which the contractor should comply. When we write 5200 mm, we do not mean 4900 mm or 5450 mm, we mean 5200 mm. Of course, we are not making a contract in our own name, but we are providing the base information, or the terms of a contract that the employer will sign.

The employer also has responsibilities and should comply with terms relating to the date of access to the site, how the contractor will be paid, and when and what happens if there are changes or delays that the parties did not anticipate. Since these terms apply to many contracts, the construction industry has **'forms of contract'** which set out *standard terms* for different types, complexities and values of contracts. Chapter 20 will describe these numerous forms of contract. However, this book concentrates on two of the most commonly used forms, the NEC3 Engineering and Construction Contract (often called NEC3 ECC or simply NEC3 or ECC) as published by the Institution of Civil Engineers or the JCT Standard Building Contract (JCT SBC05) published by the Joint Contracts Tribunal. Chapters 21 to 27 deal with these two forms of contract.

> This book concentrates on the NEC3 Engineering and Construction Contract (NEC3 ECC) and the JCT Standard Building Contract (JCT SBC05).

All these items written down in the contract become what the lawyers call **express terms**, those that have been agreed between the contracting parties over and above their basic legal rights and obligations. It is usual for these express terms to be written down, particularly in a complex contract such as one for construction. Terms can also be verbal, chiefly on more simple contracts than ours, which may have no written terms at all.

If a problem arises that is serious enough to take to court, then the court will always start by looking at the intention of the parties when they made the contract; the court will use the express terms. However, where the court considers that an important express term is missing, then the court will imply one; it is sensibly called an **implied term**. However the court will only do this where necessary to give a contract 'business efficacy'. So for example, if the contract failed to stipulate a time for completion (which does sometimes happen as a serious oversight), then the court will imply a term and decide on what it considers to be a 'reasonable time'. Similar terms would be implied by the court if necessary for work of reasonable quality and so on where the contract did not set these down in the specification or other documents.

A university example of contract terms

Many of us have rented accommodation from a university at some stage of our lives. What terms (these are express terms) might we expect in a contract with the university for a room in a hall of residence? Most contracts begin with definitions, so both parties (here the university and the student) understand the meaning of the words used. We usually expect to find a definition of time, so here 'residential year' (calendar year or teaching year) would need clarification, and other definitions might be the meaning of 'room' and 'licensee' (the student). Terms of payment are common to almost all contracts so we would expect to see details of how much we need to pay and when. Many contracts also give parties the right to cancel them (termination) but only for certain listed reasons.

Another important consideration is guests and visitors. Are visitors and overnight guests allowed, and under what circumstances? Guests will usually have to be signed in to meet fire regulations, if nothing else. We also need to define what services will be provided, such as heating, cleaning and internet access, and how they are to be paid for.

Then we come to procedures to meet likely difficulties. What might we expect to see? We might see terms relating to lost keys, changing a room, general behaviour and tidiness, noise and use of shared areas such as student lounges and storage areas. So a list of contract terms for this sort of contract might include:

1. Definitions of room, licensee (the student), residential year
2. Payments
3. Lost keys (a fee may be charged)
4. Changing your room (a fee may be charged)
5. Termination of the licence to occupy
6. Use of room in vacation periods
7. Damage to the common areas
8. Overnight guests
9. Personal property
10. General conduct, especially health and safety and fire
11. Responsibilities such as tidiness, cleanliness and departure procedure.

An example of terms in a simple construction contract

Now let us take a simple construction example. Suppose you have just set up your own business as a small builder and you have won the contract to build a large house extension for a local family. Your tender price for the contract is £100,000 and you have six months to build it, while the family move out. You have put up £20,000 of your own money and borrowed £30,000 from the bank. The contract is an NEC3 ECC contract option B, with a bill of quantities which you have priced. It has monthly payment terms so you do not need to finance the whole venture as you would in a lump sum contract (this is a contract where a single payment is made on completion). This is why you need £50,000 and not

£100,000 as working capital. You obviously want to see clear terms in the contract, so you understand your risk, and you want to know what will happen if things go wrong. You could lose all your savings and owe the bank money for years to come. Your questions and concerns might be (the answer from NEC3 ECC is given in parentheses):

- When can I start on-site? (Contract data, access date)
- When must I complete the contract work? (Contract data, completion date)
- If something goes wrong, do I get more time to complete? (Clause 60.1 compensation event)
- Can we agree a programme? (Clause 31 programme)
- What happens if the family are late moving out and this delays me? (Compensation event 60.2)
- How and when will I be paid? (At the assessment interval, usually monthly as section five, payments)
- What happens if the **project manager** orders changes? (Clause 60 compensation event)
- How do we assess the payment for changes? (Clause 61 quotation from contractor)
- What if the family move back early, before I have finished? (Clause 35 takeover)
- What happens if the project manager is dissatisfied with my workmanship? (Section four, defects, and clause 11.2(5) define a defect)
- After completion, for how long can I be called back to remedy defects? (Contract data – until the defects date)
- What about weather conditions that I could not expect? (Clause 60.13 compensation event)
- Who pays for any tests? (Section four, testing and defects)
- Is there a disputes procedure? (Option W1 or W2)
- What about insurance? (Section eight, risks and insurance)
- What happens if I find any archaeological remains? (Clause 60.7 compensation event)
- Can the project manager stop or suspend the work? (Yes under clause 34, but this is a compensation event)

What we see above is not the mystery of a form of contract, but the logic of it. A similar list but quoting different clause numbers could be written for a JCT contract such as the Standard Building Contract. Many people see construction contracts as daunting and impenetrable. They are neither. They set down measures to deal with day-to-day events such as payment, and they attempt to foresee difficulties such that they can be resolved without going to court. The project manager is the name given to the contract administrator in an NEC3 ECC contract. This person has powers under the contract to issue changes and certify payments for example.

> Contracts protect you. They cover all the main procedures and eventualities so you don't have to go to court.

'Forms of contract'

Forms of contract for construction are written following a similar process to a more general contract. They set down definitions and procedures for things that are likely to happen or need clarification on a particular type of project, such as a major project, a small project or a project using the **design and build** contract strategy for example.

A form of contract is simply a pre-agreed and published set of terms and conditions for the activities connected with a type of project, in our case within the construction industry. We have seen in Chapter 2 how our projects differ from those in for example manufacturing. Forms of contract for the construction industry address its particular needs. Most forms of contract are written for the construction stage of the project both for the contractor and sub-contractors, because this contains the most difficulties and risks and represents the main investment. Specific forms of contract have also been written expressly for the design process, suppliers and specialist functions such as adjudication. More detail of the many forms of contract is given in Chapter 20.

> *Forms of contract* are simply standardised sets of terms to suit different types of project and the various types of procurement arrangements.

If we look at forms of contract for the construction of the project we see terms relating to payment, time and progress, roles and powers, variations, defects, completion and so on. Hence we have specific standard forms of contract to use for different values, complexities and features of construction projects. Some will be better for **traditional contracts**, some for design and build, some for complex projects and some for simple and straightforward ones. Traditional contracts are where the design and construction are awarded as separate contracts to two different organisations, whereas design and build is where the contractor designs the work as well as constructing it.

> Contracts protect you. They cover all the main procedures and eventualities so you don't have to go to court.

A contract can be very simple with very few terms, but construction contracts usually have many terms because construction is a complex activity with many risks which could occur over a long period of time. So it is usual for construction contracts to start with a standard form of contract, selecting the most appropriate for the project in question, and then add further specific terms to it. In this way the parties to the contract are clear about what is required, time, payment, disputes, access, variations and so on.

Advantages of forms of contract

Thus forms of contract are simply collections of standard terms, designed for particular types of design or construction. Most of the major professional bodies publish a suite of forms of contract. Within each suite are contracts written specially for particular contract strategies, payment mechanisms, project types or a mixture of the three. Having standard forms of contract has the following advantages:

- They do not have to be written specifically for every contract.
- They are available 'off the shelf'.
- After a time, parties become familiar with them, and understand the risk.
- This makes tendering easier and increases 'certainty'.
- They are usually 'agreed' between professional bodies, so they seek to distribute risk in a fair way.
- Where difficulties cannot be resolved, cases may go to court. This establishes case precedent and clarifies issues for the future. They become tried and tested in the courts.

Despite the fact that there is a wide choice of standard forms of contract, most major employers amend them in some way, often to redistribute risk. Contractors need to read such amendments very carefully to understand the risks they are taking and the way they will be paid. This does rather defeat the idea of *standard* forms of contract, and it certainly makes life more difficult for tendering contractors, but at least standard forms of contract constitute the basis for the amended versions.

Major publishers of standard forms of contract

The New Engineering Contract (NEC3)

The contract which dominates the civil engineering side of the industry is the third edition of the New Engineering Contract (NEC3). It is published by NEC, which is a division of Thomas Telford Ltd, which in turn is a wholly owned subsidiary of the Institution of Civil Engineers. The NEC contract was first published in the early 1990s and was initially used by a number of forward thinking organisations seeking change and a better way of managing contracts. NEC2 was published in 1995 and increasingly became the contract of choice of many organisations in the United Kingdom, although it is also designed for world-wide use. NEC3 was published in 2005.

NEC3 comes with a full endorsement from the Office of Government Commerce (OGC), through the Construction Clients' Board, which recommends NEC3 for use on all public sector construction projects. There is no doubt that use of the NEC3 contract will continue to increase.

The NEC panel, which is a specialist panel of the Institution of Civil Engineers, oversees the continuing evolution of the NEC family of contracts. It is composed of a number of experts, both in the NEC and in their relative fields, who meet regularly to decide on future priorities for the NEC suite of contracts. There are currently seven panel members. Most of the panel are chartered engineers, surveyors and builders together with a solicitor. The panel also takes advice from a sub-group, who are largely legal professionals.

The NEC also produces a wide range of guidance documents and other publications and has an active user group and training organisation. They also arrange conferences and other events. There is a great deal of useful information on their website (www.neccontract.com).

Joint Contracts Tribunal (JCT) Contracts

Most standard forms of contract are published by professional bodies and the panel of contributors and commentators will usually include representatives from affected industry organisations. So for example the Joint Contracts Tribunal (JCT) which publishes contracts which are generally used by the building side of the industry has the following members:

- Association of Consulting Engineers
- British Property Federation
- Contractors Legal Group Limited
- Local Government Association
- National Specialist Contractors Council
- Royal Institute of British Architects
- Royal Institution of Chartered Surveyors
- Scottish Building Contract Committee Limited.

The JCT have been publishing contracts for over 100 years and their website www.jctltd.co.uk contains a wealth of information and downloads.

ICE contracts

The Institution of Civil Engineers also publishes the ICE range of contracts, dominated by ICE7, which is the seventh edition of a contract designed for construction using the traditional strategy, with separate design and construction organisations. The ICE contracts are produced by the Conditions of Contract Standing Joint Committee (CCSJC). The ICE conditions of contract are jointly sponsored by ICE, the Civil Engineering Contractors Association (CECA) and the Association of Consulting Engineers (ACE). The ICE conditions of contract have been in use for over fifty years and were designed to standardise the duties of contractors, employers and engineers and to distribute the risks inherent in civil engineering to those best able to manage them. In recent years ICE contracts have been supplanted by NEC contracts, but ICE contracts form the basis for **FIDIC** contracts and they are often used in their own right on contracts overseas.

Other contracts

There are many other publishers of contracts designed for specific purposes such as:

- The International Federation of Consulting Engineers (FIDIC) who publishes a wide range of contracts designed for international use.
- The Institution of Chemical Engineers (IChemE) is another publisher. Their contracts are generally designed for chemical engineering and process contracts and are named by their colour. So we have the green, red, burgundy and yellow

books designed for different payment mechanisms such as target cost or lump sum, and for international use.

- GC/Works construction contracts were designed for use on government contracts such as hospitals and schools. Their most recent main editions are dated 1998 and 1999. Their use has reduced markedly with the predominance of the NEC3 contracts.
- The Institution of Mechanical Engineers (IMechE) publishes contracts designed for use with mechanical and electrical work such as MF/1, MF/2 and MF/3. The most recent editions of these contracts were published between 1999 and 2001.

In the UK, the two most commonly used suites of contracts are NEC3 and JCT, and hence this book concentrates on these contracts.

Which edition of these contracts do employers use?

The major publishers of suites of contracts such as JCT and NEC sweep up changes in the law, and changes in industry practice in wholesale new revisions of their range of contract forms. JCT is a good example, with major revisions in 1980, 1998 and 2005. The revised versions become known by their year, such as JCT SBC05, from the 2005 range of JCT contracts. Revision and reissue are clearly a major and costly exercise for the publishing bodies. These publishing bodies will continue to support an earlier edition for a few years.

Once a major revised suite of contracts is published, it will be updated as the law changes with amendments. So for example, JCT have issued amendments to their standard building contract JCT SBC05, in 2007 and 2009, as revision one and two, but the contract is still known as JCT SBC05. Eventually, after a number of changes in the law, or changes in industry aspirations and practice, a new suite will be published. JCT have announced a 2011 suite of their contracts as this book is going to press.

Why many employers do not use the latest edition

It may come as a surprise to many readers that use of a new edition of these standard forms of contract may be delayed for some years by many employers. So, for example, some employers still use NEC2 (published in 1995) rather than NEC3 (published in 2005). Some civil engineering employers still use ICE7 rather than NEC3. The penultimate major revision of JCT was in 1998 and a few employers still use this version rather than JCT 2005. The reasons are as follows:

- Many employers will have existing projects under design, construction or defects correction. These phases can represent two to five years on larger contracts. Hence there is a long period of application for any edition, once a contract using it has been awarded.
- Employers usually modify standard forms of contract, and will need professional advice on suitable modifications to a newer version which will cost them and takes time, so they might therefore delay doing this as long as they can.

- Large employers usually publish specifications and other documents which interrelate to the conditions of contract. Amendment of these other documents is time-consuming and costly.
- Employer's staff and design organisations are usually familiar with the current edition. Conversion to a newer edition produces a training need with consequent time and cost considerations for the employer.
- Employers may be unsure of the change in balance of risk, and may adopt a 'wait and see' philosophy, until there are published comments from the industry.
- Software to support procedures generally becomes available some time after a new edition is published.
- Textbooks and other supporting documents are written some time, often years, after new contract forms are published. Having said this, NEC and JCT themselves now publish extensive guidance notes at the same time as new editions are released.
- Earlier editions of forms of contract will almost certainly have supporting case law. There may be aspects of newer editions that require testing in court to ascertain their precise application and meaning. Many employers would rather others did the testing!
- Tendering contractors may be less sure of the risk balance and may price up the unknown.
- Contractors' cash flow and other accounting systems may be linked to certain editions of contract and changing them may cause contractors to be obliged to increase tender prices.

Chapter summary

- All construction contracts are subject to contract law, and have 'parties' to those contracts, usually the employer and the contractor.
- The law sees agreements between the parties as 'bargains' which have offer, acceptance and consideration (usually the payment of money).
- The parties must also have an intention to create a legal relationship and capacity to make a contract. Formalities may also be required.
- Words and phrases in the contract itself are called terms of the contract.
- Terms must be distinguished from representations, which are documents and words used to persuade someone to enter into a contract.
- Express terms are written in the contract whereas an implied term may be inserted by a court if an important express term is missing.
- It is usual to standardise construction contracts into agreed forms of contract.
- There are different forms of contract specifically designed for different contract strategies and types of project.
- The two sets of forms of contract most commonly used in the construction industry are NEC3 contracts and JCT contracts.

5

Rights and obligations

Aim

This chapter aims to explain the main rights and obligations of the employer and the contractor in the NEC3 Engineering and Construction Contract **(NEC3 ECC)**, developed by the Institution of Civil Engineers, and the Standard Building Contract 2005 **(JCT SBC05)** published by the Joint Contracts Tribunal. NEC3 ECC is used for almost all civil engineering contracts and JCT SBC05 is one of the most widely used in the building side of the industry.

Learning outcomes

On completion of this chapter you will be able to:

>> Describe the main rights and obligations of the employer.
>> Describe the main rights and obligations of the contractor.
>> Discuss the meaning of newer terms in modern contracts relating to cooperation.

This chapter *outlines* the main rights and obligations of the employer and contractor. Full details can be found in Chapter 17 on contract law, Chapters 21 to 26 on NEC3 contracts and Chapter 27 on JCT contracts.

Contract rights and obligations generally

Contracts generally are concerned with the rights and obligations of the **parties**. A party is simply a legal word to describe the people or organisations that make the contract. People enter into contracts for many reasons in their everyday lives, from the important such as buying a house, or perhaps a car, to the less significant of buying a kettle. In construction, the employer enters into a contract or a number of contracts in order to get a new or replacement asset designed and built, usually to meet a business need or opportunity.

There are many different forms of construction contracts. They can be arranged in different ways as part of a contract strategy, and payment can be made in various ways using payment mechanisms which will be described in Chapter 14. The employer however has *rights* to specified quality, to construction within the specified time, and to communication in prescribed ways for example. Similarly the employer has *duties* or *obligations* to provide a site (called access or possession), to pay as agreed, and to provide the contractor with necessary health and safety information about the site. These are the basic rights and obligations. Many more are added in standard forms of contract to facilitate construction (such as changes or variations), to provide further drawings and information, and to attempt to deal with the many eventualities that arise in a construction project such as unforeseen ground conditions and adverse weather.

There are many interconnecting contracts in a construction project because there may be a design contract, a construction contract, supply contracts for materials and so on. However, the three main parties to a construction project are:

- **The employer** who is the person or organisation that wants the project and pays for it.
- **The designer** who is the person or organisation that designs the project and produces the drawings and specification and other contract documents.
- **The contractor** (or builder) is the organisation that constructs the project.

In the traditional contract strategy the employer has separate contracts with the designer and the contractor, shown by solid lines in Figure 5.1. The broken line shows an administrative link since a member of the design organisation will often (but not always) be the contract administrator. Whilst the day-to-day dealings will be between the contract administrator and the contractor, the contract which is being administered is the one between the *employer and the contractor*.

Working in a cooperative manner

The obligation **to work in a cooperative manner** on parties to the contract and on other people involved is a relatively new addition to forms of construction contract and reflects some of the main intentions of partnering. Partnering is explained in Chapter 11. Working cooperatively is not the same as the obligation 'not to prevent', which is discussed below. Clauses describing cooperative behaviour are a laudable intention and may well improve the actions of the people involved in the contract to everyone's benefit. The difficulty may be that such ideas have not really been tested in court. However in the author's experience such phrases can actually be very useful in practice. They are frequently found in partnering agreements which are often not contractually binding, but supplement the binding contract terms. What these phrases do is to set down in writing a behavioural aspiration of both parties. What happens in a future difficult or disagreement situation is that one party may point out that the other party's behaviour is not in the spirit of the agreement by referring to these phrases. This will often have the desired effect of producing more cooperation and reducing conflict.

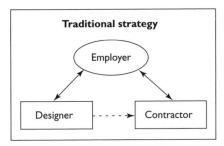

Figure 5.1 The traditional strategy.

- The NEC3 Engineering and Construction Contract (NEC3 ECC) begins with the famous phrase in clause 10.1 that says 'The employer, the contractor, the project manager [the NEC3 ECC word for contract administrator] and the supervisor shall act as stated in this contract and in a spirit of mutual trust and co-operation'.
- The 2009 revision 2 of the Joint Contracts Tribunal Standard Building Contract 2005 (JCT SBC05) now includes a similar phrase in Schedule 8 (supplemental provisions) which applies unless the contract expressly excludes it. Typical of the way that JCT SBC05 contracts are worded, the JCT SBC05 version is more thorough, yet more verbose. Clause 8.1, entitled 'collaborative working' says 'the parties shall work with each other and with other project team members in a co-operative and collaborative manner, in good faith and in a spirit of trust and respect. To that end each shall support collaborative behaviour and address behaviour that is not collaborative.'

This aim to oblige the contracting parties and their agents to cooperate is very commendable. Hopefully such aspirations are applied by the parties and should result in more harmonious and efficient contracts. Such phrases of course could be difficult to define in a court. Eggleston in his book *The NEC3 Engineering and Construction Contract: a Commentary* (Blackwell, 2006) suggests that the insertion of the word 'and' in NEC3 ECC clause 10.1 produces two distinct obligations; one a contractual obligation, and one outside the contract provisions. The JCT SBC05 (2009 revision) version of the phrase does not refer to contract obligations at all, but simply to cooperative working and supportive behaviour.

NEC3 ECC does however include other clauses relating to the cooperation of the contractor, such as clause 25.1 requiring the contractor to cooperate with others in the provision of information and the sharing of any working areas. It also requires all those attending a risk reduction meeting to cooperate. A risk reduction meeting (described in NEC3 ECC clause 16) is the result of an 'early warning' given by the contractor or project manager. Such early warnings are a very good concept and relate to matters that could result in an increase in cost or time or impair the performance of the works in use.

The employer – contract rights and obligations

Whilst most forms of contract use the word 'employer', in a few forms of contract the term *purchaser* is used instead. Most consulting engineers and architects and other design organisations call the employer their *client*. Indeed the Construction (Design and Management) Regulations 2007 (CDM2007) also refer to the employer as the client. It is worth noting that all of these roles are usually carried out by companies or partnerships, not individuals. The four most important things an employer must do are:

---

1. Decide what they want, when they want it, and how much they can afford to pay.
2. Provide a site, on time, with adequate and safe access.
3. Pay for the work in accordance with the contract.
4. Comply with the health and safety requirements of the CDM regulations.

It is an essential role of design organisations to help their clients (the employer) with items 1 and 4 above. This is because most employers are not construction professionals, and rely on the advice of their designer, architect, project manager or building surveyor to guide them through the design and construction process.

The importance of item 1 cannot be overemphasised, for it is item 1 that determines and shapes the drawings, specification and other contract documents that define the project. The construction project is the *answer to a need* of the employer. It often takes many weeks, and a lot of work, to truly understand that need, the likely cost and timescales. This is a very important part of the role of the design organisation or project manager.

If the employer's 'need' is not fully understood, documented, designed for and properly communicated in the contract documents, then the construction project *will fail* however well it is designed and built.

Employer's responsibility for health and safety

The employer has a number of important duties on projects falling within the Construction (Design and Management) Regulations 2007. These regulations apply to most projects and are covered in detail in Chapter 8. One of the employer's CDM duties is to provide pre-construction information relating to the site. This information will often refer to site investigation and ground surveys, and it is intended to help designers and contractors identify hazards and risks. The CDM regulations do not apply to certain projects such as domestic projects but, in the author's opinion, employers should follow the general principles of CDM wherever appropriate, since the regulations represent good health and safety practice.

The 2009 revision of JCT SBC05 (revision 2) now includes clause 2.1 in Schedule 8 (supplemental provisions) which encourages the parties to 'endeavour to establish and maintain a culture and working environment in which health and safety is of paramount concern to everybody involved in the project.'

Employer's obligation not to hinder or prevent

The employer's obligation to cooperate with the contractor has been covered in the section above where the word cooperate is used in the sense of collaborative team work. It is an established legal principle that a party should not hinder or prevent another party carrying out their contractual obligations, which is the other end of the scale from cooperation. Some general principles arising from legal judgments are:

- The employer must not *prevent* the contractor from carrying out work. Not only does the contractor have a duty to construct the works, but also a *right* to do so without hindrance.
- 'The employer will take all steps reasonably necessary to enable the contractor to discharge its obligations and to execute the works in a regular and orderly manner.' This is part of the judgment of the case of *Merton LBC* v. *Stanley Hugh Leach Ltd* (1985).
- However there is some doubt as to whether the employer has the obligation to provide information to a *much shorter* contractor's programme than was envisaged at tender. It has to be remembered that information is often supplied as a contractor proceeds on-site, that design organisations also have staff and other resource constraints, and that design and drawing take a finite time to complete. In the well-known 'Glenlion' case (*Glenlion Construction Ltd* v. *The Guinness Trust* (1987)) the judge decided that it was not open to one party (here the contractor) to unilaterally change the obligations of the other party *after* the contract is made.

Employer's obligation to provide the site

It is assumed in construction contracts that the employer *owns the site*, or has made arrangements for the contractor to have access. This access to site is called '**access**' in NEC3 ECC and it is called '**possession**' in contracts such as JCT and the ICE conditions of contract, seventh edition (2003), known as ICE7. There is a subtle difference, in that possession implies sole possession, whereas access in NEC3 ECC provides for other contractors to be on-site at the same time. ICE contracts were generally used in civil engineering before the advent of NEC contracts, which are now used almost exclusively.

Many construction projects such as offices, schools, hospitals, roads and bridges will be on land owned by the employer. However linear projects such as gas or water mains or sewers go beneath land owned by others. Here the employer negotiates and pays for a **way leave** from the landowner (often a farmer). This is in two parts, a working width of say 15 to 30 metres for the construction itself and a permanent way leave for future repair of the main or sewer. This permanent way leave is much narrower, perhaps five metres, and there will be legal restrictions on the landowner to prevent building anything on the way leave. The terms of the temporary way leave (or working width, or easement as it is often known) will provide for an access date for the contractor. The contractor must not enter the way leave before this date.

The date of access to the site, whether it be an area of land or a way leave, will be inserted in the contract documents, and if the employer does not provide access on this date they will normally be in breach of contract, for it is a very serious matter to prevent the contractor starting the work on-site on the date the contractor expected. The JCT Standard Building Contract (JCT SBC05) has an interesting provision, called '**deferment of possession**', which allows possession to be delayed by the employer for a period stated in the contract, but not exceeding six weeks. The employer is thus not in breach of contract, but the contract administrator must award an extension of time for completion, and the contractor will be entitled to any resulting costs.

On the date of access (possession), the site technically becomes the contractor's, and the contractor is responsible for fencing, guarding and preventing access to unauthorised persons. The contract should state who has authorised access to the site, such as the contract administrator. Strictly speaking, the employer does not usually have access to the site without the contractor's consent. So if any of the employer's staff such as operational staff have to be on-site their access times, areas and any restrictions must be clearly stated in the contract, so that demarcation and safety procedures can be agreed and they become authorised persons. This is a very important health and safety consideration and should always be considered carefully, before the contractor starts on-site. Similar arrangements, possibly involving gates and temporary traffic lights, will be necessary where landowners have to cross the working width on a pipeline or sewer.

Employer's obligation to appoint a contract administrator

Under most forms of contract the employer must appoint a **contract administrator** (called the project manager, engineer or architect/contract administrator depending on the form of contract). This contract administrator has powers under the contract and effectively acts in place of the employer. The construction contract of course is between the employer and the contractor, but most employers do not have the time, experience or expertise to administer a construction contract. This is the role of the contract administrator. A competent and experienced contract administrator is usually a major feature in a successful contract. In traditional contracts, the contract administrator will usually be a senior person from the design organisation. There are a few general obligations on the employer as a result of this appointment of a contract administrator, which are:

- To clearly notify the contractor of the contract administrator's name and any future change in contract administrator.
- The contract administrator's powers should have already been set out clearly in the contract.
- Having appointed a contract administrator, the employer is responsible for the contract administrator's actions, provided that the contract administrator is acting within delegated powers under the contract.
- The employer must not seek to unfairly influence any contract administrator (especially where the contract administrator is an employee).
- The employer should not attempt to by-pass the contract administrator, by dealing directly with the contractor.

The role of contract administrator does not apply in the same way to the design and build contract strategy. This is because the contractor designs *and* builds the works, and so there is no interface to manage between design and construction, at least as far as the employer is concerned. The relevant JCT contract is the Design and Build Contract (2005) referred to below as JCT DB05. Here the contract comprises the employer's requirements and the contractor's proposals. JCT DB05 is written naming 'the employer' in most clauses where the term 'contract administrator' would be used in JCT SBC05. However, In JCT DB05 the employer appoints an **'employer's agent'** in article 3. This employer's agent acts for the employer in terms of instructions, consents, notices, requests or statements as if it were the employer, except where the employer chooses to limit these powers in writing.

Employer's obligation to pay the contractor

All forms of contract contain payment terms, setting out when and how the employer will pay the contractor. Most contracts (except for those on very small or simple projects) provide for monthly payments to be made to the contractor. This assists the contractor's cash flow, and enables a lower initial tender price to be bid. The amount to be paid is usually determined by the contract administrator who completes a certificate in the prescribed format and sends a copy to the contractor and the employer. In JCT and ICE7 contracts this is called an interim certificate. NEC3 does not directly refer to a certificate but expects the contractor to be given details of how the payment is made up.

The contract itself should define how long the employer has to pay the contractor which is usually seven or fourteen days from the date of the certificate. If the employer is late in making payment then interest is usually paid at a rate stated in the contract. The employer is obliged to pay the amount certified, and not some lesser amount that the employer may consider more appropriate.

An amount is usually retained by the employer, in a separate account, for later payment back to the contractor. This amount is often between three and five per cent of the certified amount and is called a **'retention'** in JCT and ICE7 and in NEC3 ECC. In NEC3 ECC retention is not an automatic feature as it is in many of the traditional forms of contract, but has to be included as option X16. The purpose of retention is to give the employer a sum of money to use to pay another contractor in the unlikely event that defects are not corrected under the contract.

Employer's right to vary the contract (via the contract administrator)

Construction projects are usually very complex and require extensive foresight and coordination. It is difficult for the design organisation to think of every eventuality and provide for it in the drawings and other contract documents at the tender stage. Hence most contracts give the contract administrator the right to issue new drawings, instructions and other requirements. In JCT and ICE contracts these are called **'variations'**, but NEC3 ECC uses the word **'changes'** instead. So the right is really the employer's, but the variation has to be issued by the contract administrator under the terms of the contract. These terms may limit the power of the contract administrator to give variations, or the type of variations or even the total value of variations.

Contracts are trying to strike a balance here between the rights of the employer (through the contract administrator) to make reasonable changes, and the rights of the contractor to know what might be expected once construction starts. Furthermore most contracts provide for a way of valuing and paying for such variations. Sometimes this is through a contractor's quotation, but traditional contracts with a bill of quantities will usually use this as a basis for determining the cost of a variation. Variations normally give the contractor a right to more time to complete the project as well, which is only reasonable.

Employers and insurance

The principle of insurance is very simple although its application can be very complicated. You pay a premium to the insurer for the risk of a specified event occurring, or 'peril' as it may be called. If this specified event occurs then you are compensated by the insurer. The amount of compensation depends on the policy. It may be the value of the article at the time of the event, or it may be its replacement value.

As a simple example, you insure your house against fire for a premium of £100 per year, and thousands of other people insure their houses in a similar way. The likelihood of a house burning down is very small, but its consequence to the owner is enormous. Possibly one person suffers from a fire, and is compensated with the replacement value of the house. Thus many people pay a relatively small premium against the very unlikely serious event that could cost one of them hundreds of thousands of pounds. The insurer calculates the likelihood of the event, and a typical replacement cost, and sets the premium accordingly.

Some insurance is legally required of employers. In a construction project, the risks are typically injury or loss to the construction works, employees, the public and the completed project. Some insurance is taken out jointly by the employer and the contractor. Most contracts also insist that the contractor takes out certain insurances. This is intended to protect the employer, in case a risk event occurs and the contractor cannot pay, perhaps because they have insufficient funds or have gone into liquidation.

There are two main types of insurance relevant to the construction industry and they are liability insurance and loss insurance:

- **Liability insurance** – here the insurer **indemnifies** the insured person or company against damages and legal costs which become payable as a result of the specified event. 'Indemnify' means that the insurer will pay these costs for the insured person or company.
  An example of this type of policy is Third Party car insurance – you crash into another car and your insurance company pays out to the owner of the car you have hit, but doesn't cover damage to your car, so you are being covered for demands on you by another party.
- **Loss insurance** – here the insurer compensates the insured person or company for loss or damage which the insured person or company has suffered as a result of the specified event, whether the event happened accidentally or as a result of somebody else's negligence. Sometimes this type of policy also covers the insured for their own negligence.
  An example of this type of policy is Comprehensive car insurance – you crash into another car and the insurer covers your costs (often car repairs and medical costs, depending on the policy) as well, even though it was your fault.

Employers and regulations

In the UK, employers are bound by many laws and regulations. Lack of compliance can lead to fines, or even prison sentences. A few such regulations are:

- Health and safety legislation (especially CDM)
- Planning legislation
- Building control
- Highways legislation
- Environmental legislation and protected species
- The Competition Act
- Fire regulations
- Waste management and disposal regulations
- The EU procurement directives as transferred into English law.

The contractor – contract rights and obligations

The contractor will usually be a firm of builders or contractors who have submitted a tender price to the employer which the employer has accepted. It is normal for the resulting contract to incorporate one of the many standard forms of construction contract such as a JCT form or NEC3 ECC. These standard forms list many obligations for the contractor and have much similarity. Again as explained in Chapter 4, the courts will, if necessary, *imply* terms if express terms of importance are missing from the contract. If standard forms of contract are used properly, this is unlikely to be necessary.

Contractor's main rights

Curiously, most forms of contract are relatively silent on the contractor's *rights*, possibly because they are largely the opposite of the employer's *obligations*. Hence by deduction the main rights of the contractor are to:

- Be given access (called possession in JCT) of the site at the access date, or access to those parts of the site that are prescribed in the contract.
- Be paid in accordance with the contract at the times and in the way stipulated.
- Proceed with the construction works without interference or prevention.
- Receive reasonably prompt replies to any letters, emails or other queries made to the contract administrator. ICE7 and JCT are silent on this matter, although a term of a reasonable time would probably be implied by the courts if necessary. However, in NEC3 ECC a 'period for reply' is inserted in the contract data. This is stated in 'weeks' and two weeks is suggested as usually being suitable in the NEC3 ECC Guidance Notes.
- Receive at a suitable time to suit the progress of the works such further instructions and drawings that are reasonably necessary.
- Receive any goods or materials specified as being supplied by the employer at the times stated in the contract.

It might be added that the contractor does not have the 'right' to make a profit, as some contractors may claim. It is to be hoped that the tender prices are such that the contractor does make a profit by carrying out the work efficiently. However contractors do have loss-making contracts. The contractor does of course have a right to be paid properly and promptly in accordance with the contract.

Contractor's main obligations

Under most standard forms of contract, the **contractor's main obligations** are to:

- Comply with the contract.
- Construct and complete the works in accordance with the contract; this of course includes the specification and all the drawings.
- Ensure that any sub-contractor also complies with the contract.
- Provide competent supervision and labour. In many forms of contract, key members of staff and sub-contractors have to be 'approved' by the contract administrator.
- Complete the contract works to the standard specified, by the completion date (or any extension to that date that the contract administrator may grant).
- Obey the instructions of the contract administrator provided the contract administrator is acting within its powers under the contract.
- Comply in full with all health and safety legislation, particularly the CDM regulations.
- Comply with local authority bylaws, statutory undertakers' regulations and other acts and regulations relating to the control of building works.
- Comply with all regulations relating to the disposal of waste and the production of site waste management plans.
- Submit a programme to the contract administrator, and update it as necessary or when so instructed. This is required under most forms of contract and is a very sensible action in any case.
- Provide insurance as specified (see section below).
- Return to correct any defects in the works after they are taken over by the employer at completion (see section below).

Further obligations of the contractor under many forms of contract are to:

- Work in a spirit of mutual cooperation.
- Submit regular applications for payment, or 'attend' while measurement takes place.
- Proceed with 'due expedition' (ICE7 clause 41), or 'regularly and diligently proceed' (JCT SBC05 clause 2.4). NEC3 ECC has no similar terms, possibly on the basis that it is in the contractor's interests to do this anyway.
- Give notice of any ambiguities or inconsistencies in the contract documents to the contract administrator, so that they can be investigated and if necessary corrected. Under most forms of contract, the contractor will be reimbursed for any loss that subsequent correction causes.
- Pay any fees, charges, rates or taxes in respect of the work if this is specified in the contract. The contractor is usually reimbursed for this in the monthly payments.

- Protect and preserve fossils, antiquities and other objects of value that may be found on the site.
- Design all temporary works (equipment needed in the construction process but removed on completion).
- Design any permanent work (goods and materials that will form part of the finished project) *if so directed in the contract*. This is a complex issue and is explored in Chapter 10.

Contractor's obligation for health and safety

This is a very important obligation. Construction is a dangerous activity and causes many unnecessary deaths and injuries every year. Whilst much can be done at the design stage, the contractor has the responsibility for the work on-site where most accidents occur. The contractor has extensive duties under the CDM regulations, particularly the preparation of a construction phase plan, risk assessments and method statements. These duties are covered in Chapter 8.

Additionally, many forms of contract have specific clauses relating to health and safety such as ICE7 clause 8(3). JCT SBC05 clause 3.23 largely reiterates the main obligations under CDM. However, Schedule 8 of the 2009 edition of JCT SBC05, referred to above, goes on to put four further requirements on the contractor in clause 2.2 which are summarised below:

- Comply with any approved codes of practice (ACOP) issued by the HSE.
- Implement health and safety site inductions for all contractor's personnel and members of their supply chain.
- Provide health and safety advice.
- Ensure health and safety consultation.

This is quite interesting in that whilst the CDM regulations are legally binding, the ACOP are not. The ACOP do of course represent good practice and any contractor not following them may have to explain why to a court. So here, JCT is making adherence to the ACOP a contractual requirement in the contract between the employer and the contractor.

NEC3 ECC says in clause 27.4 that the contractor acts in accordance with health and safety information stated in the works information which of course is information provided by the employer. Such information need not include requirements already set down in the CDM regulations, but may include safety information and procedures relating to an occupied site such as a factory, specific site information relating to safety or perhaps demarcation requirements.

Contractor's obligation for security and control of the site

Contractors will ensure the safety and security of the site. This is for three main reasons:

- The contractor is usually contractually responsible for any loss or damage to materials on-site or the partly completed construction works.
- The contractor could be liable for any loss or injury to members of the public, and anyone else who entered the site.

- The contractor will wish to act in a professional manner and maintain its reputation.

So today we see high fences around construction sites, and frequently gatekeepers and access procedures and often security guards or similar arrangements.

The contractor is also responsible for 'control' of the site. Under CDM regulation 22 the principal contractor shall draw up 'site rules' and prevent access to site by unauthorised persons. The principal contractor (PC) also has general duties to plan, manage and coordinate the construction in discussion with any other contractors involved. These duties make it very important for the principal contractor to have effective planning and control arrangements in place for all site activities. Whilst the principal contractor is not responsible for detailed supervision of the work of other contractors, the PC must know what they intend to do, where, when and how, in order to ensure the overall safety and control of the site.

A particularly difficult and important area of control for the contractor is that of materials suppliers who may make infrequent deliveries to the site, and statutory undertakers (or 'stats' such as representatives of gas, electricity or water companies) working on the site. Materials suppliers will need specific site induction and careful direction to their place of offloading, together with possible attendance and lifting arrangements. Statutory undertakers frequently need access to areas of the site to remove, move or install cables, pipes and other equipment. Again they will need specific site induction. Additionally the contractor will need to ensure that the contractor's work or workforce does not impact on the statutory undertakers' work, and that the work of the stat does not place the contractor's workforce in any danger.

Contractor's obligation for workmanship and materials

Many construction problems relate to either design, workmanship, unsuitable materials or all three.

In forms of contract based on traditional procurement, the design organisation is responsible for *design and specification of materials*, and the *contractor for workmanship and using compliant materials*.

The primary role of the contractor of course is to construct the 'works'. The contractor does this by buying materials and other goods and by supplying labour, plant and equipment. It is important to distinguish between the 'permanent works', that is goods and materials that will form part of the finished project, and 'temporary works'. Temporary works are items such as trench supports, any other temporary supports, settling ponds and lagoons, groundwater pumping equipment, formwork, scaffolding, tower cranes and so on. Provided that they are safe and do not damage or have an adverse effect on the permanent works, the employer and the contract administrator usually have little *direct* interest in the quality or design of temporary works. This is the contractor's responsibility.

The contract between the employer and the contractor is all about the quality of the *permanent works*. The quality of materials and any particular workmanship requirements are usually set out in the specification and any related standards such as British Standards or European Standards. In NEC3 ECC contracts the specification will form part of the works information and is separate from the pricing mechanism. This is the usual civil engineering practice. In 'building' contracts such as JCT SBC05, the specification information is usually combined with the measurement items in the contract bills. Hence JCT SBC05 refers to materials in clause 2.3.1 and workmanship in clause 2.3.2 as being 'as described in the contract bills.' NEC3 ECC simply says 'the contractor provides the works in accordance with the works information.'

The contractor is usually required to construct the permanent works to the drawings which give levels, sizes and dimensional information for setting out the works. Under most forms of contract the contractor is responsible for the proper setting out of the works and any checks made by the contract administrator do not remove this responsibility (ICE7 clause 17 and JCT SBC05 clause 2.10). NEC3 ECC is silent on this matter, possibly because it is now unusual for the employer to pay for extensive site supervision, reporting to the contract administrator.

It was once normal for the contract administrator to have engineering staff who would check the contractor's dimensions and setting out before concrete was poured. Whilst incorrect setting out was still the contractor's responsibility, unfortunate errors such as incorrect setting out on the first floor of a tower block could be identified by the site supervisor before the second and third floors were constructed. The absence of site supervision on more modern projects may mean that such mistakes are identified much too late, when little can realistically be done about mistakes and a compromise may have to be reached that is not always in the employer's interests.

In a traditional contract, the works are designed by the design organisation and all materials should be clearly specified. Where insufficient information is given for the contractor to order materials, for example where the specification gives no statement as to the strength of bricks, then the contractor should ask for further directions from the contract administrator. If the contractor proceeds to make assumptions particularly about strengths or material properties, and orders materials on this basis, then the contractor may inadvertently take on design responsibility for which it may neither be competent nor insured. JCT SBC05 in clause 2.3.3 says that where materials are not specified or subject to the Architect/Contract Administrator's opinion then they shall be of a standard 'appropriate to the works.' It is suggested that the contractor may take the risk of deciding what is appropriate for such items as doors or skirting boards, but never for items that require design knowledge, such as the brick strength mentioned above. In a tall building for example, the brick strength is important and should have been determined by a structural engineer.

Conversely, the contract administrator should not respond to requests from the contractor as to *methods of work*, unless these have been specified by the designer. This is occasionally necessary where the design assumes a sequence or method of construction. Under most contracts, materials are specified as are any final quality or other tests. How the contractor achieves this is the contractor's own obligation, as the contractor should be the one with this expertise.

In a design and build contract by contrast the contractor *designs and builds* the works to meet whatever performance or final quality criteria were specified. This simplifies the situation from the viewpoint of the employer, because only one organisation is responsible: the contractor. However, if clear quality requirements and a good performance specification are not provided to the contractor, then the employer may well get much less than they might expect.

Contractor's obligation for defects

In many ways this obligation follows on from the contractor's responsibility for workmanship and materials. A defect is not defined in ICE7 or JCT SBC05 but it is defined in NEC3 ECC. Clause 11.2(5) of NEC3 ECC defines a defect as 'a part of the works which is not in accordance with the works information.' There is a difficulty here, in that in normal language, a 'defect' is something unsatisfactory or deficient. In NEC3 ECC, work can be unsatisfactory, but *not* be defined as a defect under the contract if it is in accordance with the works information. This particularly applies to work that has been incorrectly or poorly specified by the design organisation. On reflection, readers will see that this is fair. Contractor's should only be *obliged* to provide what they are contracted to provide; nothing more and nothing less. Contractors may be *willing* to provide more, often to please a valued client, but they are not contractually obliged to do so.

In most forms of contract there is a period of time allowed for correcting defects, usually twelve months in ICE7 and NEC3 ECC contracts, and a default of six months in JCT SBC05, which can of course be increased when the tender documents are prepared. This period is called the rectification period in JCT SBC05. NEC3 ECC uses a different method by stating a defects date as being a number of weeks after completion of the works. The effect is the same. It is a little like the twelve month warranty period on a new washing machine. The contractor's obligation is to return to site to correct all defects within this period as directed by the contract administrator in JCT SBC05; or whether or not so directed in NEC3 ECC. If the contractor fails to return and due notice is given, there are sanctions which vary according to the contract but they usually allow the employer to deduct money or bring in an alternative contractor to correct the defect.

Contractor's obligation for insurance

The contractor has an obligation for insurance, just as the employer has. Some insurances are legally required of all companies such as employer's liability insurance, and this requirement is extended in construction contracts to include the valuable work and materials which constitute the contract works. The usual insurance situation is abstracted from NEC3 ECC clause 84 and is as follows:

'The contractor insures in the joint names of the contractor and the employer for events which are the contractor's risk (see below for exclusions) from the starting date until the defects correction certificate is issued (the end of the period for correcting defects)

- loss of or damage to the works, plant or materials
- loss of or damage to equipment (temporary works)
- liability for loss of or damage to property
- liability for bodily injury to or death of a person (not an employee of the contractor) caused by activity in connection with this contract
- liability for death or bodily injury to employees of the contractor arising out of and in the course of their employment in connection with this contract'

Exclusions to the contractor's risk usually include:

- The unavoidable results of constructing the works
- Negligence, breach of statutory duty or interference with any legal right by the employer
- A fault in the design of the design organisation
- War, rebellion, riot and similar events.

JCT SBC05 has a similar list of requirements and exclusions, but they are not written in such simple language. JCT SBC05 also has three choices of insurance options relating to different types of buildings such as new build or alterations and a number of other provisions to suit the more complex nature of building work.

Chapter summary

- Contracts generally are concerned with the rights and obligations of the parties.
- Parties are the people or more usually the organisations that make the contract.
- Increasingly, contracts stipulate that the parties work in a cooperative manner, where 'cooperative' is used in a behavioural and team work sense. NEC3 ECC began this trend, and the JCT suite of contracts is being amended to include it.
- The main obligations of the employer are to provide access to the site, to pay the contractor in the manner and at the times stipulated in the contract, and to provide specified insurances.
- There are also legal requirements for the employer not to hinder or prevent the contractor from carrying out the works.
- Under most contracts, the employer must appoint a contract administrator, and this contract administrator usually has powers under the contract to vary (change) the contract works, within prescribed limits.
- Both the employer and contractor have a crucial responsibility for health and safety. Their roles and obligations are clearly set out in the CDM regulations (2007).
- Additionally the contractor has a responsibility for the security and control of the site, which largely impacts on safety considerations.
- The contractor's main obligation is to construct the contract works using workmanship and materials to the contract requirements.

- The contractor is required to supply and update a programme under most forms of contract.
- The contractor usually has a responsibility to correct defects in the constructed works usually for a period of time after completion. This period is usually 12 months.
- The contractor is obliged to take out insurance to cover a number of risks, mainly to the contract works and materials but also to persons and property.

6

Tendering and contractor selection

Aim

This chapter explains the process of tendering and contractor selection. This is the formation of a legal contract.

Learning outcomes

On completion of this chapter you will be able to:

>> Describe the process of tendering and contractor selection.
>> Explain the main tendering options.
>> Assess which option would be most appropriate in different circumstances.

IMPORTANT

This chapter describes 'traditional tendering'. In traditional tendering, an employer can negotiate or tender in any way that the employer chooses provided it is within the law. The law gives few restrictions on tendering options and processes provided there is no fraud or misrepresentation.

However, a very different situation exists where an employer is subject to the Public Contract Regulations or the Utilities Contracts Regulations (called **'the regulations'** from now on). These regulations set out in great detail rules and procedures which apply to tendering and contractor selection for many employers, and which must be followed. Because they are so important, the regulations are described in detail in Chapter 13, the EU Procurement Directives. Employers who break the regulations can be subject to massive fines.

Tendering

What is tendering?

As we saw in Chapter 4, a legally binding contract requires offer, acceptance and consideration. **Tendering** is the process by which bids are invited from suitable companies, based on the employer's **tender documents** in order to establish that legal contract. There are a number of different **tendering options** which are described later in this chapter. The tender process is shown in Figure 6.1.

> Tendering is the industry name for obtaining an offer from a contractor, which can then be accepted by the employer to form a binding contract.

The period of time that a tenderer has to decide how much to bid and submit a tender is called the **tender period**. Upon receipt of the tenders, the employer, or their advisors, must assess the tenders.

Employers may produce their own contract documents and assess tenders themselves, but this is very unlikely. The employer usually employs a design organisation such as a firm of consulting engineers, architects or building surveyors to prepare all the tender documents and then to advise which tender to accept. This advice is usually incorporated in a tender report, which is written after tenders have been assessed. The employer considers the tender report and then signs a contract with the successful tenderer who then becomes 'the contractor' under the contract. This contract between the contractor and the employer is the construction contract, which can be enforced in English law. The employer and contractor are the 'parties' to that contract. At this stage the tender documents become the contract documents. This version of the contract documents is very important and was traditionally kept as a hard copy in safe storage. It is very important because it constitutes the *terms* of the contract. The storage period depends on the employer's procedures but may be either six or twelve years depending on the type of contract.

> Once the contract is awarded, the tender documents become the basis of the contract. They become the *contract documents*.

Under most forms of contract, additions and changes can then be made and so the contract documents are constantly changing as construction work proceeds. On a large contract, hundreds and possibly thousands of changes, additions and clarifications will be made. Under most forms of contract, the contractor is paid to make these changes to the construction work, and is given more time to do so. However, the *basis* for a change in time allowed or cost is what was said in those *original contract documents*. That is why it so important that they are labelled and stored carefully.

So in simple terms, the employer wishes to find the 'right contractor at the right price' and the contractor wishes to 'win the contract' but on the right terms and at the right price for him or her. From both perspectives, the right contractor

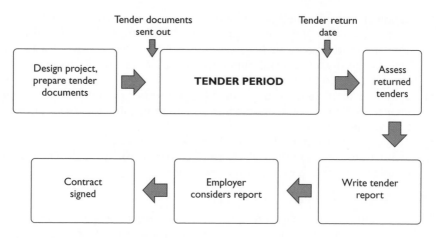

Figure 6.1   The tender process.

is largely dependent on the tendering option and process used, and the right price is largely dependent on the quality and comprehensiveness of the tender documents, and the use of a form of contract that distributes risk in a fair and transparent way. Some very good advice on tendering is given in Practice Note 6, published by the Joint Contracts Tribunal.

Types of tender

There are two types of offer resulting from two different types of tender option:

- A **single offer (bid)** is a tender for one project. This is the normal tender situation, where a contractor offers to carry out a particular contract, on particular terms, for a price.
- A **framework agreement** (sometimes called a standing offer or serial contract) is a tender to construct a number of projects over a period of time, as described below.

The tender process for single-bid tendering in detail

Going out for tender

1. The employer decides on a **tendering option** in order to attract several tenderers. This process is described in detail below.
2. This decision creates a list of tenderers.
3. The employer determines a suitable **tender period**.
4. Before sending **tender documents** to these tenderers it is advisable for the employer to reconfirm the tenderers' interest in submitting a tender.
5. The employer then sends 'invitation to tender' letters to the tenderers, accompanied by two sets of tender documents (these tender documents will become the contract documents when the contract is signed), and usually a tender return envelope.
6. Tenderers visit site, and make tender enquiries of suppliers and sub-contractors.

7. Tenderers consider their potential programme and available resources.
8. Tenderers make any further enquiries or clarifications from the employer.
9. Tenderers should raise such enquiries in writing. The employer (or usually their design organisation) should respond with a copy to all tenderers. This ensures good communication and fair dealing.
10. Tenderers then complete any schedules or information in the tender documents, price up any bills, or schedules, often using quotations from sub-contractors and suppliers.
11. Tenderers must submit their tenders by the due date, at the place and in the manner described in the instructions to tenderers. Many employers will not open a tender that is even five minutes late.

After the employer receives the tenders

12. The employer (usually through the design organisation) evaluates the tenders, particularly in respect of prices and pricing structure.
13. Any qualifications by the tenderer are reviewed – qualifications should generally not occur if queries are raised before the tender is submitted.
14. Withdrawal or rejection of tenders may result because of the inclusion of unacceptable qualifications or if major errors are identified in the tender.
15. Finalisation of tender – this stage may allow minor errors to be corrected. Many employers have rules regarding errors. All corrections must be recorded in writing.
16. Occasionally a negotiation process takes place if savings have to be made because the employer's budget is exceeded. This cannot be done with tenders under the EU Directives.
17. A tender report is usually produced by the design organisation.
18. The employer selects one of the tenders on the basis of the tender report.
19. The employer writes to the successful tenderer accepting the tender (the offer). Usually the acceptance is 'subject to contract'.
20. Signature of the formal contract (and possibly signing it as a 'deed', which is a legal device to give a longer period for breach of contract claims) confirms the agreement, and a binding contract is formed between the tenderer and the employer.
21. The successful tenderer is now called the 'contractor' and the tender documents become the contract documents.
22. Once the contract is awarded every unsuccessful tenderer should be notified. This notification should be made as soon as possible, to allow tenderers to reallocate resources to other potential work.
23. Every unsuccessful tenderer should also be issued with a list of tender prices, and the names of other tenderers, so that they can assess their commercial position in the market for future tendering opportunities. For example, if a contractor's tender was twenty per cent more expensive than the average tender, then they have misunderstood the tender, or have a serious problem with competitiveness.

The tender process is very important to employer and contractor alike. It is the way 'work is won', and the way contractors stay in business.

Good communication, based on clear and comprehensive tender documents, is vital to ensure a successful project.

Tender documents

The employer's tender documents are usually prepared by the design organisation or 'designer' for short. The designer can be a firm of consulting engineers, architects or building surveyors. However, some employers, particularly the larger ones, employ construction professionals as part of their own staff where they are known as 'in-house' staff. These in-house professionals will prepare the tender documents themselves, or supervise their production by outside organisations such as consulting engineers or firms of architects or building surveyors.

The tender documents describe what is to be built and how the work will be paid for. Tender documents are usually very detailed and many months or even years will be spent on producing them.

The tender documents should include all information that the tenderer needs to understand the project and price it up. The tender documents usually include each of the items in the box below, which are then explained in turn beneath it.

The tender documents
1. Invitation to tender and instructions to tenderers
2. Any articles of agreement
3. Conditions of contract and contract data
4. The pricing mechanism
5. The payment mechanism (which may be the same as the pricing mechanism)
6. Works information including the specification, tender drawings and health and safety information
7. Site information

1. The invitation to tender and instructions to tenderers will normally list the main documents, and will give instructions of what tenderers must return, to what place and when. Information should also be provided to tenderers on the criteria for tender assessment, how queries will be dealt with and whether alternative proposals are acceptable. Tenderers should also be given details of how to visit the site and any health and safety matters connected with such a visit.
2. Articles of agreement set out in brief terms the contract agreement, for later signature by the employer and successful tenderer.

3. The contract data will include all important information about starting and completion dates, names of people such as the project manager (architect/contract administrator in JCT contracts), the conditions of contract and any amendments made to them by the employer. In NEC3 ECC contracts it will include detail of all the options selected.
4. The pricing mechanism may be schedules of rates, bills of quantities (an itemised list of estimated costs – see Chapters 14 and 15) or directions as to the completion of activity schedules by the contractor. The contractor will complete this pricing mechanism in the returned tender with rates and prices.
5. The payment mechanism is usually specified in the conditions of contract, and is often the same as the pricing mechanism. These terms are explained in Chapter 14.
6. The works information will include the specification, the drawings, all other details, constraints or directions relating to the works, and the pre-construction health and safety information to comply with the Construction (Design and Management) Regulations 2007, known as the 'CDM regulations'.
7. The site information will include the location of underground services and details of land ownership, accesses and way leaves. It will also include the ground investigation report. This usually comprises borehole reports and laboratory and other tests on ground parameters. The *interpretation* of this ground investigation report is up to the contractor. The next chapter focuses on Site Investigation.

Ground investigation reports

It is normal for the employer to arrange for a comprehensive ground investigation at an early stage of the design process as part of the overall site investigation. The site investigation process is very important and is covered in detail in the next chapter. Ground investigation results are necessary to produce a safe and efficient design as well as to determine appropriate construction methods.

Ground investigation reports typically have two main parts: a factual part and an interpretive part. The factual part contains sections through boreholes, groundwater measurements and details of all *in situ* and laboratory tests. The interpretive part will contain calculated design parameters such as bearing capacity (based on assumed formulae and factors of safety). Importantly, it will frequently advise on suitable construction methods and techniques and on methods of dewatering and excavation support. This construction advice can be useful to the designer, but it should normally *not be included* in the tender documents. The contractor is employed for expertise in construction and so the contractor should determine the most appropriate methods and techniques.

The danger is that if *interpretive* information is included with the tender documents it usually forms part of the contract (unless the contract clearly says that it does not). If such interpretive information later proves to be incorrect or unsuitable, then the contractor will have a claim that they were misled and that the prices and the assumptions about activity durations or construction methods are therefore inappropriate. Clearly, any information relating to health and safety must be considered properly by the designer, and must be included in the pre-construction information provided to the tenderers.

Tendering options

It is important to distinguish between *'tendering options'* which are ways of selecting contractors for different *types of projects*, and the *'tendering process'* which describes the administrative *mechanics* of sending out tender documents, receiving tenders, evaluating them and appointing a contractor in a fair, robust and legal way. There are six main tendering options, described below.

> *Tendering options* are different ways of selecting the best contractor. The *tendering process* is the administrative mechanism of inviting and assessing tenders.

The six main tendering options

The six main tendering options are all restricted and controlled by the regulations, for employers affected by those regulations.

There are six main tendering options for obtaining tenders. They can be further divided into single-bid options, and two-bid options:

> **SINGLE-BID Options**
> Open tendering
> Single-bid selective tendering
> Pre-qualification
> Negotiated
>
> **TWO-BID Options**
> Two-stage selective tendering
> Framework agreements

SINGLE-BID options

1 Open tendering – in this option, the employer places an advertisement in a newspaper or journal for any interested party to bid. This option was once used frequently on the basis that more tenderers should result in the lowest price because of the increased number of offers being made. It arguably led to some of the historical problems in the construction industry, which were highlighted in the Latham and Egan reports. This option is unusual today.

- It is unpredictable.
- Many tenderers incur the expense of tendering where only one will be successful. This simply increases contractors' overheads, which is an expense that will be passed on to future employers.
- Tenderers may not have the expertise or resources required to carry out the contract in question.
- The employer has numerous tenders to consider, possibly dozens.

- A new 'team' of employer, contractor and designer must usually be formed, with all the associated relationship and communication issues.
- The lowest bidder has often misunderstood the contract owing to the poor communication inherent in this method, and the relatively short timescale for tendering.
- It is completely unsuitable for 'design and build' contracts, where tender documentation is 'light' (because here the contractor carries out the design as well as construction and the design has not been completed at this stage). Hence misunderstandings over what is required are almost certain.

2 Single-bid selective tendering – here the employer compiles lists of 'suitable contractors', often on the basis of the nature of the contract work and its geographical location. Such a suitable list might be for laying water mains in Gloucestershire, or road maintenance work in Derbyshire. Larger employers who use this method may have a number of different lists perhaps split by county and work type. Each list could have ten or even twenty or more potential tenderers on it.

For any particular project, a number of tenderers, usually six, are selected from the relevant list to form a specific short list for that particular tender. A tender list of six tenderers may often contain the previous successful tenderer for that particular list plus five others in rotation from the list. Tenderers on the specific tender list then submit their tenders or bids as they are often known.

- Whilst better than open tendering, this is still a fairly haphazard way of selecting the best contractor, with the right skills and resources.
- If the overall lists are long, new teams are frequently formed which does not help understanding or communication.
- However misunderstandings are less likely than in open tendering, particularly if the overall lists are not too long.
- There is some incentive for tenderers to gain 'repeat order work', if the successful tenderer in any category knows it will be allowed to bid for the next project in that category.

3 Pre-qualification – in this method, contractors fill in a pre-qualification questionnaire as a basis for selection onto a tender list. This allows targeted selection, if the questionnaire is constructed carefully to reflect all the important requirements for a specific project. It can therefore be a good selection system if the questionnaire is well designed. A pre-qualification questionnaire would include such items as:

- Health and safety record and procedures
- Previous experience of similar work
- Referees
- Resources, both plant and people
- Financial matters such as turnover
- Location of head office and other office bases.

Once selected in this way, normally four to six contractors will be asked to tender.

- If the questionnaire was well constructed, only appropriate contractors will be asked to tender. This is more likely to result in the 'right contractor at the right price' than option 1 or option 2.
- However there is no incentive for repeat order work as in option 2 above.

A modern variant of option 3 is the **electronic auction**, where a tender list is formed as above. The difference lies in the bid process. Tenderers submit their bids on-line during a fixed period of time, and where they can see the other bids. The idea is that tenderers can compete and reduce their initial price to undercut other bidders. Hence the method is sometimes called an 'electronic reverse auction'. Exactly what tenderers see is determined by the way the process is set up by the employer. Tenderers would usually see the current lowest bid (if selection is on price alone) or the most advantageous bid (to the employer) where price is not the only criterion. Tenderers can usually submit as many bids as they wish and, at the set closing time, the lowest tenderer becomes the appointed contractor. The process has attracted much criticism largely because the 'successful' tenderer has probably gone below its lowest fair price in order to win the contract. If this is the case it can lead to a difficult construction contract, with the contractor looking for every opportunity to make a claim for losses under the terms of the contract. This is likely to lead to deteriorating relationships and is the very opposite of what much modern procurement seeks to achieve.

4 Negotiated – here the employer negotiates the price, usually with a known and trusted contractor. This method is not usually possible with contracts falling under the regulations and, where it is allowed, it is subject to administrative limitations. Where the regulations do not apply, this method is often used with 'strategic partnering', where an employer and contractor build up a long-term relationship based on cooperation, trust and repeat order work. Contractors will often take a loss on one particular contract in the expectation of a reasonable profit on future contracts. This requires an acceptance on both sides that if the employer gains from a contractor's loss on one project, then the contractor will need to recover this loss on future projects.

The big drawback with negotiated contracts is that after a time the employer never really knows what the 'market price' is. Many employers take an occasional competitive bid on a series of negotiated contracts to 'test the market'.

It is a very good method for building long-term relationships, mutual understanding and good communication.

Contractors appointed regularly under this method build up a thorough knowledge of what the employer really wants, their systems and their people.

TWO-BID options

5 Two-stage selective tendering. In *stage one* the tenderers bid on a competitive basis on a 'basic version' of the project since it has not been designed at this stage. This could be a similar project, a scenario or a schedule of work items. The successful tenderer becomes the 'preferred contractor'. This method is often used to involve the contractor at an early stage, thereby reducing one of the main drawbacks of 'traditional contracts' (that the design is usually completed before the contractor is appointed). Hence in two-stage tendering the contractor can make an input to the design by advising on likely construction methods and 'buildability', can start to form team relationships, and is unlikely to misunderstand the nature of the project when the final version of the project is priced up. Contractor misunderstandings usually just lead to contract claims of various kinds, as the contractor attempts to recover any lost money.

In *stage two* when the design is complete the contractor submits a price for constructing the complete design, and there is usually an element of justification and negotiation. There are few drawbacks to two-stage selective tendering and many advantages.

Drawbacks
- The main drawback for the employer occurs if the contractor submits an unacceptable bid at stage two. At this stage there is probably insufficient time to go through the whole tender process again with alternative tenderers. This can give the employer little choice but to accept what may seem to be a high price.
- The second drawback is the additional time and cost to both the employer and contractor of two tender stages instead of one.

Advantages
- The main advantage is early contractor involvement and team formation.
- The contractor can advise on programme and phasing.
- Misunderstandings are minimised.
- Extra investigations (often site investigations of various kinds) can be made at the contractor's request.
- Meaningful discussions can take place with suppliers during the detailed design period, which is often many months.
- It is possible to order 'long delivery' materials (the contractor may want some financial protection here, in case the contract is cancelled before stage two).
- The method is particularly appropriate *where contractor input to the design is essential,* such as complex projects, those with specialist plant and equipment of which the contractor has knowledge, those requiring extensive coordination of specialist suppliers or projects where the methods of construction have a significant influence on the design.

6 **Framework agreements** – this form of tendering is usually used by employers who must comply with the regulations because it is relatively low risk and efficient for employers subject to the regulations. However, frameworks could be considered by any employer with an ongoing need for a similar type of project. It is important to note that framework agreements are also a contract strategy in themselves. They satisfy both definitions for they are initially a selection method, and, once selected, the contractor is subject to a contract strategy of call-off orders based on the overall framework.

> Framework agreements are both a tendering option and a contract strategy.

When considering frameworks for tendering purposes, we see that they have all the advantages of two-stage selective tendering, and also minimise the main drawback. Hence if a contractor consistently 'takes commercial advantage' or produces work of a poor quality, the framework can usually be cancelled and rebid (most have an annual cancellation clause). Equally, under long-term frameworks, many employers become clearer in their programmes, requirements and processes, which benefit the framework contractor. Frameworks are a marvellous way of concentrating everyone's minds on being sensible.

When setting up frameworks, there are many advantages to the employer in minimising the number of framework contractors, such as economies of scale and simplification of procedures. However, many employers will have at least two framework contractors for any type of work, so that workload can be adjusted to reflect contractor resources and performance. This is the other way in which employers can minimise any problems of 'overpricing' by framework contractors.

Frameworks are often used with partnering agreements (either formal or informal), which is a further advantage.

- Frameworks are a very good way of satisfying the regulations in an efficient manner.
- Frameworks have all the other advantages of two-stage selective tendering as set out above.
- Frameworks facilitate long-term relationships, and drive real efficiency, provided there is a robust price-setting mechanism that can be audited.
- Frameworks are expensive to set up both for the employer and the contractor, but once established they remove the costs of tendering (although there will still be some costs associated with producing prices for each individual order).
- They provide a much more consistent and visible workflow for contractors, which helps them reduce their prices, so the employer benefits as well.
- Setting fair prices for each contract is the main difficulty inherent in frameworks. How do you set and agree fair prices, in the absence of competition over a period of many years, when market conditions are bound to fluctuate? It is rather like taking out a fixed rate mortgage for five years. However, good ways of satisfying an employer that they are getting a fair price for the work is adjusting workload as described above, and occasionally taking a 'check price' from another framework contractor, or even one outside the framework.

Chapter 12 covers Frameworks in detail.

Appointing designers and other specific professions

Tendering generally concentrates on appointing a suitable contractor, because this is the major area of investment in a project. However appointing a suitable design organisation or professional advisors is also very important. It is worth noting that a slightly different form of tendering is often used for professional services such as engineering consultancy, building surveying, quantity surveying or architectural services. Again, the main determinants are price (fees) and quality (the complexity and extent of professional service provided).

The 'tender documents' here will describe the professional services required, and the type and scope of the commission, together with any project requirements. It is very important to describe clearly all the services required, especially if it is not simply a commission for design or appraisal work. Obtaining land, giving legal advice, preparing documents for planning approval and so forth are usually listed as additional services in these contracts for professional services.

Tendering for professional services is often called 'inviting fee bids', or inviting 'fee proposals'. The word 'commissioned' is often used to describe the appointment of the provider of these professional services. The main families of contract, published by the Joint Contracts Tribunal (JCT) and the Institution of Civil Engineers (the New Engineering Contract), both have specific forms of contract for the appointment of such professional designers and advisors.

The remainder of this chapter will concentrate on tenders for contractors.

The tender period

The tender period is the time period from when the tender is sent out by the employer, until the closing date (and time) stated in the Instructions to Tenderers. It is usually four to eight weeks (for projects in the tens of millions of pounds even up to twelve weeks), depending on the size and complexity of the project. The most important time and date for a tenderer is the *tender return date*. Most employers have a formal tender registration and opening procedure involving staff from various departments. These arrangements are usually made to open tenders shortly after the due date and time.

Most employers will reject a tender that is late. This procedure was endorsed where rejection of a tender that was 26 minutes late was upheld by the High Court. The case was *J B Leadbitter and Co* v. *Devon County Council* (2009) and related to a four-year framework agreement falling under the Public Contracts Regulations 2006. Devon County Council had been very clear with tenderers on a number of occasions that twelve noon was a strict deadline. The judge commented: 'fundamentally, Devon CC relies on the simple proposition that a procurement process requires a deadline for the submission of tenders and that a deadline is a deadline.'

Once the design is complete and tender documents are ready, many employers will be impatient to have their project started on-site, and may want as short a tender period as possible. This can be a mistake.

If the tender period is too short for tenderers to make a thorough job of all the work described below, then tenderers will face extra risk, and will almost certainly 'load their tender' (increase the price) to reflect this. Tenderers are also likely to make mistakes in pricing, or misunderstand requirements. If the mistake is to overprice a tender, then the employer has fewer realistic tenders to choose from. If the mistake results in an artificially low price, then at first sight the employer may be pleased. Unfortunately this situation generally creates difficult working relationships on-site, as the contractor attempts to recover losses and expenses (in JCT contracts) by means of claims, or by the use of compensation events in NEC3 ECC contracts.

Once a tender list has been drawn up, it is of course possible for the designer to send a 'scope of work' letter to the tenderers when confirming their interest, and actually ask them what a suitable tender period would be. This sensible procedure should be done 'without obligation' on the employer.

How tendering contractors price the tender

What is involved for the tenderers?

Tenderers have a great deal of work to do during the tender period. Such work will include:

- Assessing the scope of the tender documents to understand fully what must be provided.
- Assessing the risk, particularly any risk inherent in the form of contract used and amendments made by the employer (such amendments are very common).
- Assessing any requirements for contractor design, and making suitable draft arrangements to carry this out.
- Visiting the site and assessing all site information provided by the employer, or publicly available.
- Drafting a programme showing all the construction activities.
- Producing a health and safety plan.
- Assessing resource requirements and availability for supervision and labour.
- Assessing funding, banking facilities and potential cash flow.
- Pricing the documents which will involve obtaining supplier and sub-contractor quotations and materials prices.
- Considering environmental issues and waste disposal.
- Considering access requirements and any off-site offices or storage requirements.
- Completing any information required by the employer, such as safety, quality and environmental performance statistics, plans and policies.
- Completing any contractor proposals required by the contract.
- Completing the schedule of cost components and contract data if an NEC3 ECC contract is used.
- Holding a **tender settlement meeting** (or final review meeting).

> The tender settlement meeting is when senior members of the contractor's staff review the tender, the prices, sub-contractor arrangements and of course the risk, together with the contractor's resources and 'need for work'. Final pricing decisions are made at this meeting, which can have a significant effect on the tender price.

How tenderers check all documents

- Tenderers normally check all documents for consistency. If documents are not 'mutually explanatory' tenderers can either inform the employer (which is likely to please the employer), or price the cheapest interpretation, and expect to renegotiate the price for the inconsistently described items after contract award when there are no competitors.
- Tenderers may submit an alternative tender, with terms or materials that are more favourable to them, but look similar to the original ones.
- Tenderers may look for 'design alternatives' that save money, wait until they are awarded the contract, and then offer them to the employer, at a price that guarantees them an extra margin.

- It was once common for tenderers to look for potential claims (possibly measurement or description claims), price the possible outcome (accounting for the probability of success), and then reduce the tender price somewhat to be more likely to win the bid (or up-price other items, to take more margin). This practice was fairly common in the 1990s when competition was fierce and tender prices were low. However with the advent of newer forms of contract such as NEC3 ECC, and team working and partnering philosophies, this practice is now very unusual.

How tenderers check contracts with bills of quantities

- Tenderers look for incorrect under-measure and over-measure by checking quantities, and where they see an error, they may decide to adjust their prices. So if there is 1000 cubic metres of excavation, and only 500 cubic metres in the bill of quantities, tenderers could increase their rate and balance the money off elsewhere in the tender (or decrease the rate for the reverse).
- Tenderers might overprice any *single* item, perhaps one 500 mm valve, in the hope that there might be more than one in the final contract as constructed. In simple terms, in traditional forms of contract the price of increased quantities is based on the price for that item in the tender. Hence a high initial price produces a good financial return if more items are later 'ordered' by the contract administrator at that price by a variation or change order. It is essential that this possibility is taken into account when assessing tenders, so that the employer fully understands any risks or advantages. It can be difficult for new professionals to appreciate that because a cubic metre of concrete is priced at £100 in a tender, its true cost could be very different.
- 'Back-end loading' of the tender means that work items are overpriced and the front section of a bill of quantities is underpriced to compensate. On measurement contracts all initial quantities are 'estimated' and are then remeasured after the construction work is completed. So back-end loading is done where tenderers think the designer has generally under-measured the work items, so that the payment for work items will increase when remeasured.

Factors considered by tenderers when producing their tender price

Tenderers will also account for a whole range of other commercial factors in fixing their final tender price, such as:

- Their 'need for work', to keep their resources and people busy.
- Market conditions for this type of work in this area.
- Other tenders that they are currently pricing and may win.
- Their current financial situation and borrowing requirement.
- The increase in reputation they may get from the contract in question, particularly if it is unusual or prestigious.
- The possibility of future work of a similar kind arising from this contract.
- Knowledge of the employer, especially the speed of payment they can expect, and whether the employer has a team-working philosophy.
- Risk factors, particularly in the conditions of contract.
- Conditions of contract with many amendments to the standard forms.
- The location of the project, particularly considerations as to roads and other communication routes.

- Difficult ground conditions or dewatering issues for which there may be no reimbursement in the contract with the employer.
- The complexity of the contract documents.
- The level of liquidated damages (see Chapter 18).
- The timescale for the project; short timescales being generally much more risky.
- The likelihood of price increases that are not reimbursed under the contract.
- The likelihood of difficulties with other stakeholders such as the public or residents. This can be a real problem on high-profile defence contracts.
- Any special guarantees or warranties.
- Insurance issues such as the likelihood of theft or vandalism in the area.
- The availability of local or specialist labour.
- Whether to distribute prices to adjust their **cash flow** (see *Adjusting the contractor's cash flow* below).

All these factors can have a significant effect on the tender price, because a contractor bidding for design or construction work has to cover their 'risk'. Any construction project involves risk and during the tendering process judgments must be made as to how well risks can be assessed prior to the award of a contract and who will take the risk, the employer or the contractor. Tenderers do this by increasing their prices using their estimate of the balance between potential likelihood of the risk event occurring and its probable cost.

Many forms of contract will distribute these risks between the contractor and the employer. If a contractor risk remains with the contractor and has a likelihood of occurring of one in ten, and a likely cost of £20,000, then the tenderer ought to increase the tender by £2,000. Of course, tenderers will be limited on the extent of such a price increase if there are other companies bidding as well, which is the usual case. To make matters slightly more confusing, tenderers may not load the risky item itself with an increased price, but they may decide to put this 'risk money' elsewhere in their tender prices.

Adjusting the contractor's cash flow
By adjusting the price of work done earlier or later in the contract and balancing it off elsewhere, tenderers can radically adjust contract cash flow, and hence their funding requirements and the associated cost of them. Contracts with bills of quantities usually have a front section with general items such as site establishment, site clearance, connecting office services and so on, followed by many sections of work items where the excavation, concrete, formwork and other engineering items are measured. This front section is often called 'preliminaries' or 'prelims', particularly in building contracts. It is usually a significant part of the tender total: thirty per cent is not unusual. It is possible to deliberately increase the overall price of the prelims and balance this increase against a reduction in the price of the work items. This increases cash flow early in a contract. There are significant risks in adjusting prices to optimise cash flow if quantities change or additional similar work is added in as a variation.

In contracts with activity schedules, such as NEC3 ECC option A, contractors may adjust their chosen activities and the prices of those activities to achieve a similar effect. This can be done by including a lot of early activities, or by increasing the price of them relative to the true cost, and balancing the money by reducing the price of later activities.

In summary, the two main options for adjusting the contractor's cash flow are:

- *Front-end loading* of the tender prices to generate early cash flow. This is done by overpricing early items such as site establishment items (often called 'prelims'). The drawback is that tenderers have to balance this extra by underpricing the work items which will be undertaken later in the contract. If these work items increase when the construction takes place, then the contractor may lose money.
- *Overpricing* early work items in bills of quantities or early activities on the activity schedules; and underpricing later ones. This has the same effect as front-end loading.

'Tendermanship' and anti-competitive behaviour in tendering practices

Some of the practices described above come under the name 'tendermanship' – the way in which tenderers sometimes attempt to give themselves a commercial advantage over other tenderers, or to produce alternative cash flow models which may suit their business better.

There is no suggestion that tendermanship is illegal, but it does have some ethical and commercial implications. What is certainly illegal under the Competition Act 1998 are any agreements between businesses that prevent, restrict or distort competition, or which are intended to. The Act provides a list of such anti-competitive agreements in section 2, as follows:

'Agreements, decisions or practices which –

(a) Directly or indirectly fix purchase or selling prices or any other trading conditions.
(b) Limit or control production, markets, technical development or investment.
(c) Share markets or sources of supply.
(d) Apply dissimilar conditions to equivalent transactions with other trading parties, thereby placing them at a competitive disadvantage.
(e) Make the conclusion of contracts subject to acceptance by the other parties of supplementary obligations which, by their nature or according to commercial usage, have no connection with the subject of such contracts.'

There are a number of cases in the courts regarding collusion of tenderers, price-fixing agreements and 'cover-pricing' for which the penalties are very severe with potential fines running to many millions of pounds. Collusion or price-fixing is when tenderers agree in advance who will win the contract, and they all adjust their bids upward. Cover-pricing is when a tenderer, who does not wish to submit a proper bid (perhaps because it has too much work on at the time), obtains a 'high price' from another tenderer to ensure that the 'cover tender' is not successful. This illegal practice is largely the fault of employers who do not accept the fact that contractors may not always want to tender for their work. Simply saying 'No, we are too busy at present to tender' is apparently an unsatisfactory response for some employers. These practices are not common, but unfortunately they do exist. Any tenderer following these practices runs a high risk of prosecution.

Ethical and commercial issues in tendering practices

So, let us move away from illegal practices to commercial practices. Tenderers wish to submit a winning tender, and so will account for the likely competition. This is not the same as deliberate agreements to fix prices, or not to compete, which are illegal. Tenderers will often have some idea of the names of other tenderers, particularly if the work is specialist. They will also have some idea about the competitiveness of these other tenderers for the particular project in question based on feedback they have had from employers on previous tenders or other commercial information.

We must remember that all parties in contracts usually try to minimise their risk and maximise their gain, or obtain a cheaper price. The employer will usually have commissioned a design organisation to produce the tender documents. This process takes weeks or months, or sometimes years. As a result the employer should be very familiar with the tender documents. Frequently, on advice from their designers or commercial staff, employers modify standard conditions of contract to rebalance risk in their favour; possibly by making the contractor responsible for the effects of any weather conditions that may occur, however exceptional. It is often quite difficult for tenderers to fully assess the implications of these changes to conditions of contract. The tenderer often has only four to eight weeks to digest all the tender documents, understand the project and assess the risk; in addition to obtaining prices, programmes and all the other activities listed above.

Furthermore, the tenderer is in competition with other tenderers and will incur the cost of producing the tender. It can cost from thousands of pounds, to hundreds of thousands of pounds to submit a tender. If there are six tenderers, then each has only a one in six chance of success. The costs of tendering have to be carried by contractors in their company overhead, which is a cost percentage applied to items in all tenders, but that is only recovered on tenders that are successful and become construction contracts.

So when we take all the above factors into account, it is hardly surprising that tenderers attempt to 'level the playing field' a little, by adjusting the balance of their prices. Any such adjustment carries its own risks as well as potential rewards. One drawback with such adjustments is that they may upset the employer if discovered. This could damage relationships or reduce the possibility of further work with that employer. So the unfortunate tenderer has to take this into account as well as trying to compete with all the other tenderers.

It is up to the organisation which is assessing tenders to examine tenders carefully and include appropriate checks on 'tendermanship'. The organisation carrying out the assessment is often the design organisation, but it may be the employer's own staff. There is computer software available which compares tendered prices and highlights unusual prices or anomalies. The tender assessor can then consider the risks in accepting any given tender and advise the employer accordingly.

Provision of programmes with tenders

A construction programme describes how the contractor proposes to carry out the contract and will show sequences of activities and will frequently be based on a critical path network. Some employers prefer to see a draft construction programme included with the contractor's tender showing the main activities and durations and their interrelationship. Programmes can be very useful, but there is some debate on whether programmes should be requested at the tender stage. The programme allows the employer and designer to gain a preliminary idea of the contractor's plans, and when access to the site and further information may be required. Hence it can be very useful as an *indication* of the contractor's intentions.

The normal contract requirement for programmes is for them to be submitted within a period *after* contract award. JCT SBC05 in clause 2.9.2 requires a programme 'as soon as possible after execution of this contract, if not previously provided.' ICE7 clause 14.1(a) requires a programme within 21 days of contract award. NEC3 ECC clause 31.1 requires a programme within the period stated in the contract data (if a programme is not identified in the contract) and actually has financial sanctions if the first programme after contract award is late. All these contracts have specific provisions for the approval or rejection of programmes, and for revised programmes to be submitted as the construction work proceeds. The point is that the programme is supposed to be a flexible device that is revised to show actual progress and changing circumstances. It is not intended to be 'cast in stone' as a contract document. However both the JCT SBC05 and NEC3 ECC contracts accept the possibility of a programme being submitted with the tender, presumably with the intention of the normal programme revision procedure applying from then on.

There was a famous 'programme' case on an ICE 5th edition contract, *Yorkshire Water Authority* v. *Sir Alfred McAlpine (Northern) Limited* (June 1985). There were some special circumstances surrounding this case but a method statement and programme submitted with the tender were held to have been *included* in the contract, and thereby had the status of contract documents. A method statement is a *written* description of *how the work will be carried out*, and the people and plant envisaged. A programme depicts information graphically; in particular the activities, their durations and their logical linkage. Admittedly, the ICE contracts contain a provision for the variation of a 'specified sequence method or timing of construction' in clause 51, and the judge in this case held that the method statement was such a 'specified sequence'. However it does indicate the danger of giving a *tender* programme an unrealistic *contractual* status. Possibly the safest answer is to make it clear in the tender documents that a programme submitted with the tender documents will not be incorporated in the contract itself.

Deciding on the most appropriate tender

Employers have always tried to achieve the optimum balance between cost, quality and time. *Price* is still the main criterion for deciding which tenderer will be successful. Price will include not just the total tender price but also other financial and risk assessments that the employer needs to carry out, depending on the payment mechanism chosen. Payment and pricing mechanisms are described further in Chapter 14. Increasingly, employers look for other attributes as shown below. These other attributes are often grouped under the term 'quality'.

Where tender selection is based on price and quality, it is usual to have two submissions from tenderers in separate 'envelopes'. Wherever possible, 'quality' attributes should be assessed on objective criteria such as statistics, or by the use of selection matrices or some other means of allocating numeric values to each attribute, rather than a subjective assessment. So for example under health and safety, the number of reportable incidents could be assessed on a scale of one to five, and so on. Where tenders fall under the EU Directives, the use of justifiable *objective criteria* is a legal requirement. Objective criteria should always be determined and set *before* tenders are received. Some typical price and quality requirements for tender assessment are listed below:

- **Price**
  - Tender total
  - Pricing of major items and distribution of prices (see 'tendermanship' above)
  - Fees for target contracts, which are based on reimbursing the contractor's actual costs and then adding a fee percentage to them
  - The pricing of items in the schedule of cost components (for relevant NEC3 ECC contracts).

- **'Quality'** considerations such as
  - Health and safety records and statistics
  - Environmental records and statistics
  - Accreditation to established quality or environmental standards
  - Experience of similar projects
  - Reference projects (previous projects used as references), possibly including site visits by the organisation assessing the tender
  - References from other employers
  - Outstanding major claims against other employers
  - Liquidated damages paid over the previous three years (or other appropriate period)
  - Personnel with relevant expertise
  - Resources, particularly any specialist construction plant
  - Design expertise and resources, where this is required under the contract
  - Demonstration of an understanding of the project
  - Teamwork ethic
  - **Cultural fit** (see below).

Employers will also check
  - Health and safety policies and procedures to ensure *competence* in terms of the CDM regulations
  - Financial stability of tenderers and their current workload.

Cultural fit

Perhaps a brief explanation of cultural fit is needed here. All organisations have a culture – in other words, what it feels like to work in that organisation. So an organisation may be formal and expect staff to wear a collar and tie, or even a suit; or informal, where jeans and trainers are acceptable. Some organisations have a lot of rules and procedures, whereas others encourage staff to make their own decisions. In some organisations, a manager would be called 'Mr Smith', whereas in others he would be called 'John'. Many organisations today are concerned with their 'image' in one form or another, and may have a particular stance on the environment, ethical behaviour or some other issue. When two organisations have a similar culture, it makes it much easier for their staff to work harmoniously together. Perhaps more importantly, an employer with a particular brand or image does not want contractors who will tarnish that brand or reputation by their activities. This is especially important when there is an interface with the public, such as in highway works or building renovation. For all these reasons, and many others, employers increasingly look for cultural fit in their contractors.

The relevance of project size on selection criteria

Smaller projects

For smaller projects, price is still one of the main selection criteria, and may be the only one. At its most simple, the tender process may ask for a lump sum price from each tenderer. If we take a simple example, tender prices for a new garage might be £7,800, £7,900 and £8,000. Clearly the tender at £7,800 is the lowest price.

Larger projects

For larger or more complex projects, it is worth going into more detail in tenderer selection and assessment. This will inevitably result in greater expense both for tenderers and the employer, and this extra expense has to be worthwhile. For larger or more complex projects the tenders are likely to be assessed on a mixture of price and quality.

Additionally, the selection process may involve a number of stages. Some of the above list of quality considerations may be checked at the tenderer selection stage. Some may be checked twice with the second check going into much more detail after tenders are received. Such checks can be very expensive and time-consuming for both tenderers and employers, and so an appropriate balance should be struck between the time and cost of initial selection for tender lists, and the selection of the successful tenderer after tenders are received.

Very large projects and framework agreements

Framework agreements are described in detail in Chapter 12, but in simple terms they comprise a number of projects awarded over a period of time under an overall framework contract. Such frameworks or very large projects may be worth many tens or hundreds of millions of pounds, and justify extensive selection processes. Such selection processes might include:

- Written submissions including experience, resources, processes (safety, environmental and quality), and examples of innovation, supply chain management, carbon reduction policies and financial references.

- Interviews including presentations by tenderers and possibly scenarios for them to react to and solve.
- Presentations about how the tenderer will contribute to the employer's goals and fit with the employer's organisational requirements.
- Business plan submissions for joint development.
- Proposals as to how the tenderer will integrate the construction processes into the employer's business without disruption.
- Reference sites for visits by the employer, and possibly examples of alliances or other ventures in which the tenderer is engaged.
- Pricing exercises.

Problems associated with tendering

If we include frameworks in the term 'tendering', then tendering is by far the most common way of contractor selection in the United Kingdom. This situation has been strengthened by the EU Directives, which impose very strict tendering rules and procedures for affected employers. However, there are a number of problems associated with tendering as outlined below. These problems can be minimised by following the advice given earlier in this chapter.

- Large tender lists – the longer the list the fewer the number of contractors who will be willing to submit a realistic tender, because of the balance between their tendering costs and the probability of success. Tender lists of more than six tenderers are inadvisable for this reason.
- Tender lists for design and build contracts, or other contract strategies where tenderers carry out some design work should be restricted to about three tenders, or contractors may be unwilling to tender at all.
- Short tender periods – too short a period for tendering will force tenderers to account for greater risks in pricing and thus may increase the overall cost of the project.
- Poor tender documentation – lack of information or conflicting information between documents will cause problems for the tenderer when pricing work.
- Inviting tenders when the project is unlikely to proceed is bad practice – tenderers bear the cost of tendering and will be very reluctant to submit tenders to employers who frequently fail to place a contract.
- Failure to notify results is poor practice since tenderers need this information to assess their competitiveness. It is good practice to notify all tenderers of the other tender prices.
- Sending tender documents out late without adjusting the tender period to compensate can cause contractors' estimators problems with programming work and obtaining quotations from suppliers and sub-contractors.
- All tenderers must be treated equally and informed of any clarifications to documents or changes to the tender submission date.
- Additional or incorrect information – if it is apparent that some information is incorrect, or missing, the employer should inform all tenderers at *the same time*, so that none have an advantage.
- Qualified tenders (and options offered) submitted by tenderers cause problems for comparison between tenderers. It is better if a compliant bid (one that complies exactly with the tender documents and instructions to tenderers, and

contains no amendments or qualifications) is submitted together with an alternative incorporating changes. Many employers insist on a compliant bid as the first stage to considering alternatives.

- Fairness – contractors do not like to bid for employers who either disregard properly submitted tenders or behave in an unprofessional manner.
- Behaviour – if employers behave badly in any of the above respects, they can expect that *contractors will be reluctant to tender in future, or will load their prices. Good communication and ethical behaviour really do guarantee the lowest prices.*

A clear tender process, honesty and absolute fairness, good communication and an adequate tender period will all help the employer obtain the best prices for the contract.

Chapter summary

- The tender process is the mechanism of sending out tender documents, receiving tenders, evaluating them and appointing a contractor in a fair, robust and legal way. It needs very careful administration.
- The Public Contract Regulations or the Utilities Contracts Regulations (called 'the regulations') have a significant impact on tendering options and processes for any employer affected by them.
- The tender documents sent out by the employer usually include the invitation to tender and Instructions to Tenderers, any articles of agreement, contract data, the drawings, the pricing mechanism, works information and site information.
- Good communication, based on clear and comprehensive tender documents, is vital to ensure a successful project.
- The tender period is the time period from when the employer sends out the tender documents until the closing date (and time), stated in the Instructions to Tenderers. It is usually four to eight weeks depending on the size and complexity of the project.
- There are six main options for obtaining tenders. They are open tendering, single-bid selective tendering, pre-qualification, negotiated, two-bid selective tendering and framework agreements.
- There are two types of offer resulting from two different types of tender option. A single offer is a tender for one project. In contrast a framework agreement is a tender to construct a number of projects over a period of time. Framework agreements are frequently used by employers who are subject to the regulations.
- 'Tendermanship' is the way in which tenderers sometimes attempt to give themselves a commercial advantage over other tenderers, or to produce alternative cash flow models which may suit their business better. It is important for people assessing tenders to check for tendermanship and draw the employer's attention to any commercial implications.

- Traditionally tenders were assessed on price and the lowest price was usually successful. Increasingly employers are selecting on price together with an assessment of quality. The assessment of quality can bring a diverse range of factors into consideration.
- There are a number of problems associated with tendering which can be avoided by a fair and transparent process, comprehensive tender documents, an adequate tender period, open communication and feedback to unsuccessful tenderers.

7

Site investigation

Aim

This chapter emphasises the need for thorough site investigations, and describes the main types.

Learning outcomes

On completion of this chapter you will be able to:

>> Describe the ranges and types of site investigation required for a particular project.
>> Explain the environmental and contractual necessity for a thorough site investigation.
>> Describe in detail the objectives of a site investigation.
>> Propose a suitable preliminary site investigation and site survey for a project.

Site investigation and ground investigation

These terms are often confused, especially by civil engineers from an earlier generation, who will often refer to 'the SI' when they actually mean the ground investigation. The reason may be that in the previous century, ground investigation was the largest part of site investigation. In those days there were far fewer laws and regulations controlling the activities of those who wished to build.

Ground investigation is concerned with the properties of the ground in which a structure is designed and constructed. Site investigation is concerned with all aspects of the *location* of a structure and its interrelationship with its surroundings both during construction and in future use.

It is worth emphasising that *interrelationship* is a very important consideration. The environment and surroundings will affect the construction of the asset, and the asset will affect its surroundings and local environment. Both aspects need thorough investigation by the construction professional.

Ground investigation is a part of the overall site investigation. It is a very important part, for the ground investigation determines the parameters needed to design structures and buildings, and provides much of the information for safe construction in the ground by the contractor. Ground investigation is usually carried out by shell and auger rigs (which drive a hollow casing into the ground to remove soil for sampling), or rotary boreholes (these extract cores from hard materials or rock for analysis). Both of these methods have facilities for *in-situ* tests but are mainly designed to send suitable undisturbed samples for analysis in a laboratory to determine their physical and mechanical properties. This is the province of soil mechanics and geotechnology, and not the subject of this book.

However, in recent years the wider aspects of site investigation have come to the fore with concern for the environment, and ever-increasing legislation relating to safety, waste, protected species and the influence of a new project on the road network and local infrastructure. It is this wider aspect of site investigation that will be discussed below.

The contractual significance of the site investigation

Not only is site investigation an important part of the investment process, it also has a serious *contractual* significance. On most projects, the employer (or a design organisation) will carry out or organise an initial site investigation to determine the suitability of the site and any major issues or constraints. Additionally, it is usual for the employer to arrange a ground investigation. This is because the results of these two investigations will provide ground parameters, ground water levels and constraints on the contractor's freedom of access or working hours and arrangements which may affect their tender price (see Chapter 6 for more on tendering). Whichever of the contract strategies is used, the design organisation will need the results of the ground investigation to carry out the structural design.

Furthermore, it is not efficient and may not be possible for a number of tenderers to each carry out their own separate ground investigations. Ground investigation often requires access to private land or the highway, and clearly nobody will want a ground investigation by each of six tenderers. Having said this, on smaller projects, by an employer who knows little of construction, ground investigation may be left to tenderers. This is often the case with small projects that use the design and build strategy.

Another consideration is the possibility of legal action for misrepresentation by the contractor against the employer, where the employer does not disclose or misrepresents any relevant information about the site. Such legal action is unlikely since it would sour future business relations, but it could be a necessary step for a contractor who has lost a lot of money.

> The adequacy and outcome of the site investigation have an important contractual significance.

The results of the *site investigation* should be referred to or written into the contract where relevant, either directly or as part of specification clauses. It is likely that these results will have contractual implications if they constrain the contractor's access to site, or use of the site, or working hours and arrangements. Issues such as these are likely to have programming implications, or may affect the sequence or method of work, or the size and type of plant used. For example, an access bridge with restricted load capacity, or a narrow country road leading to the site, will limit the size and weight of plant that can be used in the construction. In many urban locations, the site investigation will reveal areas where construction noise will be limited and where working hours may be constrained.

> The results of the *ground investigation* will form a base line of assumptions about the ground itself when tenderers plan the excavation work and price up the tender.

The ground investigation is fundamental to the planning of the construction work, particularly excavation, and the price required in order to carry it out. These results will influence the tenderers' assumptions on suitable plant, cut and fill, ease of excavation, reuse of materials, acceptable side slopes to excavations, dewatering plant needed together with health and safety considerations.

There are two significant contractual issues resulting from the ground investigation:

1. *Actual* ground conditions which could not reasonably have been predicted from the ground information supplied to tenderers.
2. Incorrect quantities for items in the ground, where contracts using a bill of quantities are used. A bill of quantities is a list of quantities abstracted from the drawings, which the contractor prices as part of the returned tender.

The various forms of contract treat these two issues in similar ways but with important differences. The detail of NEC3 ECC and JCT SBC05 contracts will be explained in later chapters. However it is worth touching on the issues in a chapter concentrating on site investigation.

NEC3 ECC contracts and ground investigation

In NEC3 ECC it is assumed that the ground investigation information will be inserted (or referred to) in the 'contract data part one, provided by the employer.' NEC3 ECC uses **compensation event** 60.1(12) which talks of the contractor encountering physical conditions which 'an experienced contractor would have judged at the contract date to have such a small chance of occurring that it would have been unreasonable for him to have allowed for them.' If proved by the contractor, a compensation event entitles the contractor to more time and money. Here a thorough ground investigation will facilitate this judgment of risk on the contractor's part. So we see that the better the ground investigation, the smaller region of unpredictability that will be subject to the phrase 'small chance of occurring.' In other words, the conditions predicted by a good ground investigation ought to be those encountered during construction. Most of these physical conditions will be found in the ground investigation, but matters such as the risk of the site flooding from nearby watercourses should be dealt with in the site investigation as a whole.

NEC3 ECC covers the second issue of incorrect quantities in clause 60.4 of options B and D. These are the NEC3 ECC options which use bills of quantity.

JCT contracts and ground investigation

A different assumption forms the basis of most JCT contracts 'with quantities'. These 'with quantities' versions assume that the project has been measured and **billed** correctly by the employer (or the design organisation working on the employer's behalf). Billed is a term meaning that the quantities have been abstracted following a standard method of measurement and inserted in the bill of quantities for the tenderer to price.

Thus JCT SBC05 has a clause 2.14.1 which covers issue two by referring to the correction of any errors in quantities in the bills and entitles the contractor to extra time and payment of their financial loss and expense, if these can be justified. Presumably a claim relating to 'unforeseen' ground conditions in JCT SBC05 would be made on the basis that there were no quantities relating to the matter in the tender documents. So if rock were discovered in an excavation, the JCT argument would be that it should have been measured in the bill of quantities. By contrast, the argument in an NEC3 ECC contract hinges on the 'forseeability' argument, determined by reference to the ground information supplied to tenderers.

Inspecting the site

Most forms of contract allow the contractor to claim financial loss, expense and more time if the actual ground conditions differ from those which could reasonably have been assumed at tender. This baseline of assumptions made when the tender is priced is founded on the documents provided by the employer, but it often includes the *assumption that a tenderer inspects the site* whether or not he or she has done so. Any future claims for 'unforeseen conditions' are based on this assumption of an inspection during the tender period. Thus NEC3 ECC contains a provision in clause 60.2 which says that in determining a compensation event, the contractor is assumed to have accounted for 'information obtainable from a visual inspection of the site.'

So on a project of any magnitude or potential high risk, tenderers will visit the site and carry out their own assessment of site access and other conditions. In this way, at least for tenderers, there is usually a 'site investigation' made by tenderers, however brief, in stage seven of the investment process, and this is shown as the smaller arrow in Figure 7.1.

When is the site investigation carried out?

Site investigation is an important and integral component of the investment process and forms a part of stages three to seven. There are *four* main site investigations, which are shown in Figure 7.1, and described below. The smaller arrow indicates a possible site investigation (usually only a site inspection) by tenderers.

At an early stage of the project it is necessary to conduct an investigation that is sufficient to allow the sensible evaluation of options. Access, land ownership and the local infrastructure and environment are important considerations that may rule out some options, even at this early stage of the process of taking a

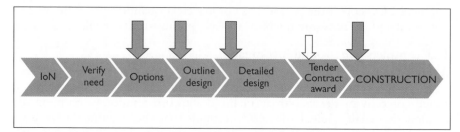

Figure 7.1 Timing of site investigation during the investment process.

project from concept to completion. The most important concern is to ensure (as far as possible) that the site is suitable for the project in the first place. More detailed investigation of all the items above will be necessary in the detailed design stage to *confirm* this suitability and explore more detailed aspects of the site and its *interrelationship* with its surroundings. Clearly, at the option assessment stage, the cost and time taken by the site investigation must be taken into account, particularly if the project may not proceed. Where a number of sites are under consideration, or where the project is linear such as a road, railway, pipe or sewer, with different potential routes, many site investigations may be indicated. However carrying out extensive site investigations at a number of locations is rarely possible or practical. Other inhibiting factors are the difficulties of entry to private land, and permissions under the Traffic Management Act 2004. This act controls access to the highway, for site investigation as well as construction itself, and its notice procedures are fairly onerous and time-consuming. For all these reasons site investigation at the option stage is fairly outline, and rarely includes the ground investigation itself.

Once we reach the beginning of the detailed design stage, comprehensive site investigation, and with it the ground investigation, really come to the fore. This is because of the importance of planning approval and design requirements. Most projects require planning approval from the relevant authority. Planning approval will probably require consideration of visual impact, noise assessment, traffic effects and particularly environmental impact such as the disposal of construction waste and any effect on protected species.

The design organisation will need clear parameters to enable the structural design to proceed. These parameters will be the output from the main ground *investigation*. These will be such items as active and passive pressures, ground water table levels, material properties, bearing capacity and so on. During the detailed design stage the need for further site investigations may become apparent as more information is gathered.

On most projects the employer arranges site investigations and provides information about the site to tenderers.

At an early stage of the investment process, employers may arrange site investigations themselves. However once a design organisation or other advisor is appointed, the employer usually relies on them to organise the necessary investigations, which will determine design parameters as well as information for construction itself. Of course the employer pays the cost of all investigations.

The information provided to tendering contractors will usually include the ground investigation report, so that they understand the ground parameters of the site. Contractors tend to be particularly interested in material properties, possible reuse of as-dug material, slope stability and groundwater ingress and control, so they can plan their excavation techniques in a safe and economic manner. The employer has a duty under CDM 2007 to provide '*pre-construction information*' both to the designer and the contractor in accordance with regulation 10. Other very important information will relate to buried services, which are the cause of many accidents and contract delays.

Service providers (such as gas, electricity and water companies) will need consulting about the location of their plant and equipment, some of which may require diverting to enable construction work to proceed efficiently and safely. In this context, plant and equipment refers to cables, sewers, gas and water mains, oil pipelines, television cables and telecommunication cables. Large-diameter pipes and important cables or telecommunication equipment can take many months to have moved and relocated, because alternative supplies must be made available.

> Early identification of service providers and full consultation are vital to avoid delays to the project.

Once the contract is awarded, the successful tenderer, now the 'contractor', will generally dig specific trial holes to *actually locate* any buried services. This is usually done after a visit by the relevant service provider, such as gas, electricity, water and so on. Prior to this plans should have been obtained from these service providers. The trial holes are dug to confirm the actual position and level of services, since service providers' plans are often lacking in accuracy. The contractor may also dig trial holes for a visual inspection of buried strata and to establish actual slope stability, rates of water ingress and material features and properties. Excavation of trial holes will also give some indication of suitable excavation plant and the 'actual digability' of the ground in question.

Throughout stages three to five, site investigations will establish the need to consult and inform regulators and stakeholders. For many projects statutory permissions such as planning consent will be needed.

> If any important regulator is forgotten, delays of many weeks or months are not uncommon whilst approvals are gained.

Key objectives of the site investigation

The objectives of site investigation are to:

- Establish the suitability of the site.
- Determine any site constraints.

- Identify landowners both of the site and adjacent land that may be affected.
- Identify all relevant regulators and stakeholders.
- Ensure a safe and economic design.*
- Attempt to foresee and resolve problems before they arise.
- Ensure safety and efficiency during construction.*
- Ensure safety during later operation of the asset.
- Help to minimise environmental impact.

Establish the suitability of the site

Because of the wide variety of construction projects, it is impossible to give other than very general guidance on site suitability. Suitability will be determined by all the factors above. Additionally, new housing developments will need to take into account the local development plan, demand, local housing patterns, existing housing stock and the potential positioning of the development in the housing market. Factory and office developments will need to consider either moving existing staff to the new location or recruiting in the local area, or both. Most residential, factory or office developments will have an impact on the local highway system, and ease of access and the vicinity of the motorway network are important considerations. Increasingly, developments take account of the possibility of 'green travel', by bicycle, on foot or by public transport.

Engineering issues and information which is largely provided by the ground investigation may indicate expensive foundations such as piles, ground remediation and the possibility of contaminated land. Additionally, the opportunity to reuse excavated material and minimise waste is becoming an increasingly important site-selection factor for environmental reasons.

Likely stakeholders to any site need identification and evaluation. This can sometimes be difficult. Information can often be obtained from the local authority, and a visit to the site will yield information on neighbours and any obvious historic or environmental features.

> The support or objection of stakeholders can be a major decision factor in the selection of a site.

The likely difficulty or ease of obtaining planning consent will often be a major consideration, and this consent will certainly evaluate the impact of the development on the local environment. The presence of protected species such as newts, badgers, bats and owls and areas such as **SSSIs** (described below) can have a major effect on the time taken to allow site access, and even determine whether any development of the site is allowed at all.

Determine any site constraints

Site constraints will usually relate to access to the site, and the size and location of the site itself. Access to the site will need to be considered, particularly any

---

* These two items are the main focus of the *ground* investigation.

narrow lanes, access through small villages or over narrow bridges or those with weight restrictions. Some rural highways are unsuitable for the movement of heavy construction plant, and local improvements may be necessary.

The size of the site itself relative to the completed asset will require careful consideration. Contractors need adequate areas for their offices and storage of equipment and materials. Safe access around and within the site, and the location of suitable welfare facilities, are requirements of CDM 2007. If the completed asset takes up much of the site area, it may be necessary to identify and rent adjacent land for use during construction.

If the site is in a city centre it may create noise, dust and mud issues during construction, particularly if there is extensive piling or spoil removal (often called 'muck shifting') to be done. This may necessitate wheel-washing facilities and special traffic movements or restricted working hours to reduce potential complaints. Again in towns and cities, parking arrangements both during construction and in subsequent use of the asset can now be a major hurdle in the planning process. It is essential that any constraints on contractors' freedom to determine their site arrangements, method and timing of work are identified before tender, and stated clearly in the tender documents.

Some locations will have easier and cheaper access and connection to local services such as gas, electricity, water and drainage. Whilst electricity, water and drainage are a construction issue, the requirements of the future development must be evaluated carefully. Similarly there may be issues with the delivery of essential materials, such as the location of the nearest concrete supplier. Labour will also be an issue for contractors, particularly in the more remote areas of the United Kingdom such as the Scottish Highlands.

Identify landowners that may be affected

Many employers will own the land of the site itself, ready for their new school, office or factory. Where land purchase is necessary, some option consideration and access and environmental factors will require investigation before the purchase is completed. For linear projects such as pipes and sewers, a way leave or easement will be needed to allow access across land, and then to allow permanent access to the constructed asset. Legal restrictions on the landowner's rights to construct buildings or other structures over the pipe or sewer will also be essential. Identifying the many landowners on linear projects is called 'land referencing' and can be particularly time-consuming and difficult. Local land agents frequently carry out this process for the employer. Farmers in particular are always concerned about the timing of projects because of potential crop loss.

Nearby landowners can also be affected by the construction of the asset, particularly if buildings are adjacent to the site. Here a condition survey is advisable to clearly establish the pre-condition of their buildings in case cracks or other damage appear during the construction work or soon after. Building surveyors are specially trained to deal with issues under the Party Wall Act 1996 which lays down the procedural requirements where building works or adjacent excavations may affect an adjoining owner. This is often the case, particularly in UK cities which often have many properties with party walls (the shared wall between two terraced or semi-detached properties). However the act is more general in its scope and provides rights and responsibilities for both building owners and adjoining owners, not simply those with party walls. The party proposing the work is called the 'building owner' and the owner of the affected property is called the 'adjoining owner'. If an adjoining owner does not 'consent' to any proposed works under the scope of the act there is a dispute procedure laid down for the appointment and payment of surveyors, who will agree a Party Wall Award.

Identify all relevant regulators and stakeholders

This can sometimes be a difficult process. Many employers with *programmes* of work will have databases of all the potential regulators and service providers in their areas so that they can be contacted, and so that regulators are unlikely to be forgotten.

Regulators

Typical regulators that may be affected are shown below. They influence different aspects of the investment process and are discussed as relevant in later sections of this chapter:

- The local authority for noise, air quality, traffic, the planning process and contaminated land.
- Environmental regulators for discharges to controlled waters, waste and water abstraction control.
- Nature conservation organisations for designated ecological sites and protected species.
- Designated archaeological and heritage sites can be the concern of both the local authority and heritage bodies.
- The health and safety executive for health and safety matters which may include vibration and asbestos.
- The sewerage provider (usually a water company) for permission to discharge to the public sewer (this is, the sewerage provider performing a regulatory function).

Local councils (county or city) will have information about their areas, and many of the regulators such as planning and building control are employed by such bodies. Local parish councils will have very specific information, and will almost certainly need informing and consulting, since they can be a very important stakeholder, particularly in country areas.

Service providers

Service providers are *stakeholders*, since if they have plant and equipment in the area, they do not want it to be damaged. Neither do they want the project to disrupt their service provision to customers. Service providers will be gas, electricity, water, telecommunications and cable television. It is up to the construction professional to determine which service providers may be affected by the project and to contact them for plans and further information.

> The construction professional must identify and consult with all service providers who are affected by the project.

This identification and consultation are carried out at a suitable stage of the investment process, usually stage four or five (the design stages). Service providers often make a small charge for the provision of plans. These plans are usually sufficient to enable the design to proceed, and to determine what, if any, service may need moving. The relevant service provider will arrange for services to be moved or made dead, and will charge for this. Large or important services can be very expensive to move. A large sewer or power cable may cost over £100,000 to move, and take many months, so early identification of significant services is essential. It is possible and safe to carry out work of a restricted nature underneath overhead power lines, and the relevant service provider should always be consulted to determine access restrictions and so that 'goal posts' of a suitable height are erected by the contractor before work commences.

For individual projects some further 'detective work' is needed. Site inspection should reveal overhead power lines, and the manholes, boxes and covers of many service providers such as gas, water, telecommunications, electricity and cable television. Unfortunately many of these covers will be cast iron, and engraved with the original name of the provider. Some covers may be a hundred years old, and relate to historic names of water or gas companies. However, they are important because the services below them are probably still in use. Another difficulty is that some covers may be located over removed or redundant services, or dead cables. For obvious reasons assumptions about disuse should always be discussed with the relevant service provider, and actual tests should be made to confirm this.

In addition to the more usual service providers such as gas, water, electricity and telecommunications, there are a number of major oil pipelines crossing the UK. These run at high pressure and are clearly a major problem if damaged. The relevant owners should always be consulted for information on such pipelines in the vicinity of any proposed construction.

Work in the highway or in footpaths

Work in the highway, and this includes ground investigation as well as construction, will usually fall under the Traffic Management Act 2004, and the relevant authorities should be consulted and informed. At a later stage a formal procedure of notices and permissions must be followed to avoid fines and possible prosecution.

If any public footpaths or bridle paths need to be closed or diverted, the local authority should be informed. Most closures take many months because they need to be advertised. If the project affects any bus routes, then the relevant provider will need consultation to arrange any diversions.

Ensure a safe and economic design

The overall design will need to take into account all the factors from the site investigation such as accesses, limits of land ownership, buried services, environmental issues and any contaminated land, planning constraints or conditions and so on.

Safety

Safety must be considered both generally, and particularly under CDM 2007. Here the designer is a named duty holder, and has specified duties under regulation 11. These duties can be summarised as avoiding foreseeable risks to health and safety, and communicating with other duty holders such as the client and contractors. It is a legal requirement under CDM 2007 for the designer to prepare design risk assessments.

Waste

Waste must also be considered and minimised during the design. There are now many UK regulations concerned with waste management. It is important to have a waste management plan, so that material that can be reused is never 'discarded'. The abstract below is taken from the explanatory memorandum (number 314) to the Site Waste Management Plans Regulations 2008. The client will normally rely on the design organisation to prepare the plan for them.

'2.1 These regulations require a site waste management plan to be prepared and implemented by clients and principal contractors for all construction projects with an estimated cost greater than £300,000 excluding VAT. The plans must record details of the construction project, estimates of the types and quantities of waste that will be produced, and confirmation of the actual waste types generated and how they have been managed. More detailed reporting requirements apply to projects exceeding £500,000.'

When a designer or contractor is considering how to deal with waste, they should follow the waste hierarchy, shown below, beginning at the top by trying to design out waste in the first place. Disposal of waste is the last resort, and should be avoided as far as possible by careful thought and design.

ELIMINATE	Design out waste
REDUCE	Minimise waste generation
REUSE	Reuse materials on-site wherever possible
RECYCLE	Reprocess materials for off-site use
RECOVER	Recover energy from waste sent off-site
DISPOSE	This is the least desirable option – the last resort

Further useful information on excavated material and waste is obtainable from www.netregs.gov.uk

Structural design issues

The structural design will rely on information from the ground investigation, particularly parameters relating to bearing capacity, sliding resistance, active and passive pressures, settlement and material properties. The effects of groundwater will also affect horizontal loadings on buried structures and may induce flotation unless precautions are taken.

The design organisation will usually carry out or commission a full topographical survey to determine ground levels and the position of all relevant features and visible structures such as manhole covers.

In areas prone to mining subsidence, information can be obtained from the coal authority. Where mining subsidence is a serious possibility, it is usual to engage the services of specialist consultants to give expert advice on the design and construction of the project.

Asbestos

Where there is work in any older buildings, it may be necessary to carry out a survey for asbestos. Asbestos is now recognised as a very dangerous, carcinogenic substance, but it was formerly used for many purposes, particularly insulation. Hence roof spaces of buildings more than a few decades old may have asbestos insulation. Pipes and boilers in particular used asbestos for insulation in boiler surrounds and pipe lagging. Finally a number of manufactured goods used asbestos such as asbestos cement (AC) pipes and roofs. Many grey corrugated roofs are actually made of asbestos cement. They can be found on older industrial buildings and in house extensions and garden sheds and garages. Whilst AC is less dangerous than asbestos itself, it still requires handling with care. There are many regulations concerned with asbestos, and expert advice should always be sought, where the presence of asbestos is suspected. The main regulations applicable here are the Control of Asbestos Regulations 2006 which bring together three former sets of regulations.

Flood risk

With changing weather patterns it is becoming increasingly necessary to consider flood risk, both during construction and once the asset is completed. The Environment Agency (EA) reports that over five million people in England and Wales live in properties that are at risk from flooding from rivers or the sea. The EA website gives indications of likely flooding, and the EA are always ready to give project-specific advice.

Archaeology

Buried archaeological remains and any historic buildings or structures in the vicinity will also need consideration during design. The local authority will usually have information on these. A further source of information is English Heritage (www.english-heritage.org.uk). It may be necessary to carry out a full condition survey of nearby historic or listed buildings both to ensure support where necessary, and to establish a baseline in case there are future claims of damage from their owners. Under most forms of contract, if contractors encounter archaeological remains they are obliged to stop work and report the matter. Again under most forms of contract the contractor is entitled to payment for any costs or delay arising. The employer therefore has a vested interest in a proper investigation during the design phase so that there are no surprises once construction begins.

Economic design

Economic design does not simply include all the factors above. It should also balance the long-term running costs of buildings and similar structures with their initial capital cost. This is called a 'whole life cost' analysis. A well-known technique for comparing capital costs with long-term revenue or operational costs is the discounted cash flow technique (called DCF). We may be accustomed to predicting the value of our savings twenty years in the future at a given rate of interest. What DCF does is the opposite of this process, by converting each future revenue cost to a present value. We can then compare the present value of future revenue costs with the initial capital cost of the asset. As an example, a very approximate rule of thumb at a 5 per cent discount rate is that £1,000 per annum is equivalent to £12,400 in capital cost. If we apply this to a project with two options, one costing £20,000 and saving £1,000 per annum, with a second one costing £30,000 but saving £3,000 per annum, we see that option one is not economic, whereas the more expensive option two is financially beneficial.

The existing infrastructure

Finally, the design may need to take account of the size or capacity of existing services adjacent or near to the site and often the local road network. Whilst some civil engineering projects do not need services, most building assets require electricity, water, waste disposal, telecommunications, gas supplies and so on. The future requirement for all these services needs consideration during the design, and enquiries should be made of all the relevant service providers to determine likely costs and timescales.

Attempt to foresee and resolve problems before they arise

Clearly this is a duty of any competent designer. The site investigation will reveal many potential problems which will need to be overcome in design and construction. It is usually much cheaper to address problems and attempt to solve them during the design process, rather than leaving them until the construction stage, where delay becomes very expensive. Site investigation may even reveal problems and issues that are so serious that the project needs to be changed significantly or even cancelled. This is one purpose of the 'gateways' discussed in the investment process, all as described in Chapter 2.

Problems and issues that still remain after the design process should be brought to the attention of tendering contractors, by describing them in the contract documents. In NEC3 ECC contracts they should form part of the 'site information'. In this way, tenderers can attempt to make financial and programme allowances for them. Problems that arise during construction, particularly those relating to ground conditions or site conditions, will usually result in a claim for financial loss and expense by the contractor (JCT SBC05) or an early warning leading to a compensation event under clause 60 of NEC3 ECC. Such problems will usually have extensive contractual repercussions.

Ensure safety and efficiency during construction

For the construction work itself, the contractor needs to plan site accesses, the site layout, storage areas, location of offices and welfare facilities, temporary power supplies and other facilities. Access to and from the site can be an issue which needs discussion with the local authority and local police. This is particularly true where large and slow vehicles (such as lorries and concrete delivery vehicles) will exit the site onto public roads.

The contractor will need to plan safe and efficient construction techniques and will use information from the site investigation, particularly the ground investigation. Predictions of ground water levels and permeability derived from the ground investigation will have a major impact on the planning of dewatering techniques for excavations and below ground structures. The contractor will also use information from the ground investigation to plan excavation support, and a safe angle for any slopes.

Most projects involve the need for cranes; often the use of fixed tower cranes. Ground conditions and access restrictions will assist the contractor in planning these activities such that lifting plans can be properly and safely prepared.

On many sites there can be potential issues from neighbours relating to dust, noise and vibration created during construction, and there may well be specific conditions of planning consents relating to these items. In urban areas, a contractor's working hours may be restricted, and noise levels may be monitored, making the selection of plant very important. Piling plant in particular can be very noisy. During excavation work, the contractor will need to consider the control of mud, possibly by wheel washers and mobile road-cleaning vehicles to avoid problems from users of local roads.

Ensure safety during later operation of the asset

If all the above results of the site investigation, including future implications, are taken into account during design and construction then safety in future operation of the asset should be assured. Of course, proper handover and training should take place and all relevant documentation should be given to the employer for future use, including the health and safety file which the employer is required to keep and make available under regulation 17 of CDM 2007.

Minimise environmental impact

Most projects have an impact on the environment. The extent of environmental legislation is increasing, and many planning consents contain conditions relating to the environment. Larger projects may require an Environmental Impact Assessment (EIA), which is a major undertaking, often carried out by specialist consultants. EIA is a subject in itself and there are many excellent books on the subject. In addition to the issues already discussed, the site investigation must also consider the following:

- Protected species
- Hedgerows and trees
- Noxious or invasive plants
- Designated sites.

Protected species

Many species of wildlife are now protected in the UK. The joint nature conservation committee (JNCC) is a statutory advisor to the UK government and contains very useful information on its website (www.jncc.gov.uk). The construction professional cannot be expected to be an expert on the *details* of protected species, or on the many laws governing this area. In most cases the construction professional will carry out a desk study by using any available databases or information from the local authority or similar bodies. The Multi-Agency Geographic Information for the Countryside publishes a good database (www.magic.gov.uk). It is important to remember that where planning permission is required, as it is for many projects, the planning authority will consult with the relevant nature conservation bodies as to the likely impacts and effects of the project.

After completing a desk study, the construction professional will visit the site after carrying out a personal risk assessment and gaining any permission from landowners. The point of the site survey is to verify and extend the desk study and establish whether further work by experts is needed. Planning and carrying out a site survey are described in the section below. For general information, this section simply highlights some features of the more widely encountered protected species and information on hedgerows and trees.

Bats

Bats roost in a variety of localities in both urban and rural areas such as in holes and cracks in trees, in roofs and walls of buildings, under bridges as well as the normal assumed places such as underground caves. Bats hibernate between October and April and breed between May and September. If bats are likely to be encountered, a survey by a licensed specialist worker should be arranged in order to establish the location and size of the roost.

Badgers

Badgers are widespread throughout the UK and their presence can affect many construction projects, particularly those in the countryside or in quieter areas away from much human activity. Badgers are protected under the Protection of Badgers Act 1992. Badgers generally live in setts which are a network of underground chambers and tunnels. It is an offence to directly disturb a badger sett or to carry out works close to a badger sett, which could cause a disturbance, without a licence from the relevant nature conservation body. If a badger sett is likely to be disturbed by construction work it is usual to have the badgers moved, and for entrances to the existing sett to be temporarily blocked to prevent the badgers returning during the course of the construction work.

As well as living in woodland, badgers can be found in road and railway embankments, under buildings, in hollow trees and in quiet undisturbed areas of industrial sites and water and sewage works. The sett can usually be identified by its entrances. Entrances are usually rounded or oval-shaped holes and the presence of large spoil mounds is usually apparent.

Many construction activities can affect badgers and a licence must be obtained for the following activities around active badger setts:

- use of heavy machinery within 30 metres
- use of light machinery within 20 metres
- use of hand tools within 10 metres.

Water voles

Water voles are often confused with rats and are usually 150 to 200 mm in length with a long tail. 'Ratty', in Kenneth Grahame's book *The Wind in the Willows*, was actually a water vole. Water voles can be found in burrows in the banks of slow-flowing rivers, streams, ditches, ponds and lakes. It is the burrow that often betrays the presence of water voles although experts will also look for faecal latrines located near the burrow. Water voles tend to be more active during the day. Whilst they are active all year round, they are usually less evident in winter. Water voles and their burrows are legally protected against damage and disturbance.

Otters

Otters may be found on coasts and estuaries and in fresh water habitats such as rivers, streams, ditches and lakes. Otters construct holts (or dens) within their home ranges. These may include woods and wet areas adjacent to the water itself. Holts take a variety of forms including cavities in the roots of bank-side trees, piles of logs or flood debris, drains and caves. Otters also have a number of resting sites (sometimes called couches or hovers), above ground. These may occur in reed beds or scrub.

Red squirrels

Red squirrels are becoming increasingly rare as they succumb to a squirrel pox carried by grey squirrels, to which the latter are immune but reds are not. Red squirrels are generally found in woodland.

Reptiles

All reptiles in the UK are protected and should not be killed or injured. In the UK there are six native species of land-dwelling reptiles:

- common lizard
- sand lizard
- slow-worm
- adder
- grass snake
- smooth snake.

Reptiles hibernate in winter but are active during the rest of the year. This can make locating reptiles difficult for the construction professional. The author has experience of a large project surveyed in winter, when no grass snakes were evident, only to have them appear once construction work had begun in the spring. The project was stopped for many months at a cost of over half a million pounds.

Reptiles breed between April and June and the young are born between July and October. Reptiles can be found throughout the UK and are typically found in dense grassland, or scrub with open areas where they can bask. They may also be found in vegetation on railway embankments and in hedgerows, and sheltering under rocks and logs.

Amphibians

Great crested newts and natterjack toads are fully protected by law. Natterjack toads are extremely rare in Britain and unlikely to be found on potential construction sites. This is not the case with great crested newts unfortunately, and they can be a great problem for the construction professional.

Great crested newts can be found in ponds in rural, urban and suburban areas. Great crested newts are nocturnal and spend most of their time within grass, scrub, woodland and under logs within 500 metres of a water body. They spawn in the ponds between March and June. It is an offence to intentionally kill, injure or take a great crested newt; or to recklessly damage or destroy access to both its breeding sites and its terrestrial habitat; or to disturb a great crested newt while it is occupying a place which it uses for that purpose. What this means in practice is that newt surveys have to be undertaken where there are likely ponds or other habitats within a radius of about 500 metres of the potential construction site. These surveys can be difficult and take a long time on linear projects across open countryside, such as gas or water pipelines.

Once found, great crested newts can be moved by a registered newt handler to another location (called translocation). 'Newt-proof fences' are then erected around the construction site or around the newt habitat. Such fences are usually about 500 mm high, relatively inexpensive, buried 200 mm in the ground and constructed of plastic mesh or similar material.

Nesting birds

United Kingdom legislation provides general protection for all wild birds, and prohibits the killing, injuring, taking or selling of any wild birds or their nests or eggs. All birds and their nests are protected under the Wildlife and Countryside Act, while some rare species, such as the barn owl, birds of prey and kingfishers, carry further protection against disturbance.

Where construction work involves tree or hedge removal, it should be arranged to fall outside the nesting season, which is from March to the end July. This can be a major consideration in planning a construction project.

Birds can also be found nesting in unusual places on-site, including on scaffolding or machinery. If this occurs the equipment cannot be used until the birds have finished nesting and the area may need to be sealed off to prevent disturbance.

Hedgerows and trees

Hedgerows

Hedgerows are protected under the 1997 Hedgerow Regulations, which make it a legal requirement to notify the relevant district or borough council before removing a hedgerow or part thereof. Under regulations the removal of any hedge longer than 20 metres requires planning permission. If the hedge is shown to be significant in terms of its age, environmental or historical importance, then the planning authority can refuse such permission.

Tree protection

A tree preservation order (TPO) protects a single tree, group of trees or woodland. It protects trees from being topped, lopped or removed. Consent must be obtained before any work on a tree with a TPO is undertaken. Failure to gain permission is an offence and could result in a fine. Trees subject to a TPO are usually mature trees of particular variety, stature or interest. All trees located within a conservation area are protected. The designation of **conservation area** ensures that all new development is sympathetic to the special architectural and aesthetic qualities of the area, particularly in terms of scale, design, materials and space between buildings. The local authority's planning department should be contacted to determine trees subject to TPOs and before any construction work commences in the vicinity of any protected trees. It should be remembered that tree roots usually spread for a distance which is greater than the height of the tree.

Noxious and invasive plants

Ecology is adversely affected through the spreading of noxious and invasive plants, and it is an offence to cause their spread. These plants are covered by Schedule 9 of the Wildlife and Countryside Act 1981 or under the Weeds Act 1959:

- Japanese knotweed (grows densely, shades other plants)
- Giant hogweed (poisonous sap)
- Ragwort (poisonous to sheep, horses and similar animals)
- Thistle (shades out other plants, reduces biodiversity)
- Himalayan balsam (grows densely, shades out other plants).

It can be very expensive to remove Japanese knotweed through removal and disposal of surrounding soil because its roots can spread for many metres. If the construction professional comes across what appear to be noxious or invasive plants on-site then it would be wise to take expert advice.

Designated sites

Sites with important ecological attributes (both plant and animal species) or natural landforms can be given special protection that can be applied at a regional, national or international level according to their importance or rarity. Examples of designated sites are:

- Area of Outstanding Natural Beauty (AONB)
- County Wildlife Sites (CWS)
- Local Nature Reserve (LNR)
- National Nature Reserve (NNR)
- Special Area of Conservation (SAC)
- Special Protection Area (SPA)
- Sites of Importance for Nature Conservation (SINCs)
- Sites of Nature Conservation Interest (SNCIs)
- Site of Special Scientific Interest (SSSI)
- RAMSAR (Wetlands of international importance)
- Regionally Important Geological Sites (RIGS)
- World Heritage Sites.

It is an offence to carry out works on a designated site without the permission of the nature conservation bodies. They usually require several months' notice before works begin, assuming permission is given. Similarly, planning consent is much more difficult to obtain in any designated site.

The construction professional is most likely to come across Conservation Areas, SSSI and ASSI categories. The JNCC website states that 'Legislation in the United Kingdom makes provision for Sites of Special Scientific Interest (SSSIs) designated for their biological or geological features. By March 2005, there were

6,569 SSSIs in England, Scotland and Wales, and a further 225 Areas of Special Scientific Interest in Northern Ireland (ASSIs), covering between them over 2.4 million hectares.'

A useful contact is Natural England which was established in 2006 by the Natural Environment and Rural Communities Act 2006. It was formed by the amalgamation of three founder bodies, the Countryside agency, English nature and the Rural development service. Natural England has power to designate AONBs and SSSIs. Natural England publishes interactive maps showing SSSIs and other features on its website (www.naturalengland.org.uk).

Conservation areas

These are very common. A conservation area is defined by the Planning Act 1990 as 'an area of special architectural or historical interest, the character and appearance of which it is desirable to preserve or enhance.' They are therefore designated areas where building works, particularly to existing buildings, may be controlled and even minor alterations may require planning consent. Other items which may be controlled are street lighting levels and types and kiosks for electricity or other services. Conservation areas can usually be found by accessing the website of the relevant local authority. They will also be found in the development plan for that area. A conservation area (relating mainly to buildings) should not be confused with a Special Area of Conservation (SAC) which relates to the protection and improvement of habitats.

Carrying out an environmental investigation

When carrying out an environmental investigation, it is important to consider all stages of the investment process, not just construction. Environmental risks should be considered for the following work stages:

- Surveying work
- Site investigation (particularly ground investigation and boreholes)
- Construction work, including
  - Location of site offices and cabins
  - Location of storage areas
  - Access and parking arrangements
  - Temporary works
  - Excavation and ground work, including dewatering and disposal of groundwater
  - The main construction work itself
- Site emergencies
  - Such as spillage of fuel
- Future operation of the constructed asset.

Step 1 – Identify the environmental obligations of the project

Identify any potential legal obligations that may affect the project such as contaminated land and waste disposal.

- Identify environmental or heritage issues.

- Review carefully any environmental reports or other information provided by the employer.

Step 2 – Identify the environmental risks particular to the site

Carry out a desk study
- Review all relevant documentation.
- Check any available maps, such as Ordnance Survey maps.
  - Look particularly for watercourses, footpaths, woodland and historic or archaeological items shown on these maps.
- Search all relevant databases (examples are given above) to determine the location of SSSIs or designated sites.
- Contact environmental regulators at an early stage to determine any concerns that they may have.
- Discuss the project with the local authority.
- Obtain all relevant details in the employer's possession, such as information on previous local projects, databases, lists of personal contacts and so on.
- Liaise with the employer and any other designers to establish how they can help identify and overcome potential environmental difficulties.
- Where the project is on an existing site, talk to current operational personnel, who will often have extensive knowledge of previous projects or operational issues.
- Document your findings in an environmental risk register for the site. This will be developed as the investigation proceeds.

Carry out a site survey
- Before visiting the site, ensure that the permission of any relevant landowners has been obtained, preferably in writing.
- Carry out a personal health and safety risk assessment for your site survey.
- It is usual for at least two people to carry out the site survey. This avoids 'lone working' (a Health and Safety issue to be avoided where possible).
- Another advantage is that two or more people are more likely to identify issues and are able to discuss them later on return to the office or other base.
- Survey the site in order to:
  - Confirm the findings of the desk study and add to them as appropriate.
  - Assess the size and location of any watercourses or bodies of water.
  - Note footpaths or bridle paths.
  - Look for trees that may have preservation orders on them.
  - Look for hedgerows that could need removal.
  - Identify the location of possible protected species.
  - Identify possible invasive plants.
  - Identify any suspected contaminated land.
    - This may be the result of operations within any disused industrial building. Local street names like 'gas street' or 'mill street' may yield further clues of former industrial sites.
    - Topography such as unusual mounds may indicate previous site clearance or filling.
    - Former industrial tips may be indicated on old maps of the area.
    - Investigate the sites of former garages for petrol and oil contamination.

- Note any particular environmental features of the site, particularly undisturbed areas of grass or scrub which may hide protected species or flora.
- Note any heritage features of the site, particularly old buildings or remains of old buildings.
- Note any **environmental receptors** and assess their sensitivity.
- Determine where further investigation will be required, particularly from experts.

An **environmental receptor** is anything that receives something from the construction work or the future constructed asset. Rivers and watercourses may receive groundwater during excavation (a licence will be required), a nearby school or hotel may 'receive' noise, a badger sett may be disturbed and so on. Other receptors could be nearby houses or other buildings. Wherever there is an environmental receptor, it is wise to note its distance, its key features and its likely sensitivity. Most environmental receptors will require further investigation and assessment.

For many straightforward projects the site survey will yield little evidence of an environmental nature, in which case the designer can proceed with confidence. Wherever features are noted that are outside the area of competence of the site survey team, expert advice must be obtained. This is often the case where the location or existence of protected species is in doubt.

Step 3 – Complete documentation and take further advice

- Complete any standard site survey forms that your company may use.
- Update the environmental risk register.
- Take further advice where necessary on return to office base.
- Arrange any return visits for yourself.
- Appoint experts where necessary for detailed investigations.
- If any significant environmental risks related to the site are detected, inform the employer.

Step 4 – Prepare an environmental management plan (EMP)

- Information gathered in steps 1 to 3, particularly the risk register, can be used to form the basis of an environmental management plan (EMP).
- The EMP should be site specific, accessible, regularly revised and referred to on a regular basis.
- It should assign responsibilities for actions.
- The EMP will be used during the design stages and can later be communicated to the contractor so that it can also form the basis for the construction stage environmental management plan (CSEMP).
- The contractor can use the CSEMP to develop method statements to describe in detail how specific items of work will be carried out with due regard to environmental matters and who will be responsible for them.
- The CSEMP should include plans for any emergencies such as spillage, particularly if nearby watercourses may be affected.

Step 5 – Monitoring and use in future projects

- A monitoring system should be implemented to ensure that items on the EMP are being addressed in the design.
- All monitoring data should be retained so there is an audit trail.
- Any databases of environmental features, location of protected species or details of service providers' plant and equipment should be updated for use on future projects.

Enhancing the environment

A really successful construction project will actually *enhance* the environment. The Institution of Civil Engineers used to say 'Civil engineers tame the great forces of nature,' but the Institution now says 'Civil engineering is all about creating, improving and protecting the environment in which we live.' So not only should construction professionals 'protect', they should also 'improve'. They should try to balance environmental, social and economic factors to achieve sustainability. Some ways in which this can be achieved are:

- Careful and creative design to improve the visual aspects of the area.
- Landscaping and sympathetic planting.
- Creation of new habitats.
- Sustainable drainage systems.
- Consideration of flood risk.
- Careful transport planning with attention to 'green travel', pedestrians and cyclists.
- Minimisation of energy use and carbon footprint.
- Minimisation of waste and maximum reuse of materials.
- Use of materials from sustainable sources.
- Effective use of insulation to reduce heating and cooling requirements.
- Reuse of water and 'grey water' from drainage.
- Maximisation of natural light and heat sources.
- Consideration for the local community.

Chapter summary

- A comprehensive site investigation is essential for all projects.
- The overall site investigation will usually include a ground investigation to determine physical and mechanical properties of materials.
- Site investigation will include the identification of all relevant stakeholders, landowners and regulators.
- These bodies will all require extensive further consultation and communication.
- Regulators have legal powers with respect to construction, and their permission must be gained to avoid delays and possible prosecution.
- Service providers such as gas, electricity, water and telecommunications have buried plant and equipment throughout the UK, particularly in urban

areas. This plant and equipment will always need to be located precisely before construction begins, and may need diverting.

- The identification of protected species affected by the project has become a major concern of regulators and hence is fundamentally important to construction professionals.
- Site surveys should be carried out thoroughly and with care.
- Environmental risk registers and environmental management plans should be prepared and used.
- All projects should minimise waste and harmful effects on the environment.
- Wherever possible, construction projects should enhance the environment once completed.

8

The Construction (Design and Management) Regulations 2007 (CDM 2007) and safety

Aim

This chapter explains the requirements of the Construction (Design and Management) Regulations 2007 (CDM 2007) and their impact on the construction professional. It also introduces safety consciousness.

Learning outcomes

On completion of this chapter you will be able to:

>> Explain the importance of the CDM regulations and their purpose.
>> Describe the roles of the duty holders.
>> Describe and be able to discharge your duties under the regulations.
>> Carry out risk assessments.
>> Apply the principles of prevention.
>> Define the terms hazard, risk and control measure.
>> Describe the Reporting of Injuries, Diseases and Dangerous Occurrences Regulations 1995 (RIDDOR).
>> Appreciate why safety considerations underpin all aspects of your work.

Note on terminology

The CDM 2007 regulations refer to the employer as the 'client'. The CDM terminology will be used in this chapter.

Introduction

> Designers create many of the risks; contractors can mostly only manage the
> risks, while the workers have to endure them. Clients must learn more about
> their responsibilities and think about the risks. Construction professionals
> are involved at all stages. We need to work together to ensure consistently
> high standards.
> *'Health and Safety in Construction – Guidance for*
> *Construction Professionals'. ICE*
> *John Barber MA LLB CEng FICE MHKIE FCIArb Barrister*

In the construction industry the site is the dangerous place, not the design office
or the offices of the employer. Most accidents, injuries and deaths occur on-site.
What this quotation emphasises is that many **risks** could be eliminated by improved
design and more aware employers ('clients' in the context of the quotation). Since
accidents and deaths almost always occur on-site it is all too easy to blame the contractor, and in fact this approach used to be very common. Contractors may be at
fault of course, but they have to manage a difficult and complex construction
process, outdoors, with ground conditions that cannot always be predicted and
with a workforce that may be drawn from many trades and the self-employed, and
today from many countries. Additionally some of the larger employers seek to
impose their own views and practices on health and safety. In view of all these
factors, for the contractor, managing health and safety is a Herculean task.

The construction industry is one of the most dangerous industries in the UK.
The people killed or seriously injured on sites are generally not the managers who
organise the work and draw up the risk assessments, but the workers. Figure 8.1
shows the trends in fatal and major injuries from 1989 to 2009 and is taken from
the Construction Intelligence report published by the **Health and Safety Exec-**

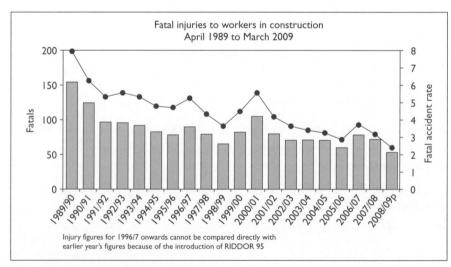

Figure 8.1   Trends in fatal injuries (chart 5).

utive, which can be accessed on www.hse.gov.uk/construction/pdf/conintreport.pdf. The bars show the number of fatal accidents by year. The line *above the bars* is the fatal accident rate (number of fatal accidents per 100,000 workers).

When we examine the causes of these fatal injuries, we see that 'falls from a height' is the major contributor, followed by being struck by a moving or flying object as shown in Figure 8.2. The '09p' means that at the time of publication, the 2009 figures were provisional.

The HSE and HSC as regulatory bodies

The Health and Safety Commission (HSC) was created under the Health and Safety at Work Act 1974. The Health and Safety Executive (HSE) functions on behalf of the commission. The purpose of these two bodies is to promote awareness of health and safety issues, to provide guidance to both organisations and individuals, to assist with the development of legislation and where necessary to take enforcement action such as prosecution in the courts. In practice, the construction professional is more likely to deal with the HSE, hopefully for advice and guidance, rather than as the subject of enforcement. The HSE have an excellent website at www.hse.gov.uk which is packed with valuable and practical information.

The development of CDM 2007

The CDM regulations aim to improve safety in the construction industry by allocating roles and duties and giving guidance. The original CDM regulations were introduced in 1994 but concerns were raised that their complexity and the

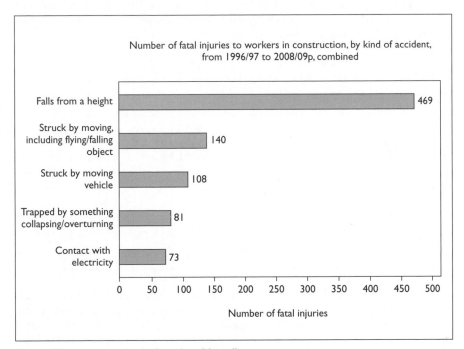

Figure 8.2 Injuries by kind of accident (chart 6).

bureaucratic approach of many duty holders frustrated the underlying health and safety objectives of the regulations. These views were supported by an industry-wide consultation in 2002 which resulted in the decision to revise the regulations. The CDM regulations were revised and republished as CDM 2007.

The Health and Safety Executive (HSE) enforces CDM 2007 in accordance with the Health and Safety Commission's (HSC) published Enforcement Policy Statement. 'Enforcement' has a wide definition, ranging from advice, to more formal sanctions and prosecution and the HSE say:

The principles of the enforcement policy are:
- Proportionality
- Targeting
- Consistency
- Transparency
- Accountability

HSE will focus on serious problems and not trivialities.

The detail of the Construction (Design and Management) Regulations 2007 (CDM 2007) and the ACOP

The CDM 2007 regulations are supported by an Approved Code of Practice (ACOP) which amplifies and explains the regulations themselves in more straightforward language. The ACOP is not legally binding, as the regulations are, but since the ACOP represents official good practice, it ought to be followed. The ACOP is available as a free download on the HSE website.

The ACOP to CDM 2007 is a practical, working document for use by construction professionals.

The regulations place a lot of emphasis on the role of the client (the 'employer') in improving safety. The client could be an individual, partnership or company and includes property developers or management companies for domestic properties.

Elsewhere in this book such a 'client' is called the 'employer' but we shall use 'client' within this chapter to avoid confusion.

However CDM 2007 does not apply to *domestic* clients, such as householders having building work or renovation done to their house because 'client' refers to persons who 'in the course or furtherance of a business', either carry out a project themselves or accept the services of someone to carry out such a project for them.

It is important to note that it is the *type of client* that matters in terms of the application of CDM 2007, not the type of work. So, for example, construction work to a shop would come under CDM 2007 because the client is a business; even though the work is similar to work we might expect to see on a domestic property such as a house.

CDM 2007 does not apply to 'domestic' clients, but only to businesses.

CDM 2007 is divided into 5 parts:

Part 1 deals with the application of the regulations and definitions:

- The regulations (except for Part 3 below) apply to all 'construction work' in the UK, carried out by or for businesses.

Part 2 covers general duties that apply to *all* construction projects:

- Competence, cooperation, coordination, prevention
- Duties of clients, designers, contractors.

Part 3 contains additional duties that *only* apply to **notifiable** construction projects (more on 'notifiable' below):

- Duties of clients, designers, contractors
- CDM coordinators, principal contractor
- **Health and safety file**
- **Construction phase plan.**

Part 4 contains Regulations 26 to 44 which are practical requirements that apply to *all* construction sites.

26. Safe places of work
27. Good order and site security
28. Stability of structures
29. Demolition or dismantling
30. Explosives
31. Excavation
32. Cofferdams and caissons
33. Reports of inspections
34. Energy distribution installations
35. Traffic routes and vehicles
36. Prevention of drowning
37. Traffic routes
38. Vehicles
39. Prevention of risk from fire etc
40. Emergency routes and exits

41. Fire detection and fire-fighting
42. Fresh air
43. Temperature and weather protection
44. Lighting

Part 5 contains the transitional arrangements and revocations for the transition to CDM 2007. These can be ignored since CDM 2007 has been in force for many years now.

Notifiable projects

> **Notifiable projects**
> Regulation 1 says that 'for the purposes of these regulations a project is notifiable if the construction phase is likely to involve more than:
>
> a) 30 days or
> b) 500 person days
>
> of construction work.'

Any day on which construction work is carried out (including holidays and weekends) should be counted, even if the work on that day is of short duration. A 'person day' is one individual, including supervisors or specialist trades, carrying out construction work for one normal working shift.

Hence for our purposes almost all of our construction projects are notifiable and Part 3 of the regulations applies in addition to the other parts. As we shall see, whether the construction professional is a designer or a contractor, CDM 2007 affects virtually everything that we do.

Who notifies and when?

Notification is carried out by the **CDM coordinator** (see below) using a standard form called an F10, which must be *signed by the client* and sent to the HSE. Clients thereby acknowledge their duties under CDM 2007. The notification should be sent to the HSE office covering the site where construction work is to take place as soon as possible after the CDM coordinator has been appointed by the client. The F10 is a straightforward form containing basic data. It is used by the HSE to collect statistics and to target inspections. Hence for example, the HSE may decide to visit sites carrying out roof work in the Midland counties.

Duty holders under CDM

CDM 2007 places legal duties on virtually *everyone* involved in construction work. Whilst the corporate body is usually named (client, contractor and so on) the performance of these duties lies with individual construction professionals who are working for these organisations. This is why the CDM 2007 is so important to us all. Those with legal duties under CDM 2007 are known as '**duty holders**' and they are:

• **Clients** as we described above.

- **CDM coordinators** – A 'CDM coordinator' has to be appointed under Regulation 14(1) to advise the client on notifiable projects. The ACOP states that the CDM coordinator's role 'is to provide the client with a key project advisor in respect of construction health and safety risk management matters. They should assist and advise the client on appointment of competent contractors and the adequacy of management arrangements; ensure proper coordination of the health and safety aspects of the design process; facilitate good communication and cooperation between project team members and prepare the health and safety file.'
- **Designers** – The term 'designer' has a broad meaning and relates to the function performed, rather than the profession or job title. Designers are those who, as part of their work, prepare design drawings, specifications, bills of quantities and the specification of articles and substances. This could include architects, engineers and quantity surveyors.
- **Principal contractors** – A 'principal contractor' has to be appointed for notifiable projects. The principal contractor's role is to plan, manage and coordinate health and safety while construction work is being undertaken. The principal contractor is usually the main or managing contractor for the project. In addition to their specific duties, the principal contractor must also comply with the duties of 'contractors'.
- **Contractors** – A 'contractor' is a business who is involved in construction, alteration, maintenance or demolition work. This could involve building, civil engineering, mechanical, electrical, demolition and maintenance companies, partnerships and the self-employed.
- **Workers** – A 'worker' is anyone who carries out work during the construction, alteration, maintenance or demolition of a building or structure. A worker could be, for example, a plumber, electrician, scaffolder, painter or decorator.

The delivery of construction projects to CDM 2007

In order to deliver safe projects and to comply with CDM 2007 there are a number of steps to be followed, important documents to be prepared and communication to be undertaken. The process and documents are described below.

Pre-construction information

Pre-construction information is all the information in the client's possession about the site and the construction work. Some will already be held by the client from previous contracts or earlier site investigations. Most clients will also have a full site investigation carried out specifically for the construction contract in question. Site investigation is described in Chapter 7. It is also usual to ensure that all service providers such as gas, water and electricity have provided relevant plans showing their services under or above the site.

It is sensible contract practice for the employer to provide such comprehensive site information. It is also a legal requirement under CDM 2007 regulation 10 which says that the client (the employer) must promptly provide designers and contractors with pre-construction information consisting of all the information in the client's possession (or which is reasonably obtainable), including:

(a) any information about or affecting the site or the construction work
(b) any information concerning the proposed use of the structure as a work-place
(c) the minimum amount of time before the construction phase which will be allowed to the contractors appointed by the client for planning and preparation for construction work
(d) any information in any existing **health and safety file**.

The health and safety file

The preparation and storage of a health and safety file is a requirement for any *notifiable project*. The file should be made available to the designer, contractor and anyone else who could benefit from it. Whilst a health and safety file does not have to be prepared for a non-notifiable project, any relevant health and safety files from previous projects should be made available to any interested parties as part of the site information.

The health and safety file is an as-built record of all features of a project that will help to reduce the risks in future construction work, including cleaning, maintenance, alterations, refurbishment and demolition. The ACOP gives a suggested list of contents, which is abbreviated below:

(a) A brief description of the work carried out.
(b) Any residual hazards which remain and how they have been dealt with (for example, surveys or other information concerning asbestos; contaminated land; water-bearing strata; buried services and so forth).
(c) Key structural principles (for example, bracing, sources of substantial stored energy – including pre- or post-tensioned members) and safe working loads for floors and roofs, particularly where these may preclude placing scaffolding or heavy machinery there.
(d) Hazardous materials used (for example, lead paint; pesticides; special coatings which should not be burnt off and so on).
(e) Information regarding the removal or dismantling of installed plant and equipment (for example, any special arrangements for lifting, order or other special instructions for dismantling and so on).
(f) Health and safety information about equipment provided for cleaning or maintaining the structure.
(g) The nature, location and markings of significant services, including underground cables; gas supply equipment; fire-fighting services and so forth.
(h) Information and as-built drawings of the structure, its plant and equipment (for example, the means of safe access to and from service voids, fire doors and compartmentalisation and so on).

The construction phase plan

The contractor prepares this important document before starting work on-site. In fact for notifiable projects Regulation 16 places a duty on the client not to allow construction phase work to start until:

- The principal contractor has prepared a construction phase plan which complies with Regulations 23(1)(a) and 23(2).
- The principal contractor is satisfied that the requirements of Regulation 22(1)(c) (provision of **welfare facilities**) will be complied with during the construction phase.

Appendix 3 of CDM 2007 gives extensive guidance on the contents of the construction phase plan. This has been summarised below:

- *A brief description of the project,* the client and other role holders and the location of existing relevant information.
- *How the work will be managed,* including management structures, arrangements for liaison, site security, induction and training, reporting, risk assessments, emergency procedures and so on.
- *Arrangements for controlling significant site risks* such as delivery and storage of materials, dealing with services, preventing falls, lifting arrangements, traffic routes and segregation arrangements, health risks such as manual handling and so on.
- *The health and safety file,* including layout and format and arrangements for gathering and storing information and so on.

The construction phase plan should be implemented, communicated by the principal contractor to any other contractors affected by it and kept up to date as the project progresses.

Welfare facilities

Schedule two of CDM 2007 lists in detail the welfare facilities that should be provided. They can be summarised as:

- Sanitary conveniences
- Washing facilities including hot and cold water, soap and towels
- Drinking water
- Changing rooms and lockers
- Facilities for rest.

Summary of duties under CDM 2007

Because the duties under CDM apply to companies and also to construction professionals themselves, they are reproduced in full from section 23 of the ACOP. They can also be found on the HSE website www.hse.gov.uk/construction/cdm/summary.htm

Summary of duties under CDM 2007

Construction (Design and Management) Regulations 2007

	All construction projects (Part 2 of the regulations)	Additional duties for notifiable projects (Part 3 of the regulations)
Clients (excluding domestic clients)	• Check competence and resources of all appointees • Ensure there are suitable management arrangements for the project welfare facilities • Allow sufficient time and resources for all stages • Provide pre-construction information to designers and contractors	• Appoint CDM coordinator* • Appoint principal contractor* • Make sure that the construction phase does not start unless there are suitable welfare facilities and a construction phase plan is in place • Provide information relating to the health and safety file to the CDM coordinator • Retain and provide access to the health and safety file (*There must be a CDM coordinator and principal contractor until the end of the construction phase)
CDM coordinators		• Advise and assist the client with his/her duties • Notify HSE • Coordinate health and safety aspects of design work and cooperate with others involved with the project • Facilitate good communication between client, designers and contractors • Liaise with principal contractor regarding ongoing design • Identify, collect and pass on pre-construction information • Prepare/update health and safety file
Designers	• Eliminate hazards and reduce risks during design • Provide information about remaining risks	• Check client is aware of duties and CDM coordinator has been appointed • Provide any information needed for the health and safety file

	All construction projects (Part 2 of the regulations)	Additional duties for notifiable projects (Part 3 of the regulations)
Principal contractors		• Plan, manage and monitor construction phase in liaison with contractor • Prepare, develop and implement a written plan and site rules (initial plan completed before the construction phase begins) • Give contractors relevant parts of the plan • Make sure suitable welfare facilities are provided from the start and maintained throughout the construction phase • Check competence of all appointees • Ensure all workers have site inductions and any further information and training needed for the work • Consult with the workers • Liaise with CDM coordinator regarding ongoing design • Secure the site
Contractors	• Plan, manage and monitor own work and that of workers • Check competence of all their appointees and workers • Train own employees • Provide information to their workers • Comply with the specific requirements in Part 4 of the Regulations • Ensure there are adequate welfare facilities for their workers	• Check client is aware of duties and a CDM coordinator has been appointed and HSE notified before starting work • Cooperate with principal contractor in planning and managing work, including reasonable directions and site rules • Provide details to the principal contractor of any contractor whom he engages in connection with carrying out the work • Provide any information needed for the health and safety file • Inform principal contractor of problems with the plan • Inform principal contractor of reportable accidents, diseases and dangerous occurrences
Workers/ everyone	• Check own competence • Cooperate with others and coordinate work so as to ensure the health and safety of construction workers and others who may be affected by the work • Report obvious risks • Comply with requirements in Schedule 3 and Part 4 of the Regulations for any work under their contract • Take account of and apply the general principles of prevention when carrying out their duties	

Hazards and risks

Hazards

The HSE define hazards in their publication *Five steps to risk assessment* at www.hse.gov.uk/pubns/indg163.pdf. They say a hazard is anything that may cause harm, such as chemicals, electricity, working from ladders, an open drawer and so on. The ACOP provides a comprehensive list in Appendix two. If we think of some typical construction hazards we could draw up a list such as the one below:

- Moving vehicles
- Buried services
- Poor lighting levels
- The edge of a high building (falls from height)
- Falling objects
- An unsecured ladder
- Collapse of formwork
- Collapse of an excavation
- Confined spaces
- Presence of groundwater
- Working with concrete
- Contaminated land.

We also need to consider less obvious hazards such as:

- Multiple occupancy of a site (many different contractors, or contractors and operational staff of the employer)
- An untidy site
- Lack of knowledge of site rules
- Using mobile phones whilst working
- Not working to the agreed method statement
- Lack of training or supervision.

In order to carry out a **risk assessment** as described below, we need to begin by identifying hazards. Start by visualising the activity taking place and write down possible hazards. If you can, visit the location of the work activity so you can picture it better. Ask colleagues what they think. Look at risk assessments for similar activities. Many employers publish lists and other guidance. If possible ask the workers who will carry out the activity what they think. Consider forming a small group to carry out a 'brain-storming exercise'. Visit the HSE website. HSE publishes practical guidance on where hazards occur and how to control them. Look at manufacturers' information or data sheets or other publications.

Remember to think about long-term hazards to health (this could be high levels of noise or exposure to harmful substances) as well as safety hazards. A long-term hazard to health is smoking, of course. A construction example is 'vibration white finger' which is a serious condition caused by long-term use of vibrating tools such as jack-hammers.

Risks

The HSE go on to say that 'a risk is the chance, high or low, that somebody could be harmed by hazards, together with an indication of how serious the harm could be.' In assessing risks, the HSE say 'a risk assessment is simply a careful examination of what, in your work, could cause harm to people, so that you can weigh up whether you have taken enough precautions or should do more to prevent harm. Workers and others have a right to be protected from harm caused by a failure to take reasonable control measures.'

Hence a risk is the likelihood of an undesired outcome multiplied by the severity of the harm.

RISK = LIKELIHOOD × SEVERITY

This is also stated as risk equals probability multiplied by consequence. The result is the same. So we see that risk has two components. Let us suppose that you are going to undertake some dangerous sports with the likelihood and severity as shown. Which sport would you prefer?

- 1/100 chance of a broken finger
- 1/10,000 chance of a broken back
- 1/10,000,000 chance of death.

We do mental risk assessments all the time. We might cross a minor road through a village without using a nearby pedestrian crossing, but would we do this on a busy dual carriageway? If the speed limit on both was 40 mph, the severity of our injuries would be the same, but the likelihood is a different matter. We might contemplate swimming in a lake, but we are unlikely to swim in a fast-flowing river since the likelihood of drowning is much greater.

The process of assessing risks is not usually particularly scientific, but a risk assessment matrix is often used. A five-point matrix could use ratings thus:

LIKELIHOOD	RATING	SEVERITY	RATING
Certain	5	Fatality	5
Highly likely	4	Major RIDDOR	4
Likely	3	RIDDOR over three-day injury	3
Unlikely	2	Not RIDDOR	2
Highly unlikely	1	Very minor injury, person can continue working	1

The term 'RIDDOR' in the table is explained below.

We can see that the maximum risk score is twenty-five (certain × fatality) and the minimum is one (highly unlikely × very minor injury). A company will have a policy on this, and possibly another matrix to determine actions and **control measures**. However if we adopt a simple approach a score of five or less would be acceptable for work to commence without further assessment. At six we could allow work to commence if all reasonable steps had already been taken to reduce risk. At a score of eight or more, work could not commence and further risk assessment must be carried out and more control measures put in place to reduce the risk. This reduced risk is called the residual risk. Our assumption here is that work cannot commence until the original risk or the residual risk is less than six.

Risk assessment

Proper risk assessment is a fundamental part of all good health and safety practice. Construction professionals *are required by law* to:

• Assess the risks inherent in designs
• Assess the risks associated with all construction activities.

The HSE publish a guide to assessing risks called *Five steps to risk assessment* at www.hse.gov.uk/pubns/indg163.pdf. They also publish many examples of risk assessment, for bricklayers for example, on their website.

Who should assess risks?

Designers

The ACOP suggests that designers should critically assess their design proposals at an early stage and then throughout the design process to ensure that health and safety issues are identified, integrated into the design process and addressed as they go along. These assessments are called 'design risk assessments', or DRAs. Of course it is not always reasonably practicable to eliminate every hazard (such as the need to work at height or in excavations) but design solutions should reduce risk to an acceptable level by reducing the likelihood and potential severity of harm, trying to reduce the number of people involved and for how long.

Contractors and all employers of people

All employers of people have a legal duty to carry out risk assessments. This is set out in the Management of Health and Safety at Work Regulations 1999 and regulation three says:

'Every employer shall make a suitable and sufficient assessment of –

(a) the risks to the health and safety of his employees to which they are exposed whilst they are at work and
(b) the risks to the health and safety of persons not in his employment arising out of or in connection with the conduct by him of his undertaking.'

The risk assessment process

Everyone connected with the investment process should assess risks. So we see that risk assessment actually begins with the client (employer), who also has clearly defined duties under CDM 2007.

The client (employer)

The risk assessment process begins with the client (employer) when considering different options and the timescale allowed for design and construction work.

The client should then ensure that all pre-construction information is provided. In doing this, the client should consider risks. Hence where there are buried power cables, the client might consider relocating a structure to avoid them or would ensure that they were identified and marked in the pre-construction information. The client should also consider having them moved by the electricity company. The client reduces potential risks further by only appointing experienced and competent people to the many roles in the investment process.

The designer

The designer should carry out design risk assessments (DRAs), using the principles of prevention, described below. The DRA should attempt to remove the risk or reduce it. A very simple example is reinforcing bar (or rebar as it is known). Rebar is heavy. A six metre long 25 mm bar weighs 24 kg. The manual handling limit for a fit male lifting in an ideal position is 25 kg. Hence a seven-metre bar exceeds the limit. Designers should consider the weights of rebar when detailing reinforced concrete. Shorter bars weigh less. Of course there is a limit to the process since all overlapping bars must use the minimum lap length for stress transfer.

As another example, a designer contemplating a pipeline crossing a highway at a depth of three metres (in order to go below an existing sewer) should consider the risks associated with working in deep trenches, traffic control and also possible injury to highway users. An alternative would be to use a 'no-dig' technique of pipe-laying such as directional drilling. This would be safer. Here the designer is 'avoiding the risk' to use the terminology of the **principles of prevention** which are set out below.

Where a hazard cannot be prevented, the designer must ensure that the nature of the hazard is properly communicated to the contractor, so that the contractor can take adequate precautions.

The contractor

The contractor plans and organises the work on-site. Before any work activity takes place the contractor must carry out a risk assessment thus:

- Obtain all relevant information and review information provided by the designer as above.
- Identify the hazards. Most contractors publish guidance for their staff on the hazards associated with different types of work activity such as scaffolding, pouring concrete, fixing formwork and so on.
- Try to organise the work to avoid the hazard if possible.
- Decide **who might be harmed and how**.
- Evaluate the risks and try to reduce them. All risks must be brought down to an acceptable level of residual risk or control measures must be put in place which have the same effect.
- Decide on precautions and **control measures**. In doing this the contractor will look to their company systems for advice on typical control measures.
- Write down the risk assessment in order to record it.
- Prepare a **method statement** based on the risk assessment to describe in straightforward terms how the work will be carried out.
- Communicate the method statement to the workers who will do the work and their supervisors. Ensure they understand it.

- Check at intervals that the method statement is being followed.
- After the work is complete, discuss any learning points with the workforce in order to improve future method statements.
- Keep records for future use.

Deciding who might be harmed and how

In deciding who might be harmed and how, the HSE give this advice:

> For each hazard you need to be clear about who might be harmed; it will help you identify the best way of managing the risk. That doesn't mean listing everyone by name, but rather identifying groups of people (such as 'people working in the storeroom' or 'passers-by').

Some workers, such as inexperienced or young workers or those whose English is not good are subject to particular requirements. The worker carrying out the activity is not the only person to consider. Activities can generate dust and noise that may affect nearby workers, other contractors on the site may be affected by activities and so might the general public.

Control measures

Control measures are procedures, processes or equipment that further reduce risks. Examples would be:

- Demarcation fencing and barriers to separate vehicles from people
- 'Traffic calming' to ensure vehicles travel at slow speeds
- Good signage and warning notices
- 'Beepers' fitted to vehicles when reversing
- The use of banksmen to direct the operation of excavators
- Training and information such as 'tool-box talks'
- Hand railing and edge protection
- Barriers
- Signs and guards
- Personal protective equipment (PPE)
- Welfare facilities.

PPE is often seen as an easy answer to reducing risks, whereas it ought to be seen as the last resort. Hence for example control measures such as 'wear gloves' and 'wash hands after working' are not the real answer to reducing the risks from handling dangerous chemicals. The risk should be reduced in the first place by purpose-designed containers, means to prevent spillage, mechanical handling gear where possible and so on.

Method statements

Once the risk assessment has reduced risks to an acceptable level, a **method statement** is usually produced. This method statement describes how the activity should take place in a safe and proper manner. It should be in plain language, not too long and easy to understand. The method statement should also set out clearly the control measures that are to be used. The method statement should be read, signed and understood by the operatives who will carry out the work.

The principles of prevention

The ACOP states in Appendix 7 (The principles of prevention):

'Duty holders should use these principles to direct their approach to identifying and implementing precautions which are necessary to control risks associated with a project.'

The general principles of prevention (from ACOP Appendix 7)
- Avoiding risks
- Evaluating the risks which cannot be avoided
- Combating the risks at source
- Adapting the work to the individual, especially as regards the design of workplaces, the choice of work equipment and the choice of working and production methods, with a view, in particular, to alleviating monotonous work and work at a pre-determined work-rate and to reducing their effect on health
- Adapting to technical progress
- Replacing the dangerous by the non-dangerous or the less dangerous
- Developing a coherent overall prevention policy which covers technology, organisation of work, working conditions, social relationships and the influence of factors relating to the working environment
- Giving collective protective measures priority over individual protective measures
- Giving appropriate instructions to employees

An example of a risk assessment

A risk assessment should be repeated until alternative methods or control measures have reduced the risk to an acceptable level such that the work activity can be undertaken. So using the principles of prevention we try to remove the risk or reduce it. Once we have done this, we can introduce control measures to reduce it further.

Let us take the example of joiners fixing floor pans for the first-floor pour of a tower block. The floor pans are five metres above ground level. This is work at height and the hazard is the edge of the floor pan area. The likelihood of falling may be one in a thousand per working day, and the severity would be broken bones and possibly death. The *designer* has already done a design risk assessment and this sort of construction is usual and unavoidable on a tower block.

The *contractor* carries out a risk assessment when planning the work. Since the contractor cannot remove the hazard itself, control measures are introduced. Until sufficient floor area is fixed to enable full edge protection (handrails and toe boards) to be installed, the contractor ensures that all joiners are wearing 'fall-arrest gear' and suitable footwear to minimise slips. The contractor produces a method statement which describes how the work will be carried out and copies it and files it for future reference. The contractor then ensures that the joiners are trained in the use of fall-arrest gear and that they have all read and understood (and possibly signed) the method statement. Once the floor pans are fixed, full edge protection will be installed to protect future trades-people, such as steel-fixers and the concrete gang, from falling.

Reporting accidents (RIDDOR)

The Reporting of Injuries, Diseases and Dangerous Occurrences Regulations 1995 (RIDDOR), place a legal duty on:

* employers
* self-employed people
* people in control of premises

to report work-related deaths, major injuries or over-three-day injuries, work-related diseases and dangerous occurrences (near miss accidents).

Reporting accidents and ill health at work is a legal requirement. The information enables the Health and Safety Executive (HSE), and local authorities, to identify where and how risks arise, and to investigate serious accidents. HSE can then help, and provide advice on how to reduce injury, and ill health in the workplace. In serious cases, the HSE may decide to prosecute.

The RIDDOR regulations are very extensive and the HSE website simplifies them into the following list:

* Deaths.
* **Major injuries.**
* Over-3-day injuries – where an employee or self-employed person is away from work or unable to perform their normal work duties for more than 3 consecutive days.
* Injuries to members of the public or people not at work where they are taken from the scene of an accident to hospital.
* Some work-related diseases.
* **Dangerous occurrences** where something happens that does not result in an injury, but could have done.

Major injuries

Some reportable major injuries are:

* Fracture, other than to fingers, thumbs and toes.
* Amputation.
* Dislocation of the shoulder, hip, knee or spine.
* Loss of sight (temporary or permanent).
* Injury resulting from an electric shock or electrical burn leading to unconsciousness, or requiring resuscitation or admittance to hospital for more than 24 hours.
* Acute illness requiring medical treatment, or loss of consciousness arising from absorption of any substance by inhalation, ingestion or through the skin.

- Acute illness requiring medical treatment where there is reason to believe that this resulted from exposure to a biological agent or its toxins or infected material.

The issue that affects most contractors is the 'over-3-day' injury. This could perhaps be a pulled muscle from lifting incorrectly. Unfortunately for the contractor this is a RIDDOR, and whilst it is clearly not in the same category as amputation, or loss of an eye, it will still go on the contractor's safety record when tendering for future work.

Dangerous occurrences

Dangerous occurrences are events on-site that may not have caused injury, but that may alert the HSE to poor or incorrect safety practices. Some selected dangerous occurrences are:

- Collapse, overturning or failure of load-bearing parts of lifts and lifting equipment.
- Explosion, collapse or bursting of any closed vessel or associated pipework.
- Plant or equipment coming into contact with overhead power lines.
- Electrical short circuit or overload causing fire or explosion.
- Any unintentional explosion, misfire etc.
- Collapse or partial collapse of a scaffold over five-metres high, or erected near water where there could be a risk of drowning after a fall.
- Dangerous occurrence at a pipeline.
- Explosion or fire causing suspension of normal work for over 24 hours.
- Accidental release of any substance which may damage health.

Chapter summary

- Considerations of health and safety are fundamental to all stages of the investment process.
- The HSC and HSC are regulatory bodies who advise on health and safety.
- The Construction (Design and Management) Regulations 2007 (CDM 2007) must be followed by clients and all construction professionals.
- Important aspects of CDM 2007 are the roles of duty holders, the health and safety file and the construction phase plan.
- Risk assessment is a key part of the design and construction process.
- A hazard is anything that can cause harm.
- A risk is the result of multiplying the likelihood of an event (one in a thousand) by its consequence (serious injury, death).
- To reduce risks we carry out risk assessments and use the principles of prevention and we install suitable control measures.
- We then produce method statements to describe the work to be done and the control measures that must be used.
- Reporting accidents and ill health at work is a legal requirement under the RIDDOR Regulations 1995.

9

Contract strategies: separated, integrated and management

Aim

This chapter aims to improve knowledge of the different contract strategies available for construction procurement.

Learning outcomes

On completion of this chapter you will be able to:

>> Appreciate the nature of construction contracts.
>> Describe the different contract strategies such as 'Traditional', 'Design and Build', 'Management Contracting' and so on.
>> Apply the best contract strategy to different cases.

Contract strategies

The contract strategy is a major part of the investment process. The contract strategy determines the contract links between the employer, the design organisation, contractors, sub-contractors and other professionals engaged in the construction process. Each of these links will also have contract documents and usually a form of contract. The contract *documents* and form of contract will determine the detail of the way in which the contracting parties engage with each other; but the contract *strategy* will determine the links between all the most important contracting parties. As an example compare the traditional strategy with the design and build strategy. The links between the employer, designer and contractor are shown in Figure 9.1.

The fundamental difference between these two strategies is who the design organisation works for: the employer or the contractor. This has very significant effects which will be described in the relevant discussion on the strategies below. The solid arrowed lines indicate legally enforceable contracts with contract doc-

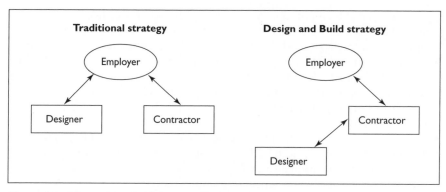

Figure 9.1   The traditional and design and build strategies.

uments and a form of contract (such as NEC3 ECC, JCT SBC05 or JCT DB05 in the case of design and build contracts).

There are a number of factors to consider when contemplating a suitable contract strategy, but the two primary considerations are the *allocation of risk* and the involvement of the contractor in the design; at what stage, and if at all. One of the major problems with some contract strategies such as the 'traditional strategy' is that the design is often almost complete before the contractor becomes involved. This can lead to communication problems, lack of **'buildability'**, unrealistic timescales, inefficiency and waste. Buildability is a term that is often used by contractors to indicate a project where real thought has been put into how it will be constructed in order to produce safe, efficient construction at minimum cost, making use of the latest techniques or those available to that particular contractor.

Factors which determine the contract strategy used

Today, for most major employers, the implications of the Public Contracts Regulations, or the Utilities Contracts Regulations (both described in Chapter 13), called **'the regulations'** below, will have a major impact on the contract strategy used and the way in which that strategy is implemented, particularly in terms of tendering and contractor selection. Other significant factors are:

The nature of the employer

- The first question is whether the employer and the contract in question are subject to 'the regulations'.
- Is the employer experienced in construction procurement and projects?
- Will the employer take an active part in the design or management of the project?
- The employer's level of risk acceptance.
- How will the employer raise the funds and is there a timing requirement as far as payments are concerned, such as cash flow considerations?
- How certain the employer needs to be of the likely final cost of the project and at what point the employer needs to know this final cost.

- How certain the employer is as to exactly what employer-needs the project should address? Any uncertainty can give rise to a long period of project development and the likelihood that the specification for the project may change.
- When does the employer require the project to be completed, and how flexible is this requirement? In other words, will the employer pay a premium for the increased cost and risk of an earlier completion date?
- How can the project maximise any employer public relations or reputation considerations?
- Can the project be used to enhance the employer's brand or advertising?
- How does this project fit in with any existing operational use of the site and the employer's operational strategy?
- For employers with longer term programmes of work:
  - How does the project fit in with the employer's longer term procurement and contract strategy?
  - Can it be incorporated into an existing EU advertised strategy, where the employer is subject to the regulations?
  - Does the employer have existing contract, purchasing or framework agreements which need incorporating?
  - How will the project fit into the employer's business plans and any commercial or customer-based considerations?

Design issues

One of the major differences in contract strategy is the matter of design. It is important to decide who does the design and who they work for. This will depend largely on resources and the difficulty or complexity of the design for any particular project. The employer should consider the following issues when determining their contract strategy:

- Does the employer wish to carry out some of the early stages of the investment process using 'in-house' resources? In-house is a term for staff who are directly employed as part of the employer's organisation.
- The employer's involvement in the design. Does the employer employ experienced investment and design staff for example? Does the employer wish to take an extensive role in design decisions, or will the employer leave this to the design organisation?
- The amount of design that can sensibly be done before the contractor is selected.
- The experience and skill of the design organisation (if one is appointed), especially in regard to construction methods, or particular items of plant or equipment.
- The benefits of integrating or separating design and construction.
- The benefit of getting contractor involvement in the design.
- The standard of design liability and who it is allocated to.

What will contractors accept in the prevailing market climate?

Wise employers always consider the likely reaction of tenderers to their contract strategy and also to the form of contract used and any amendments they make to it. These employer decisions fundamentally affect the tendering contractors' risk. Contractors are more likely to accept higher risks (or bid lower prices) at times of economic recession, or where there is a scarcity of available work. It may simply come down to a choice of bidding on conditions that a tenderer does not care for, or going out of business for lack of work. Conversely, when work is plentiful, tenderers may increase prices or talk to employers to try and influence their contract strategy and the risk profile of projects. A summary of some major factors which the contractor will take into account are:

- The amount of construction work available at that time, particularly in the sector of work that the project falls into (such as roads, bridges, office blocks, schools, prisons and so on).
- How busy the contractor is and the available resources.
- The contractor's longer term considerations as to future work and hence whether to accept a temporary overload situation if the contractor is currently busy.
- The contractor's knowledge of the employer, particularly in terms of cooperation and regular payment.
- The contractor's assessment of the project's risk profile.
- The form of contract used and the amendments to it, which affect matters such as payment and risk.
- The payment mechanism used in the contract which influences the rate and certainty of payment against the costs the contractor will incur. Payment mechanisms are described in Chapter 14.
- Whether the project will be good for the contractor's portfolio of work and future marketing opportunities. This is particularly the case with novel work, or the *start* of a large tranche of similar work, such as nuclear power or the newer forms of energy generation.

Other factors

- The cost of the project.
- The complexity of the project.
- The speed required from start to completion.
- The number of interfaces with other contracts and their complexity.
- Any significant stakeholder or public relations issues.
- Ethical issues, particularly if the employer has shares bought by 'ethical investors'.

Variations on the main contract strategies

There are many different variations on the main contract strategies. The main contract strategies are listed below.

Separated design and construction
- Traditional

Integrated design and construction
- Design and build

Management
- Management contracting
- Construction management

Framework agreements (see Chapter 12)

Partnering (see Chapter 11)

Commercial arrangements (see Chapter 11)
- Alliances
- Joint ventures

Funding arrangements such as PFI (see Chapter 11)

Of these strategies, traditional and design and build are used most often. Framework agreements and partnering arrangements are often added. Management contracting and construction management enjoyed a brief vogue in the later part of the twentieth century but are little used today. In consequence, the descriptions of these two strategies will be less comprehensive than those of the more common ones. This relatively small usage is highlighted by the Royal Institution of Chartered Surveyors (RICS) survey described below.

Every two or three years, RICS publish a very interesting survey of contracts in use. The survey is available on the internet. It must be remembered that it covers mainly 'building' contracts and so reports predominantly on contracts using the JCT suite of contract forms. The 2007 survey (the eleventh undertaken) covered £7.8 billion of contracts, with 1,370 projects reporting data. It is therefore a very large and comprehensive survey, giving useful insights into current procurement in the 'building' sector. Some of the conclusions, which have been selected from the report, were:

- The vast majority of building projects use a standard form of contract.
- The majority of building contracts in the UK continue to use 'traditional' procurement.
- Over 50 per cent of contracts in the £10,000 to £50 million value bands were procured on a design and build basis.
- There has been no apparent increase in partnering between the 2004 and 2007 surveys.
- The 2007 survey has identified a much greater use of the JCT Construction Management documentation than the 2004 survey. This time 15 instances of its use was captured. This still accounts for only a tiny proportion of contracts used but represents 9 per cent of the value of projects in the sample, largely accounted for by three very large schemes averaging £216 million each.
- Management contracts continue to be used sparingly. Only 8 instances of the use of the JCT forms were identified in the current survey, though this was more than in 2004.

This survey highlights the small current utilisation of the management contracting and construction management strategies, together accounting for only 23 projects out of the 1,370 in the survey.

Separated design and construction (traditional)

This strategy has been around since the late 1800s and is notable for the separation of design from construction. Because it has been in existence for such a long time there are numerous standard forms of contract which adopt this arrangement and it still acts as the default for many construction projects. This strategy is often called 'traditional' because of this historical perspective. It must be emphasised that the word 'traditional' does not indicate an old-fashioned or outdated strategy. Indeed the traditional strategy has a valuable place and is much used.

> The word 'traditional' does not indicate an outdated strategy.

In the traditional strategy the employer begins by appointing a design organisation, usually a firm of consulting engineers or architects. An appropriate form of *design contract* should be chosen for this appointment, such as the 'NEC3 professional services contract' or 'JCT standard conditions of appointment for an architect'. These design contracts clearly set out the services required of the designer and have terms written to cover issues that might arise during design. Many employers once had their own design departments, but this is becoming less common today; hence the need for careful appointment of external design organisations.

The design organisation will frequently assist the employer in the early stages of the investment process, such as design option selection and outline design. This early assistance to the employer is one of the benefits of the traditional strategy. Another major advantage is that statutory approvals which require early design work or discussion can be obtained. These approvals could be such items as planning approval, approvals relating to environmental matters, emissions and discharges, road closures or permissions under the Traffic Management Act. For larger projects requiring public consultation or environmental impact assessments the assistance of the design organisation at this early stage is again invaluable.

Once the employer approves the outline design, the design organisation completes the *detailed* design for the project, specifies it and produces all the drawings and details required for subsequent construction. An appropriate form of contract should be selected for the construction. This *construction contract* is between the employer and the contractor. The ICE7 and most of the JCT forms of contract are based on the traditional method, as are a number of the options in the NEC3 engineering and construction contract (NEC3 ECC). These construction contracts are designed to cover access, payment, delays and many other issues that may arise during construction such as adverse weather and unexpected ground conditions.

Priced offers, called 'tenders', are invited from interested firms of contractors and assessed, and a contract is signed. The successful tenderer, now known as '*the contractor*', builds the project. The contract is usually administered and certified by the contract administrator (called the Project manager, Engineer or Architect/contract administrator depending on the form of construction contract used). This contract administrator is usually a senior member of the organisation

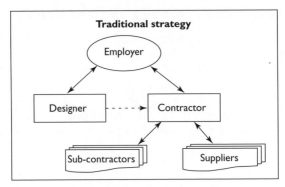

Figure 9.2 The traditional strategy showing sub-contractors and suppliers.

that designed the project. In Figure 9.2, contract lines are shown as solid lines, the contract administration link is shown as a broken line. Sub-contractors and suppliers are contracted directly to the contractor, usually under standard forms of sub-contract or terms of supply.

This contract administration link gives us another advantage which is the maintenance of design continuity from design stage and through construction. Virtually all projects require input from the designer throughout construction, for further details, clarifications, assessment of any contractor alternatives and the checking of actual conditions (especially ground conditions) against design assumptions.

Formerly, most traditional contracts used a bill of quantities (an itemised list of estimated costs – see Chapter 15), in which case the design must be complete (or almost) to enable the bill of quantities to be prepared. However traditional contracts can also be used without quantities (the contractors prepare their own) or with other payment mechanisms such as activity schedules. Payment mechanisms are described in Chapter 14.

The characteristics of the traditional strategy

- The employer selects the design organisation, and so can apply any criteria that are relevant (for employers subject to the regulations these must be *objective* criteria).
- The employer gets the benefit of taking advice and making decisions during the *design development stage* of the investment process, through their relationship with the design organisation.
- Separation of the design stages from the construction contract means that the project can be changed or stopped without incurring significant expenditure. The total cost of all the design stages is usually well below 10 per cent of the cost of the project. Hence aborting a project at this stage may be costly, but it is nowhere near as expensive as proceeding to construct an inappropriate asset.
- The design organisation can assist the employer in obtaining external approvals and consents.
- The contractor has few, if any, design obligations for the permanent works. In most forms of contract, any such contractor design has to be expressly stated.

- The design should be substantially complete at the time the contractor is selected, especially when tendering on bills of quantities. Completing the design before construction sounds like common sense, but it is not done in many of the other contract strategies.
- There are many advantages with contracts that use bills of quantities, which will be explained in Chapter 15.
- There is reasonable certainty of the final cost of a project, except for design omissions, measurement errors when preparing bills of quantity (where used) and justifiable claims under the contract.
- Matters of payment and contract claims are usually determined by the contract administrator.
- This strategy has a fair distribution of risk and is familiar to most of the construction industry.

Drawbacks with the traditional strategy

One of the main drawbacks with this strategy is that it is difficult to explore 'buildability' during the design, because the contractor has not yet been appointed. Designing and building have evolved into two separate professions in the UK, with little transfer of professional people between them. Thus most contractors know little of design, and unfortunately many designers know little about construction methods, or ways of achieving construction efficiency through design itself.

Sometimes a design can be optimised for construction generally. However a more difficult problem is that one contractor may have specific people, expertise, plant equipment or techniques which could produce real economies with an appropriate and specific design. Unfortunately, in a competitive tendering situation (which is the norm) this specific opportunity for efficiency by targeted design cannot be taken advantage of. This opportunity is one of the advantages of the design and build strategy which is described in the section below. Despite its benefits, there are drawbacks with the traditional strategy, such as:

- Misunderstandings are frequent, since the contractor is usually only introduced to the design during the tender period. There is therefore little time (often only a few weeks) for the contractor to understand the project, what is required, what the problems might be and what the risks are. These misunderstandings can lead to subsequent claims for more time, financial loss and expense.
- Designers do not always have the experience or ability to manage the interface between the design and construction.
- The strategy is slower overall because there is no concurrency between design and construction.
- Delays to design decisions, or drawings and information, after contract award, are usually the employer's liability under most forms of contract. So the employer should ensure that the designer has time and resources to complete the design and to issue any amendments or modifications once construction begins.
- The separation of design and construction into two separately appointed organisations does not assist teamwork, and may not address cultural differences between contracting and design organisations.

- Where projects are awarded on a one-off basis, this separation also makes partnering more difficult to achieve.

The traditional strategy in framework agreements

Many of the drawbacks of the traditional strategy are minimised where it is used as the selected strategy in a long-term framework agreement. Here we would have a design framework agreement and one or more construction frameworks over perhaps four or five years to undertake a programme of projects. A framework agreement, described in Chapter 12, is one in which many consecutive or concurrent separate projects are undertaken. Thus the principal drawback of the traditional strategy is removed. Construction can begin as soon as the first project is designed, and then the work can proceed on a phased design-construction basis. Additionally misunderstandings and teamwork issues are much less likely because the design organisations and contractors can build up a long-term relationship.

Design and build (D and B)

This contract strategy has existed for many years in industries outside that of construction. Much manufacturing is in effect design and build (or in this case design and manufacture). Design and build (or D and B as it is usually known) seeks to overcome the major problem with the traditional strategy, the separation of design and construction into two organisations.

The proponents of design and build hope to achieve a quicker route to contract completion by concurrent working and a quicker site start, fewer misunderstandings between designers and constructors, more harmonious working and the opportunity for 'buildability' to be fully incorporated in the design. In theory these factors produce a more efficient process and hence the possibility of a reduction in price for the employer. Unfortunately this price reduction is rarely seen because of the increased contractor risk in these projects. Having said this, design and build is a very effective strategy and is probably used more than any other in modern construction, as we saw in the RICS 2007 survey. The survey goes on to say:

'In the 1985 survey, Bills of Quantities dominated the survey with minor use of the Design and Build forms. By the 1998 survey, Design and Build was ahead of Bills of Quantities. This was the time of major shifts in procurement strategies. Clients wanted certainty of risk transfer. This survey (2007) reinforces the dominance of Design and Build as a procurement strategy, with a continued decline in Bills of Quantities. However, Bills of Quantities refuse to die.' (RICS, 2007)

Bills of quantity are a payment mechanism described in Chapter 15 and they are used in JCT contracts using the traditional strategy. So in the quotation above, RICS will be referring to the traditional strategy when they refer to bills of quantities. As we shall see in the chapters on NEC3 ECC contracts, it is also possible to use bills of quantities in the NEC3 ECC contract options B and D which are more likely to be used for civil engineering work.

Another major driver for design and build is the progressive outsourcing of many employers' in-house design departments. Hence the employers no longer have their own design capability. Without their own design capability, employers must buy it in either from a separate design organisation, or directly via the contractor in D and B. The greater simplicity of a single D and B contract is very tempting for employers in this situation. Rather than buying in design and construction separately, as in the traditional strategy, employers are often attracted by the greater ease of buying these services from one organisation as in D and B.

In the traditional strategy the contractor is tendering against a project which should have been designed completely, based on a comprehensive site investigation. By definition, in the design and build strategy the design is not complete at tender, since the contractor has yet to complete the design. The contractor usually produces an *outline* design on which to base the bid, but necessarily this design will be far from complete because of cost and time constraints at tender. Consequently there will be more unknowns and ill-defined items that are hard to price. Hence contractors tendering a D and B contract will usually incorporate a large financial allowance for risk in their tenders. In other words, the tender price of D and B contracts is often higher than those which use the traditional strategy.

The employer should always ensure that a site investigation (usually abbreviated to SI) is available for tendering contractors, so that they can better assess the risk. However, there is another difficulty here. Site investigation is not usually a 'one-shot' exercise, but may often be carried out on a number of occasions in increasing detail, as the design develops and the necessity for further investigations becomes apparent. This is described in Chapter 7. So in design and build, the later SIs may be carried out *after* the price is submitted, when of course it is too late to adjust the price. This is another major risk factor that contractors will price for in their tenders.

'Single-point responsibility'

The other major appeal of design and build to employers is 'single-point responsibility', since the employer only has a contract with the contractor. In contrast, in the traditional strategy the employer has separate contracts with the designer *and* the contractor (the constructor) as shown in Figure 9.3 which shows the contract relationships as lines with double arrows.

Under most forms of traditional contract the employer has to pay any contractor costs arising from late drawings or details, which the designer should have supplied earlier. Additionally, where a defect is found in the completed work, it may not be obvious or provable whether it is due to poor design or poor construction. If on investigation it is demonstrated to be a design error, then in the traditional strategy the employer usually has to pay the contractor for the cost of correction. The employer may have a claim in negligence against the designer of course but such claims are often difficult to sustain in court.

In design and build, neither of these problems affects the employer directly. The problem would now be a contractor issue because in D and B the *contractor* has the contract with the designer. Of course, the contractor may have to pursue the issues with the designer but for the employer, at least, this is a much simpler and clearer strategy to use. We must remember, as always, that it is the employer who wants the new asset. It is the employer who will be paying for the design

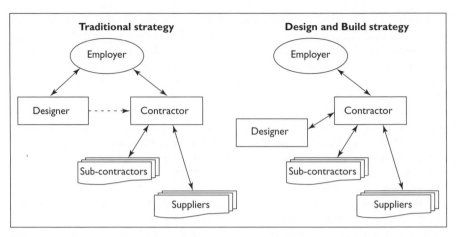

Figure 9.3   Contract relationships compared, particularly 'single point responsibility'.

and also paying for the construction, and of course 'he who pays the piper calls the tune'. So the *employer* selects the contract strategy.

Tendering issues in design and build

Another issue with D and B is the much greater tendering cost and additional time that has to be added into the tender period. This is because tendering contractors have to produce an outline design before they can price the work. For this reason it is usual to restrict the number of tenderers to three or possibly four, rather than the six tenderers that is common for the traditional strategy. Contractors rarely employ their own designers and so will use a design organisation with which they have a good relationship. One way to distribute the high cost of tendering is a 'no win, no design fee' arrangement, or possibly a 'win bonus' which the contractor pays to the designer if the tender is successful, in return for a reduced initial design fee.

A further problem with D and B is that it is difficult to compare tenders, since each tenderer will have produced a different design. However, the employer does get the benefit of the ideas of each tenderer, some of which may be very economical or original. Hence whilst tender comparison is more problematic, D and B can encourage new ideas and innovation.

A real example of employer issues in design and build

Some time ago a water company had to invest in a major extension to a water treatment works, so that treatment met new European standards for drinking water. The project cost was about £40 million in modern terms. The water company then employed 500 engineers of its own and, when extra resources were needed, they still used the traditional strategy, but through consulting engineers.

This project was very urgent and design resources were fully committed, so they used the design and build strategy. However, the water company did not appreciate the issues. Two simple examples illustrate this. Normally, as design work proceeds with a design organisation, road widths and bend radii are agreed to meet operational needs. With design and build this was not done. When the contractor's drawings arrived they showed roads of 3.5 metre width with passing bays and bends unsuitable for the large tankers required for GAC delivery (granular activated carbon, used to remove pesticides and other impurities). Modification was of course possible, at an extra cost of £700,000.

Beneath the roads were many large-diameter pipelines, some up to 1.5 metres diameter. It was normal design practice to surround pipes with granular material and then backfill with type 1 material (a standard graded stone material, used below roads) to just below road formation. This avoids the potential for road settlement in future years. This detail was not specified, and when the contractor's drawings arrived for comment they showed 'as-dug' material as backfill to road formation. Of course the use of type 1 fill was perfectly possible, but at an additional cost of £300,000. Then there were the issues around maintenance access and craneage in the buildings which had not been clearly specified, and so it went on, and on ...

The water company did not appreciate (then) the difference between the incremental design development that it was accustomed to with a design organisation, and the fundamental need to specify absolutely what is needed in a design and build contract when it goes out to tender.

The designer role in design and build contracts

The first design role is again to advise the employer on option selection, and the feasibility of the options. This will be performed by the employer's staff or external design organisation. It is particularly important that the designer explores the employer's business needs in detail, so that he or she can draw up a suitable performance specification and any design or material standards that must be communicated to the contractor, in the works information. What is crucial to the success of design and build contracts is a clear performance specification. This specification must be comprehensive and written with great care. Many employers do not appreciate the amount of consultation with the designer that goes on in the traditional design process, often over a period of many months. This consultation will take the form of meetings, correspondence, discussions and so on; all *developing* the design to meet the employer's needs. This consultation is usually absent once the design and build contract itself is awarded and this is why the initial specification and description of the project requirements are so important.

The second design role will be to produce the detailed design, but this time, on a design and build contract, *the designer is employed by the contractor*. Since the contractor carries design liability to the employer, it is crucial that the contractor has a similar design liability in the contract with the designer. This second element of design is usually greater and more time-consuming than the first. However, both are essential.

The third design role is checking contractor proposals once the D and B contract is awarded. The fourth is checking design assumptions once construction commences. Both of these are discussed below.

It is important to remember that design and build *does not reduce the total amount of design effort,* it simply changes the contractual links, and allows the contractor to have a much greater influence on the design itself.

> Design and build contracts do not reduce the *total* amount of design needed, they simply allocate the responsibility differently.

Design issues before contract award

Different designs from the tendering contractors can lead to another design issue. In the traditional contract strategy the many approvals that are needed, such as planning permission, way leaves, consents, Traffic Management Act approvals and so on, are gained during the design process at *an appropriate stage of detail and before contract award*. Planning permission in particular requires a reasonably developed design, with general arrangement and elevation drawings, and details of proposed exterior materials. So the employer could accept a D and B tender (and design), which subsequently fails to gain planning permission. There are a number of ways of overcoming this. One is to have a 'severable contract'; in other words a contract with clauses allowing cancellation if say planning permission is not obtained. The more usual way is for the employer (or the design organisation) to develop the design far enough as a preliminary design to obtain the main consents and approvals, before inviting contractors to tender for the project. The contractor then *completes* the design.

Unfortunately this brings us to a legal difficulty and that is whether the contractor is obliged to check the preliminary design that has already been carried out once the contractor's tender is accepted and work on the detailed design has commenced. Checking an earlier design adds another cost of course. The current legal position seems to be that the contractor is obliged to check the preliminary design before continuing to develop it, but it depends on the specifics of the legal case as always. The safest way forward is for the contract to specify clearly what the contractor may assume to be correct in the preliminary design, and what they need to recheck.

A further legal issue with D and B is that the *default* legal standard for design is 'fitness for purpose' rather than 'reasonable skill and care', which is applied to designers in the traditional strategy. Because of the onerous nature of the 'fitness for purpose' standard, it is usually changed to reasonable skill and care by specific clauses in the contract such as option X15 of the NEC3 ECC contract. See Chapter 10 for more on this.

Design issues after contract award

There is another 'design issue' with D and B and paradoxically it arises *after* contract award. In the traditional strategy, the design should be fully developed and completed at tender. The employer will have had the opportunity to be involved, to 'comment and approve' and to make changes at little cost. In D and B, the detailed drawings are produced *after* the tender is awarded, and are based on the employer's requirements and the contractor's proposals. Many forms of contract attempt to deal with any issues of incompatibility between these requirements and proposals which arise as the detailed design is produced by the contactor. In the event of a 'discrepancy' the employer's requirements often take precedence as they do in the JCT design and build contract 2005 (JCT DB05). This contract is written specifically for the D and B strategy, whereas NEC3 ECC does not contain the very detailed provisions for this contract strategy that are found in JCT DB05.

The NEC3 ECC contract is frequently used for D and B contracts, particularly those of a predominantly civil engineering nature, by specifying a large proportion of or complete contractor design. NEC3 ECC is typically brief on the matter, simply saying in clause 21.1 'the contractor designs the parts of the works which the works information states he is to design.'

Even very detailed employer requirements and contractor proposals can never approach the dozens or often hundreds of drawings which will have been produced as a part of the traditional design process; drawings that employers can see and comment on (should they wish), before the contract documents are finalised for tender. In D and B, these details and drawings are produced by the contractor (or the contractor's design organisation) *after* contract award. They will usually be sent to the employer for comment. This process of comment can be a massive task, and the employer would be advised to provide an ongoing resource to review the contractor's detailed drawings and proposals to ensure they meet the employer's needs. Again JCT DB05 provides for this in schedule one, which contains over a page of detailed provisions for the process of employer comment. It recognises that there may be changes requested by the employer, and disagreements as to discrepancies. When using NEC3 ECC it is very important to describe the process for employer comment in the works information, because NEC3 ECC does not prescribe a process, as JCT DB05 does.

It is important to remember that most projects need to go through all of the seven stages from concept to completion, and *someone* must carry them out. Hence despite the fact that D and B moves the responsibility for design and provision of drawings to the contractor from the design organisation, the actual work still has to be done. Similarly, if employers wish to comment on drawings and proposals, they still have to provide the time and resource, whether the strategy is traditional or design and build. Different strategies may make small efficiencies in communication or the avoidance of duplicated activities, but all the fundamental stages from concept to completion have to be undertaken and paid for.

Novation

Other items that can be lost in D and B are the ideas, knowledge and thought that the designer put into the preliminary design. Developing a design is a progressive process in which ideas are considered, weighed up and possibly discarded. So the designer of the preliminary design may have already rejected certain ideas and options for good reason. Another person picking up the preliminary design to complete it will not know this. A frequently used method to retain the preliminary designer's skill and knowledge and carry them forward to the detailed design is **novation**.

Novation is a legal mechanism whereby one contract is substituted for another. So a contract between the employer and the design organisation would become novated to a contract between the design organisation and the contractor. Of course the design organisation should be fully aware of the likelihood of novation when it undertakes its preliminary design services in the first place. The prospective contract between the employer and contractor (which the tenderers are bidding on) should contain clear terms as to the services required of the design organisation in the future and exactly what will be transferred to the contractor. So in theory the project-specific knowledge that the design organisation has used in the preliminary design is not lost.

There are two difficulties however. Firstly, the design organisation may not be the contractor's ideal choice of partner. There may be no history of cooperation or an organisational mismatch. The contractor is effectively obliged to accept the design organisation if it wants to bid for the construction contract. The second difficulty is that initially the design organisation works for the employer and is paid by the employer. So the employer has all the benefits of support, advice and a close working relationship. However, after novation, the design organisation no longer works for the employer, but instead works for the contractor. Many employers fail to appreciate the very different relationship with the design organisation that results from this. Many employers do not realise the gap in design advice that this produces, because they have been taking that advice for granted during the preliminary design stage.

Specifying design and build contracts

Let us develop this idea further. Many employers do not appreciate the amount of consultation that takes place in the design development stage of a traditional contract between the employer and the design organisation. Here the designer focuses on the employer's requirements and the options; often over a period of time, with many meetings and discussions with the employer, probing, questioning, clarifying and summarising. There is therefore a good chance that the employer's real needs are explored and understood and fully designed, and then specified in the contract. It is much more difficult to explore the employer's needs in the shorter initial stages, before a design and build contract is placed. Real attention must be given to understanding and specifying the employer's needs if success is to be achieved with D and B contracts.

> Probably the most important factor in successful D and B contracts is a comprehensive specification which truly reflects the employer's needs.

D and B contracts are relatively inflexible to change and change can be difficult and expensive, even in the contractor's design stage, largely because the contractor will have made assumptions about design delivery in the construction stage. However, it is changes and delays to construction that produce the highest costs. Once construction starts, change is even more expensive, and unlike many traditional contracts there is usually no basis for agreeing the price of changes. Traditional contracts using bills of quantity usually use the tendered rates and prices as the price (or as a basis for agreeing the price) of both added and amended work.

What can make this situation worse is the temptation with a D and B contract of an apparently quicker time from design start to construction completion, because of the integration of design with construction. It is common for a contractor to start constructing some parts of the contract work whilst other parts are still in the design stage. This minimises the design resources needed at the beginning of the project (by spreading them over a longer period) and allows a quicker start to construction activity. The drawback is that, because of this very integration, delays to the design usually have a direct and expensive effect on the construction programme itself.

Specification of any items of a required quality or aesthetic feature is very important in D and B. The contractor will, quite understandably, use the cheapest materials available to meet the construction methods and any standards specified.

So if for example the employer wants high-quality (and cost) finishing materials such as tiles, paint finishes or high-quality timbers, then these must be specified clearly. If the employer wants solid oak doors with brass door furniture then they must be specified. Again the difficulty of D and B is that this decision as to specific materials such as door quality would actually be a very late feature of the traditional design process. It cannot be a late feature of D and B, but must be clearly specified in the D and B contract, otherwise in this example the employer is likely to get chipboard or pine doors with aluminium handles.

Performance specifications and performance guarantees

A particular type of specification often used in D and B is the 'performance specification'. In a traditional contract there is usually time for the designer to fully design, specify and produce detailed drawings for all parts of the project. Of course in D and B this is not possible, since the design has only been partially completed when the contract is awarded to the successful contractor. So what are often specified by the employer (or the design organisation) are not the specifics of an item, but its required performance. The contractor is then responsible for producing a design to meet that performance requirement and any specified tests.

Let us take the example of a beer glass. In the traditional strategy the designer would consider various types of glass and their structural properties, particularly their tensile strength and Young's modulus. From a calculation of loading, the designer would calculate the resultant stresses, and hence required material thicknesses, using appropriate factors of safety. The designer would then produce CAD drawings showing the exact shape and the precise dimensions and put these on the tender drawings. By contrast, a performance specification would state the capacity (one pint), the breakage resistance, possibly the abrasion resistance (for the dishwasher), a requirement for a handle or not, any surplus capacity to prevent spillage, the clarity of the glass required and any testing requirements. Finally, and this is a very important point, the performance specification needs to 'state the obvious', in this case that the glass is cylindrical in shape. Whilst very unlikely, it would otherwise be possible to get a pyramidal or cube-shaped glass that met the specification. In real construction performance specifications, 'stating the obvious' is often overlooked. So if roads on a factory complex must be at least six metres wide, the performance specification must state this, otherwise the contractor may offer four metre roads with passing places.

If the contractor's design failed to meet the specified tests then the contractor would have to redesign and reconstruct the beer glass at his or her own expense. In real life there is a bit of a problem however. It is actually very difficult to specify performance criteria that are not subject to later argument.

Take the author's experience of a large extension to an existing water treatment works. Here the employer specified the average condition and chemical characteristics of the raw water (the river water), its range of chemical composition, turbidity (clarity) and so on. He then specified the required treated water quality standards and characteristics in great detail, and the tests that must be carried out to verify the contractor's design and to check that the constructed asset meets those final water quality parameters. A contract was awarded on a design and build basis for a new water treatment works, for £50 million, and specified three-monthly final water quality tests over two years to demonstrate that the design and construction met the specified standards. The new treatment works failed two of the tests. The employer assumed that the project would be corrected at the contractor's expense. However, the contractor proved that over two of the specified testing periods the river water was not of the average quality specified, thereby invalidating the tests. As it happened, one occasion was due to excessive turbidity (particles in the water) caused by rain and flooding, and the other by the worst UK drought in ten years. Unfortunately this example is replicated in many different ways and forms. It is actually very difficult to produce a 'water-tight' performance specification.

Another mechanism used in specifying D and B is the **'performance guarantee'**. Here the contractor guarantees certain aspects of performance for a sum of money or other specified payment. NEC3 ECC contracts incorporate this option as X17 'low performance damages'. The specified performance could perhaps be the efficiency of a pump, which will determine its power consumption against head and flow. High power consumption could result in very high long-term costs for the employer, far outweighing the initial capital cost of the pump or the impeller coating. Low friction impeller coatings are available for a few thousand pounds. Without a clear performance specification or performance guarantee, a contractor is unlikely to provide them, even though they will drastically reduce the whole life cost of the pump.

The characteristics of design and build

The employer has little control over the detailed design and lacks the design advice and support during design development which are part of the traditional strategy. This is acceptable if:

- The initial design has been thorough and a sound, comprehensive specification has been written.
- This means that the project requirements can be clearly detailed in the contractor's brief (called the 'employer's requirements' in JCT DB05).
- The employer has the staff (or employs an external organisation) to thoroughly check the contractor's design proposals at tender stage.
- Or the project is very simple, with straightforward design with few issues.
- Or conversely where the project has a high level of technical complexity, particularly involving specialist processes or manufactured plant (and the contractor will often take responsibility for this, with a performance guarantee, for example).
- Whilst D and B may not have such formal administration arrangements as traditional contracts, the employer will need help and advice with making appropriate contract payments.

Advantages of design and build

- The main advantage of D and B is a closer relationship between the contractor and designer, allowing for better planning and programming, and 'buildability'.
- Most risks are taken by the contractor, and so the employer has more certainty of price. However this price is likely to be higher than in traditional procurement because of the tenderers pricing in more risk in their initial tender prices.

The contractor may also insert provisos in the contract regarding unforeseen site conditions, since much site investigation has yet to be carried out.

- The employer can often see an early start to work on-site, since with D and B much design can proceed in parallel with construction. Hopefully this early start will result in an earlier completion, but this is not always the case.
- The employer is no longer responsible for any design delays by the design organisation.
- D and B is a good strategy where there is a need to coordinate the civil engineering design with that of mechanical and electrical plant designed by suppliers. As far as the employer is concerned the contractor is responsible for all the design, and hence is obliged to coordinate the very different design skills and methods of these various organisations.
- There is often good cost certainty (provided the employer does not make changes). Many D and B contracts are 'lump sum' rather than being remeasured as work proceeds.
- If there are defects, the employer is no longer caught in the argument as to whether it is a design problem or a workmanship problem. The employer has 'single-point responsibility' in the contractor.
- D and B is good for standard building types, such as warehouses or industrial buildings.
- Conversely it can also be a good strategy for complex process plant contracts, where the contractor may have more expertise to bring to the design than do many design organisations.

Disadvantages of design and build

- Tendering costs for contractors are much higher in D and B, and with them the abortive cost of a failed tender bid.
- Tender periods have to be longer to allow tendering contractors to produce outline designs.
- It can be difficult to compare tenders because of varying contractor proposals.
- D and B is not a good choice where architectural considerations are very important or the employer wants a building which is carefully designed to meet exact needs.
- Similarly, D and B can present problems where there are complex or difficult planning or environmental implications of a project.
- Design and build is not a good strategy where the employer wants the flexibility to make changes.
- Variations and changes can be very expensive, since they frequently disrupt both the design and construction processes.
- It is quite difficult to produce comprehensive employer requirements because of the shorter timescale to the tender stage and the lack of interaction during the process of detailed design.
- There are frequently disagreements as to whether there is really a 'discrepancy' between employer requirements and contractor proposals and who should pay.
- Employers do not have the benefit of a professional advisor during design and construction as they do in the traditional strategy (the design organisation), and may have to employ a separate organisation (at a cost) to perform this service.

Management contracting

The principle of management contracting is that the contractor is appointed early and to a contract with the employer. The usual presumption is that the management contractor does not undertake any actual construction work but concentrates on management, liaison and *coordination* of the construction work. The expectation is that the management contractor works more as a partner with the employer in developing as well as implementing the project. The main reason for this 'partnership' is the prospect that this will produce a more integrated and buildable design thereby overcoming the main problem with the traditional strategy.

A very important point is that the *contractor* has contracts with the sub-contractors, and this is what distinguishes management contracting from construction management, which is described below. Hence, unlike construction management, the **'interface risk'** is usually carried by the management contractor, since the latter places all the sub-contracts. This can be seen in Figure 9.4, since the contract lines run from the contractor to the sub-contractors and suppliers.

Interface risk

This is a very important concept. All construction involves a web of linked contracts and sub-contracts. These links all require careful planning and phasing. Construction is a very complex activity so inevitably things often go wrong, possibly because of unforeseen circumstances, delays, poor planning, late material supply and so on. Any of these factors is likely to cause someone delay and extra cost. A problem within any one contract will often also affect the contracts which *interface* with it; delaying them as well. This is clearly a risk, often called the 'interface risk'.

Many standard forms of contract and sub-contract seek to list the more common of these risks, and describe how they shall be valued, and who should pay for the cost of the problem. The interface risk varies with the method prescribed for dealing with the consequences, and who is allocated to pay for them. So if a contract prescribes that the employer pays for the effects of any serious industrial disputes or strikes (as some JCT contracts do), then the contractor is protected from this particular interface risk between itself and the suppliers. There is thus little interface risk for the eventuality of an industrial dispute; at least as far as the contractor is concerned! Of course, if the contract does not prescribe in this way (and most contracts do not), then this is a risk that the contractor must carry, consider and possibly increase the tender prices to cover; particularly in an industry or location which is prone to strike action, for example.

The characteristics of management contracting

The characteristics of management contracting are:

- The employer wants the design to be undertaken by an independent design team which the employer has chosen (in other words, not one working directly for the contractor as in D and B). The employer can thus influence the design and maintain a greater control over it.

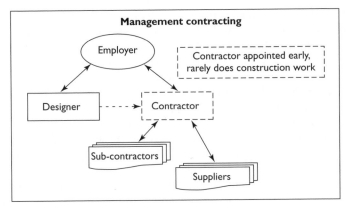

Figure 9.4 Management contracting.

- The employer wants the management contractor to make an input into the design. This is usually done to get the benefit of construction expertise.
- Where early completion is desirable, the early involvement of the contractor can produce a faster start on-site and a more buildable design. Additionally sub-contractor work packages can be awarded as the design proceeds, thereby providing the prospect of some concurrent design and construction which is not easy to achieve in the traditional strategy.
- It is generally used where the project is fairly large, since most contractors will not be willing to commit the time and resources to early involvement for a small project.
- It is a good strategy where the project requirements are complex, particularly having complication in areas where the contractor has expertise, such as complex process plant, in which a particular contractor specialises.
- It is also a useful strategy on projects where the contractor can bring experience of very similar projects. Here the contractor's programming and construction advice in the design stage can be particularly valuable.
- Good cost prediction and budgetary control can often be achieved by utilising the contractor's commercial expertise. The difficulty is that there is usually no certainty of the total price for the project, until all the sub-contracts have been awarded.
- It can be a valuable strategy where there is a high likelihood of changes during construction. Changes are a particular problem for the D and B strategy, as explained above. In contrast, in management contracting, work packages to sub-contractors can be phased as the design proceeds.
- The management contractor places all the contracts directly, and generally takes responsibility for them, thereby reducing the employer's risk. This does depend on the way the various contracts are set up of course.
- The employer wants competitive bids for works packages, and more control over them. This can be achieved by insisting that the contractor always seeks three tenders for all sub-contracts, for example.

The management contractor is paid a fee, often five to ten per cent of the construction cost. Some of this fee is usually for the provision of site offices, services, storage and common construction plant which would have to be provided anyway. However, some of the fee is for management services and is hence an 'overhead' or additional cost. This overhead can be justified if the contractor adds real value to the project. For the fee, the management contractor adds value in terms of contracting experience and coordination and, unlike the construction manager (described in the next strategy), usually takes responsibility if things go wrong, particularly in respect of interface risks.

Because of the prevalence of sub-contracting, many major contractors effectively become management contractors anyway in the sense that they do little direct building work themselves. However, it is only in the strategy of management contracting that the contractor is formally engaged at an early stage to provide an input to the design process and bring construction advice to the employer.

Many long-term framework agreements are in reality a form of management contracting in that the contractor and design organisation are appointed early, at the beginning of the framework, and thus the contractor's expertise and advice can be utilised during the design stages of the projects that form the framework.

Figure 9.4 looks very similar to the traditional strategy, the differences being that the management contractor is appointed early in the investment process, rarely does any direct construction work and is available in an advisory capacity.

Issues with management contracting

As discussed above, the prime issue with management contracting is who takes the 'interface risk' and liability for sub-contractor defects or default. A secondary issue can be problems over late payment to sub-contractors by the main contractor. In the early days of management contracting, it is reported that the employer often took the interface risks. They could be mitigated by the use of **collateral warranties**. A collateral warranty is a separate contract between the employer and a sub-contractor, which seeks to guarantee specific performance, quality or other standards. The main contract was between the employer and the management contractor. Under the doctrine of 'privity of contract' it was not possible for a third party (E, the employer) to exercise any rights under a contract between C (the contractor) and S (a specialist supplier) for example. This used to be the case even if E were affected by the default of S. Hence the need for collateral warranties to give E rights over S, particularly for performance or defects.

The law was changed in 1999 by the Contracts (Rights of Third Parties) Act. Unless a contract stipulates otherwise (and they usually do) E can now exercise rights under a contract between C and S. However, this provision is still little used, although it could replace collateral warranties. The point really is that collateral warranties and possibilities under the Contracts (Rights of Third Parties) Act simply introduce another layer of complication, risk and uncertainty; three items that most employers try to avoid wherever possible.

The importance of the regulations to many UK employers and the widespread use of long-term framework agreements arguably provide all the main benefits of management contracting, such as early contractor involvement and better team formation. There is also the possibility of awarding successive work packages without further advertising in the OJEU (Official Journal of the European Union), provided the contractor is appointed properly under the regulations, and provided the contractor makes the decisions as to sub-contractor tender lists and contract award. So, once again, we see the regulations having a major impact on the procurement of much UK construction work.

Forms of contract for management contracting

Standard forms of contract are available for management contracting from the JCT suite of contracts, or by the use of option F of NEC3 ECC. It is interesting to note that option F provides a very similar contractor risk profile to the other NEC3 ECC options, since it uses the same scheme of compensation events and requirements for time and quality. Under option F the contractor is paid the cost of sub-contractor work, plus the cost of any work the contractor does itself (without reference to a schedule of cost components) plus the fee. The work the contractor does itself, if any, should be stated in the contract data, and priced by the contractor. The contractor can be contractually required to start work (usually advisory work) at any time after the contract is signed, since ECC has the useful concept of the 'start date', which can be set at any time after the 'contract date' (when the contract is signed) and before the 'access date' (access to the site itself for construction purposes).

Construction management

Construction management was developed (along with management contracting) to address the separation of the design organisation from the contractor that we find in the traditional strategy. Again, we see the contractor appointed early in a more professional advisory role. In construction management the contractor advises the employer and design organisation on design issues and also on construction issues during the design stage of the project. These construction issues during design could include potential construction methods and risks, programming, suitable sub-contracts and so forth. Once the construction contracts are awarded, the construction manager manages and coordinates the various contractors who are carrying out the construction work itself. The construction manager is paid a fee for these services, by the employer.

The construction manager could be an individual, but is more likely to be a firm of quantity surveyors, or another contractor acting in this particular role. Figure 9.5 shows the contractual relationships as solid lines. Here the employer contracts directly with the works contractors as well as with the designer and the construction manager. The works contractors will in all probability be similar companies to the sub-contractors in previous strategies. The construction manager has a contract with the employer to organise and coordinate the works contractors, and hence an administrative link (shown as a broken line on Figure 9.5) to the works contractors, but is not contractually liable to the employer for the performance of the works contractors.

Hence construction management provides an alternative 'management' contract strategy that is broadly based on the same principles as management contracting. The important difference is that, in construction management, the employer contracts directly with the works/trade contractors, whereas in management contracting the management contractor contracts with them. So in management contracting the works/trade contractors become the main contractor's sub-contractors but in construction management they become *direct contractors*

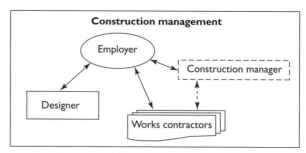

Figure 9.5 Construction management.

to the employer. This is a very significant difference and completely changes the risk profile. In simple terms, the employer now carries the 'interface risk' described in the previous section, rather than the contractor. The employer pays if things go wrong, particularly at the interface between the contracts.

On the other hand, there is a strong argument to say that a professional from the construction side of the industry is better placed to organise and coordinate many construction contracts than a project manager from the design side of the industry, which is more often the case with other contract strategies. The difference of course, yet again, is the fact that the employer carries the interface risk in construction management.

The construction manager rarely carries the interface risk. However, like any professional the construction manager must use reasonable skill and care. If the construction manager is *negligent*, then the employer would have a legal recourse under the tort of negligence, just as the employer would for any other negligent professional such as an architect or consulting engineer. However, there has to be negligence in the first place, and that negligence has to be proved in court, with all the attendant cost, risk and time implications. So the employer is very dependent on careful selection of the construction manager, and the construction manager's continuing professional dedication and competence that are expected.

Despite the attractions of construction management, this issue of risk is one of the main reasons that construction management is little used today. The RICS survey of contracts in use at 2007 points out that construction management is little used except in a few notable instances, as can be seen from this extract:

'The 2007 survey has identified a much greater use of the JCT Construction Management documentation than the 2004 survey. This time 15 instances of its use were captured (12 using the 2005 version and three the 2002 edition) compared to just three instances in the 2004 survey. This still accounts for only a tiny proportion of contracts used but represents 9% of the value of projects in the sample, largely accounted for by 3 very large schemes averaging £216 m each.' (RICS, 2007)

Construction management and employers subject to the regulations

In construction management, the employer takes the interface risk.

Another potential problem with construction management is the difficulty it presents for employers subject to the **regulations** (arising from the EU Procurement Directives, described in Chapter 13). Since the RICS survey concentrates on 'building' projects, its respondents are less likely to be affected by the regulations than are the larger employers in the civil engineering sector. The problem under the regulations for affected employers is that construction management involves placing many *individual* contracts; a contract with each works contractor. So in this strategy *each individual contract* is potentially subject to the regulations, if it exceeds the threshold value for that category of work.

This can be a very serious consideration and brings with it all the risks, costs and time constraints of advertising and appointing under the regulations, not just for one main contractor, but potentially for *many* contractors. All these advertised contracts will then have to be phased and coordinated and, of course, the employer is taking the interface risks if there are any delays. If an unwise employer were to seek to avoid the regulations by using construction management and splitting a large project into individual contracts, each below the threshold value, then the employer could be liable for 'disaggregation' which is illegal under the regulations.

Another consideration is that the construction manager is a 'service provider' under the regulations and, on any project of size, the construction manager's fee will exceed the threshold value. So right at the outset, the employer must follow the advertising and appointment precepts of the regulations, which can take a considerable time. It is said that an unsatisfactory construction manager can always be changed, but this too would require advertising under the regulations if the fee exceeds the threshold value.

The characteristics of construction management

- The employer or the employer's staff should be familiar with the construction process, since the employer has a close connection with their advisor, the construction manager, and will need to make informed decisions as the project proceeds. These decisions will often be made under time pressure since, unlike the traditional and D and B strategies, construction will be underway as subsequent contracts are placed.
- The employer has an ongoing input to the selection of all the works contractors with whom the employer has direct contracts. This can be an advantage, in that it gives the employer control over the specification of subsequent contracts and in the selection of tenderers. However it represents a workload for the employer which is carried by the contractor in other contract strategies.
- Construction management can be used when the employer requires an early start on-site, despite the design being incomplete (here time takes priority over cost).
- It can be a useful strategy where the project is technologically complex, especially where there are numerous contracts that are interdependent and where the construction manager has the expertise to coordinate and organise the many facets of work required.
- It is never a good idea for the employer to require changes during construction but in construction management, since the work packages are awarded as the project proceeds, some degree of change can often be accommodated. Hence

there is some flexibility to develop ideas and their implementation as the project proceeds.

- Because employers contract directly with the works contractors, they are more likely to be paid promptly, and they could reduce their tender prices for this. It is an unfortunate fact of UK construction that some main contractors do not always pay their sub-contractors promptly and in full.
- The reverse side of this is the increased risk of insolvency. This is because the works contractors are likely to be smaller companies than the main contractors with whom the employer contracts in the other contract strategies. Because they are smaller, the works contractors may have less financial ability to absorb the costs of problems or delays, or to pay liquidated damages to the employer if they are late in completing their contracts.
- Construction management is unsuitable for small or simple contracts since they can be awarded much more quickly and cheaply using one of the other strategies such as traditional or D and B.

The role of the construction manager

The construction manager may be an individual, but for larger projects is likely to be a firm, possibly of contractors or project managers. The construction manager's duties might include:

- Communication and team working with the employer's staff.
- Working with the design organisation in terms of construction issues and timescales.
- Breaking the contract work down into suitable packages for tender.
- Advising on suitable tenderers, tendering procedures and tender assessment.
- Producing and maintaining a master project programme.
- Advising the employer on any advance orders of materials or equipment.
- Coordinating the work of all trade contractors and suppliers.
- Advising on contract management and supervision personnel.
- Dealing with any necessary changes or variations and advising the employer of the likely construction and cost consequences before they are issued.
- Coordinating monthly payments and final account preparation.
- Advising the employer on any contract claims.

The Scottish Parliament Building – an example of construction management going wrong

Construction management is sometimes used when the employer wants a quick start on-site, before the employer has really decided what is wanted. This can be a recipe for disaster. A very good example is the Scottish Parliament building. The Scottish Parliament building went so badly wrong that it was subject to a government inquiry (Holyrood Inquiry, 2003). In 1997, early cost estimates for the project were about £30 million. Later estimates, on a more extensive brief, rose to £70 million. Bovis were awarded the job of construction manager for a *fee* of £5.7 million.

'The decision to adopt construction management as the procurement vehicle for the construction of the building was found to be "one of the most significant, if not the most significant" decisions taken during the course of the project. Construction management offers the advantage of speed but with the disadvantage of price uncertainty until the last contract has been awarded.' (Lord Fraser, Holyrood Inquiry, 2003)

The final cost was in excess of £430 million. The project was dogged by poor management and massive and late changes. It is an example of how construction can go very wrong; however the problems were not simply due to the construction management strategy, but also to constant major changes by the employer.

Chapter summary

- Contract strategies are about how the contract and administrative links are arranged between the employer, the designer, the contractor and sub-contractors (works contractors).
- Contract links confer rights and obligations under the contract used and seek to describe the main risks and who should take them.
- Administrative links can be concerned with formally administering the contract between the employer and the contractor (in the traditional strategy), or to organise and coordinate the work of the works contractors as in construction management.
- More recent contract strategies such as design and build, management contracting and construction management seek to improve the link between the designer and the works contractors, and to bring contractor expertise and coordination into the whole investment process; not just at the construction phase.
- The 'regulations' have had a major impact on many larger employers in the UK and have influenced the relative use of the various contract strategies, largely because of the expensive and time-consuming advertising and appointment procedures set down in the regulations.
- Probably as a result of the regulations, management contracting and construction management are little used today.

10

Designers and design contracts

Aim

This chapter explains the role of designers, their obligations and design liability.

Learning outcomes

On completion of this chapter you will be able to:

>> Describe the role of the designer.
>> Describe the nature of design.
>> Discuss the issues around design liability.

Design

As we have seen in the previous chapter on contract strategies, one of the major issues for the employer to determine is who designs the project. The traditional strategy has stood the test of time, but its main drawback is the separation of design and construction into two different organisations: the architect or consulting engineer, and the contractor. Many of the other contract strategies have evolved to try and address this issue, but have generated new drawbacks of their own. We see that design is very important.

Everything that is built has to be designed by someone. Even on the smallest scale in construction we see design. So two bricklayers constructing a garden wall 'design' the foundations when they settle on the dimensions of a concrete strip footing, and 'design' the wall when they decide on its thickness and mix the mortar, determining its strength and durability by the mix proportions that they choose. It is very important to understand this wide-ranging concept of design because many construction professionals take on design liability without realising it, and hence expose themselves to potential legal action if their design is negligent or inadequate.

Everything that is built has to be designed by someone. Different contract strategies simply move the accountability, but not the work effort.

The need for design and hence design responsibility is fundamental to all contracts. Traditionally, the design role was clear; the architect or consulting engineer usually working for the design organisation carried out the design, and the contractor built it. More recent projects often follow this traditional contract strategy, although many are now 'design and build', where the contractor is responsible to the employer for *all the design* as well as the construction. However, most contractors do not employ their own designers, so they go to the design organisation directly. Additionally, many contracts lie somewhere between the two extremes with *some* design by the contractor; this can result in a problem, particularly with design interfaces. In many instances, where the contractor designs some of the works, the 'contractor's design' can affect the 'designer's design', and vice versa. The other major problem is often a lack of clarity over who is designing what and when, and who is taking the design liability if things go wrong.

The designer

The designer is rarely an individual and the word really refers to the **design organisation**. So 'the designer' usually means the legal entity responsible for the design and with whom contracts are made, usually by the employer, but often with the contractor if the design and build strategy is used. Of course, design organisations are made up of individuals, who are also called designers, which can be a little confusing to the inexperienced.

Historically design organisations were groups of partners and associated staff. These organisations would usually be consulting engineers or architects, specialising in 'civil engineering' or 'building' projects respectively. Increasingly, as measurement became a more specialist function, firms of quantity surveyors were formed, largely supporting the 'building' side of the industry which usually requires more complicated methods of measurement. Building surveyors specialised in areas of work such as building renovation and repair, and other professions and organisations evolved to meet the many specific needs of the industry as it grew and became more complex.

Today many of these partnerships have developed into much larger firms, some employing thousands of staff, and the partnership structure has often changed to a company structure with boards of directors and sometimes shareholders. These new large firms are often multidisciplinary and in many cases have diversified into project management, management consultancy work and even the operation and maintenance of highway and rail networks. Having said this, it is not unusual still to find quite small professional practices or even individual specialists in the 'building' side of the industry. Much of the work on the building side of the industry is for small clients or domestic clients, who have very different needs from the larger organisational employers which predominate in the civil engineering sector.

So the designer (the design organisation) may not work directly for the employer, where the word 'employer' is used to denote the promoter and recipient of the project itself. As we have said, the designer frequently works directly for the contractor. It all depends on the contract strategy used, as is shown in Figure 10.1. So in the traditional strategy, with separated design and construction organisations, the design contract is between the employer and the designer. In contrast, in the design and build strategy the organisation that pays for the designer's services, or the designer's *client,* is the contractor. This change in con-

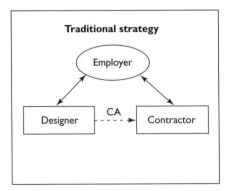 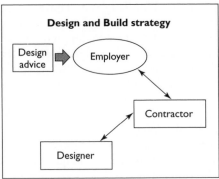

Figure 10.1 Design organisation roles and contracts.

tractual relationship has very significant effects which are discussed in Chapters 9 and 11 on contract strategies.

It is worth noting that in the design and build strategy the employer will usually need ongoing **design advice** to assess the implications of the designs being put forward by the contractor. If the employer does not have qualified staff of its own then another design organisation will have to be appointed to give the employer this continuing support. This is shown as the hollow arrow in Figure 10.1. Many of the larger employers have construction professionals on their staff to organise and assess design work being carried out by external design organisations. Organising and assessing will require much less time and resource than the main design role, but it is important nevertheless.

Another role that is frequently carried out by the design organisation in traditional contracts is **contract administration** which as shown in Figure 10.1 as 'CA'. If the design organisation is undertaking contract administration in addition to design, then they will need paying for it, by the employer, either as a separate commission or as part of their overall design commission.

What does the design organisation do?

The term 'design' can be a little confusing too, for the functions of the design organisation are wide-ranging and essential to most of the key stages in the investment processes. So, this wide-ranging work will often cover stages three to six (option selection to tendering) and frequently stage seven (construction) in a contract administration capacity. The design organisation may also be involved in stages eight and nine (see Chapter 2 for a refresher about these stages). Hence, once the employer has verified the need, the design organisation comes to the fore as the major driver and supporter until construction itself (stage seven).

> A design organisation does not simply produce calculations and drawings, a pricing mechanism and specifications.

The role of the design organisation encompasses a wide scope of 'design' services. The design organisation (usually a consulting engineer, architect or building surveyor) will often advise the employer on:

- Initial development and understanding of the employer's needs, and hence the type and scope of project required.
- The need for any feasibility studies.
- The advisability of public consultation.
- The likely risks and costs of the project.
- Appropriate timescales.
- The employer's obligations under the CDM regulations.
- Any requirement for regulatory approvals.
- The type and scope of any *preliminary investigations* (such as site investigation, noise surveys and ecological surveys).
- The need (if any) to appoint *other professional advisors* such as quantity surveyors, lawyers, accountants, financiers, insurers and specialist consultants.
- The need (if any) to engage the early services of a contractor, for construction advice.
- The need (if any) to engage the early services of a contractor, for site preparation or other preliminary construction work.
- The choice of a suitable contract strategy.
- The most appropriate form of contract to use.
- Advice on possible tendering options, and suitable contractors to be invited to tender.
- Which tender to accept.
- Contract administration.
- Claims and disputes.

Fundamentally, the design organisation is responsible, through all the stages of the investment process, for ensuring that the built project meets the employer's needs.

> The design organisation is responsible for ensuring that the requirements of the employer are properly defined and communicated in the tender documents.

Where a number of design organisations are involved, the 'lead designer' will be the design organisation responsible for managing information and coordinating the efforts of all the other designers. This coordination is essential to ensure a holistic approach to design such that the design forms a coherent and consistent entity. Coordination of the health and safety aspects of design is the role of the CDM coordinator under CDM 2007 who is likely to call design reviews to ensure that all the different aspects of design work come together. Designers, whether in the lead design organisation or a support organisation, will of course have extensive duties themselves under CDM 2007, and these are discussed further in Chapter 8.

Who is the designer?

This is an important question, but it is not easy to answer. The underlying assumption of traditional forms of contract such as ICE7 and many JCT forms is that the design organisation designs the project and the contractor builds it. Similarly in the design and build contract strategy the assumption is that the contractor designs *and* builds the project. Contract forms such as the Joint Contracts Tribunal Design and Build contract 2005 (JCT DB05) have been written specifically with this contractor design in mind. Contractor design, usually on a more limited scale, can also be specified in a general construction contract such as NEC3 ECC and JCT SBC05. NEC3 ECC relies on design information and responsibility being set out in the works information, and JCT SBC05 in a section called the 'contractor's design portion'. The issues around this are discussed below.

However, the 'designer' in law extends to far more people and activities than many construction professionals appreciate, and hence accountability, and with it legal liability, for design can be very wide. Where organisations take on design liability without realising it, they are frequently not insured for design risks and so bear the full consequence of any design failure themselves. The Approved Code of Practice (ACOP) to CDM 2007 is very comprehensive in its description of the designer and clause 116 is reproduced in full below.

Clause 116 from the ACOP to CDM 2007

Designers therefore include:
(a) Architects, civil and structural engineers, building surveyors, landscape architects, other consultants, manufacturers and design practices (of whatever discipline) contributing to, or having overall responsibility for, any part of the design, for example drainage engineers designing the drainage for a new development.
(b) Anyone who specifies or alters a design, or who specifies the use of a particular method of work or material, such as a design manager, quantity surveyor who insists on specific material or a client who stipulates a particular layout for a new building.
(c) Building service designers, engineering practices or others designing plant which forms part of the permanent structure (including lifts, heating, ventilation and electrical systems), for example a specialist provider of permanent fire extinguishing installations.
(d) Those purchasing materials where the choice has been left open, for example those purchasing building blocks and so deciding the weights that bricklayers must handle.
(e) Contractors carrying out design work as part of their contribution to a project, such as an engineering contractor providing design, procurement and construction management services.
(f) Temporary works engineers, including those designing auxiliary structures, such as formwork, falsework, facade retention schemes, scaffolding, and sheet piling.
(g) Interior designers, including shop fitters who also develop the design.
(h) Heritage organisations who specify how work is to be done in detail, for example providing detailed requirements to stabilize existing structures.
(i) Those determining how buildings and structures are altered, for example during refurbishment, where this has the potential for partial or complete collapse.

Reading this list from the ACOP can be very sobering for many construction professionals, especially contractors who often assume that they have no design responsibility or accountability in a traditional contract. Hence as far as CDM 2007 is concerned, the legal duties of the designer for health and safety extend to anyone engaged in the activities above. It is possible that the courts may not adopt such a broad definition of designer in terms of negligent design but it would be prudent to assume that they will.

A very common example of the issue in a traditional contract is where the design organisation has not specified a material fully, and the contractor needs to order it promptly so as not to hold up the construction work; the *strength* of structural blocks in a wall for example. The correct answer is for the contractor to identify the lack of information early and ask, in writing, for further instructions from the contract administrator, preferably stating the date by which it is required and the potential consequence of a late answer. If the requested design information is unreasonably delayed, then under most forms of contract the contractor will have grounds for a loss and expense claim and a revised completion date (if the work is on the critical path). However if contractors proceed to estimate the strength of the structural blocks and order them on their own account, they would be taking on design liability, even if they did not realise it. They could later face a claim of inadequate design and potentially have to take down and rebuild the wall for design issues, not construction ones. A really serious problem could occur if further floors were built above the wall, which then put the rest of the structure at risk, and where remedial work becomes almost impossible or prohibitively expensive.

Designer's obligations

Designer's general obligations

The designer, or design organisation, has a number of duties and obligations. Some arise from the design contract, some arise from professional responsibilities and some are a matter of law. The designer's main obligations are listed below:

- To act professionally at all times and in accordance with any codes published by the relevant professional body.
- To comply with the design contract.
- To ensure that employers are aware of their duties under CDM 2007 and for designers to carry out their own duties in accordance with CDM 2007.
- To take out appropriate insurance, particularly **professional indemnity** insurance.
- To carry out design work with reasonable skill and care and in accordance with published and accepted standards of professional practice.
- To ensure that all staff engaged on design work are competent and qualified as appropriate.
- Where acting as contract administrator, to act impartially and to apply the terms of the construction contract fairly between the employer and the contractor.

Designer's obligation for safety

The designer has a very important role in health and safety by identifying and eliminating hazards during the design process, and reducing risks where hazard elimination is not possible. In the first sentence about designers, the Approved Code of Practice to the CDM 2007 in section 109 says:

> 'Designers are in a unique position to reduce risks that arise during construction work, and have a key role to play in CDM 2007.'

The 'designer' is one of the five key roles identified in CDM 2007 and has a list of duties which revolve around eliminating hazards, and providing information about remaining risks, together with ensuring that clients (the employer) are aware of their CDM obligations. It is now normal for designers to carry out a Design Risk Assessment (DRA) for every aspect of design that they undertake.

Another key designer attribute is communication. Designers should always communicate their ideas and designs clearly through the tender documents which will become the contract documents after the contract is signed. However, CDM 2007 particularly requires designers to communicate and cooperate with the CDM coordinator, other designers and the contractor to ensure that once design is completed, all remaining risks are understood.

Finally, designers should only undertake work that they are competent to do and that they have the resources to undertake. This can sometimes be difficult in a commercial world. The issue is discussed in the next section.

Designer's obligation to act in an ethical manner

Let us continue with this issue of ethics and professionalism. Most designers aspire to membership of the relevant professional institution such as:

- Institution of Civil Engineers (ICE)
- Institution of Structural Engineers (IStructE)
- Royal Institute of British Architects (RIBA)
- Chartered Institute of Building (CIOB)
- Royal Institution of Chartered Surveyors (RICS).

These institutions publish codes of ethics for their members to follow, which define and regulate the way in which they carry out their work. As an example, RICS publish three simple principles:

Integrity
- Members shall act with honesty and integrity at all times.

Competence
- In the performance of their work members shall act competently, conscientiously and responsibly. Members must be able to provide the knowledge, the ability and the financial and technical resources appropriate for their work.

Relationships
- Members shall respect the relevant rights and interests of others.

The professional institutions all have entry requirements based on academic qualifications and relevant experience. They represent a body of knowledge for their members and a recognised professional qualification to reassure employers (clients). Most professional institutions also require their members to have a commitment to a code of conduct and to continuing professional development. The point of all this is to assist employers in appointing suitable professional people on whom they can rely.

Acting ethically is not simply a moral obligation, it is almost certain to lead to commercial benefits as well. This is because employers (clients) are much more likely to return with future business to organisations which act in an ethical manner and follow precepts similar to the ones above.

Probably the two most difficult ethical issues facing designers are

- To avoid **conflicts of interest**.
- To only take on work that they are competent to do and where they have the resources available.

Conflicts of interest

A conflict of interest occurs where designers may be unable to fairly and properly represent their main client due to obligations to others, or their own interests. In the interdependent construction industry, this can happen very simply since construction professionals can easily work for one client this year, and a competitor the next. Or possibly the construction professional is the contract administrator ruling against a contractor's claims for loss and expense on one contract, whilst potentially undertaking the design role in a separate design and build project for the same contractor.

The Competition Commission is one of the independent public bodies which help ensure healthy competition between companies in the UK for the benefit of companies, customers and the economy. They have this to say on conflicts of interest:

'A conflict of private interest (or duty) and public duty arises where a member has any interest which might influence, or be perceived as being capable of influencing, his or her judgment even unconsciously. While in practice a member's judgment may not be influenced by a direct pecuniary interest, in law such an interest, however small, disqualifies the member from acting.'

Here of course the Competition Commission are talking of a conflict of two 'private' interests, contracts or other professional work for example, but the principle is the same. The principle is whether the individual's proper judgment could be influenced. Most design organisations are very careful about being placed in this position, particularly if they are working on separate commissions for two different clients, the 'employer' and the contractor perhaps, but where they have to make judgments between these two organisations. It is possible if different personnel are involved with very clear and separate reporting lines and audit procedures, but it is easy in these circumstances for an accusation of bias to be made against the design organisation.

Many design organisations would not take on design work for a contractor where the designer was acting as contract administrator in a big claim or dispute situation to which that contractor was a party. The design organisation would normally wait until the matter was settled before undertaking such work. In their code of ethical conduct, the Royal Institution of Chartered Surveyors says 'Declare any potential conflicts of interest, personal or professional, to all relevant parties.'

Examples of conflicts of interest

A member of a planning committee should not vote on a project that he or she is sponsoring or paying for, because it is difficult to form an independent view. Say you were to manage some properties for a client that do not comply with recently introduced fire safety regulations, and your client has asked you to say nothing to new tenants. Professionally you must refuse to do this and assist your client in complying, before tenancies begin. Or an employer who is temporarily short of funds may ask a contract administrator (CA) to under-certify the contractor's completed work. The CA should politely refuse (although this may not be easy) whilst explaining the contractual position to the employer.

Contract administrators come under particular pressures from employers who do not really understand the nature of the construction contract they have signed. Here the contract administrator may be attempting to administer the contract fairly and under its terms by certifying claims for loss and expense perhaps, or for extensions of time. The employer will have to pay for these claims and may have difficulty in understanding the CA's actions, particularly since the *employer* is paying the CA's fees.

The best advice to follow is in the relevant profession's code of ethics, and wherever there is a potential for a conflict of interests, then declare it clearly and openly to all those affected. In cases of serious conflict of interest that compromise professional codes of ethics it may be necessary to decline a commission.

Taking on appropriate work

The second difficult area is to ensure that the design organisation only takes on appropriate work. So for example the Chartered Institute of Building says in its code of ethics for members 'Members shall not undertake work for which they knowingly lack sufficient professional or technical competence or the adequate resources to meet their obligations.'

All design organisations should have competence and expertise in certain areas of work, and the resources (usually people) to undertake this work. However design organisations rarely have 'spare staff' waiting for a new commission. If they do then the under-utilisation of these staff will add significantly to the organisation's overheads. So a new commission may be signed when there is nobody to undertake it *at that time*. Again, openness is the best policy, and the resource position should be explained clearly to the employer. For example, 'we can make a start immediately, but our main design team will not become available until April, and with your approval, we will go outside for structural services to a firm we have used before with great success.'

The NEC contract most used for design services is the NEC3 Professional Services Contract, or NEC3 PSC as it is known, which sensibly gives great importance to the consultant's (designer's) programme, in clause 31. This can either be submitted after the commission is signed, or it can be identified in the contract data; in other words it can be incorporated into the contract.

Where a design organisation is extending into newer areas it will often need to use external specialists. It should consult its design contract to ensure that this 'sub-contracting' is permitted. This is called 'sub-consulting' in clause 24 of NEC3 PSC, which requires the employer's consent to both the name of the sub-consultant and the terms of the design contract. The employer may refuse to consent if the employer considers that the appointment 'will not allow the consultant *(the main design consultant)* to provide the services *(required)*.'

Finally it must never be forgotten that competence and the provision of adequate resources are a **legal requirement** under the CDM regulations 2007. So regulation 4 states that 'no person shall appoint a CDM coordinator, a *designer*, a principal contractor or contractor unless he has taken reasonable steps to ensure that the person to be engaged is *competent*' and goes on to say 'no person shall accept such an appointment unless he is competent.' This places a legal duty on both the designer and the employer. Regulation 9 stipulates that clients 'shall take reasonable steps ... *including the allocation of sufficient time and other resources ...*'

Designer's obligation to cooperate

In NEC3 ECC clause 10.1 the 'designer' is a under a general obligation to cooperate with the employer and contractor, where the designer is named in the ECC contract as either the project manager or the supervisor. The ECC project manager role is similar to the architect/contract administrator in a JCT contract, and the supervisor role is similar to the JCT clerk of works role. Of course the project manager and the supervisor are not parties to the NEC3 ECC contract but they have defined roles within it. Additionally, the NEC3 PSC contract, which is likely to be the contract between the designer and the employer, puts an obligation on the employer and the designer in the PSC clause 10.1 to 'act as stated in this contract and in a spirit of mutual trust and co-operation.' The RIBA Standard Form of Agreement (2004) known as SFA/99 which is frequently used for works of a more building nature has a similar stipulation in clause 1.6.

Obviously, designers should not need a legal *duty* to cooperate with the employer, since they are engaged by the employer to provide professional services – usually design services plus whatever other services are required. As professionals, designers should always have the needs of their client (the employer) at the forefront of their minds. Their role frequently extends to more general advice and support to the client throughout the design and construction process.

Additionally, codes of ethical conduct will often have terms relating to behaviour towards a client such as RICS which says 'never put your own gain above the welfare of your clients or others to whom you have a professional responsibility.'

Designer's obligation for insurance

As an employer of people, the design organisation will have a legal obligation to take out certain insurances, such as employer's liability insurance. However the insurance that is of most relevance to the design organisation (and the client) is **professional indemnity insurance** (usually abbreviated to PI). This insurance covers the designer for professional negligence. So within the terms and limits of the policy, the insurer will pay the insured (the designer) the cost of a claim against it. The organisation making the claim will usually be the employer, but it could be the contractor in a contract made under the design and build strategy.

Most professional institutions insist that their members take out a minimum level of PI. However, forms of 'design contract' such as NEC3 PSC and those published by the Royal Institute of British Architects (RIBA) have specific clauses requiring PI insurance to at least the limits of liability set out in that design contract. These limits will often be several million pounds. Additionally such forms of design contract usually stipulate that the design organisation provides proof to the employer of its insurance cover.

The cost of professional indemnity insurance is very high, typically over £600,000 a year for a consultant/architect earning £15 million of fees. PI also usually carries a large **excess**, often £500,000 (an excess has to be paid first by the insured party and then the insurer pays the remainder). PI is expensive because some design errors can lead to massive financial consequences. Take the example of the remedial cost of a negligently designed piling layout for a tower block. Here negligent design could easily cost hundreds of thousands of pounds, perhaps millions.

The final point is that since their business is design, most design organisations understand the issues around design liability. Many contractors do not understand the issues around design, because their real business is construction and can be put at risk if they deliberately or inadvertently become accountable for design.

Crucially, PI insurance only indemnifies the design organisation against failure to meet a standard of **reasonable skill and care** which is explained below.

Designer's legal liability for problems with the design

Design responsibility means providing the resources and expertise to carry out the design. The cost of a civil engineering *design itself* is about four to six per cent of the construction cost (often ten per cent or over in building contracts), so it is by no means insignificant. Design responsibility is not the same as design liability.

Design liability

This means taking the legal consequences of an inappropriate, inadequate, defective or negligent design. This liability will depend on the wording of the design contract and the operation of the tort of negligence (see Chapter 19). It is quite possible for a designer to be sued in contract law and negligence. However, any damages or compensation will only be paid once.

Thus while the cost of the design itself is significant, the *cost of the result of a design* which is inappropriate, inadequate, defective or negligent can be immense. An inappropriate design will probably result in an inefficient asset that may cost more to build, operate and maintain, or one which does not meet the employer's needs. An inadequate or defective design is potentially much more serious. The resulting structure may be useless without expensive repair or modification; in an extreme case it may even collapse.

An *inappropriate design* will result in an unhappy employer, who probably has little recourse in law. It is quite likely the result of an insufficient briefing stage to determine the employer's real needs. In contrast, a problem with the *adequacy* of the design or a *defective design* is a legal matter. It may be a matter of negligence, and possibly breach of contract. If the design contract sets out clear parameters for design and these are not followed, then a **breach of contract** action might result. If the design organisation has broken a term of the design contract, then it will probably have to pay damages to the employer in contract law, to put the employer in the same position as if the term had not been broken.

However, for most design organisations a much greater risk is that their design could be found to be **negligent.** Negligence is a tort, not a breach of contract. Contract law and negligence are described in Chapters 17 and 19. Negligence is concerned with a breach of a duty of care that is owed by the designer to the employer. The test of negligence is not what the design contract says but comparison with an objective standard. The standard for a designer is usually one of exercising **reasonable skill and care.**

Design liability in design and construction contracts

However, the law normally sets a higher standard for services of a *design and construction* nature, and that standard is '**fitness for purpose**'. In simple terms 'it has to work'. As an example, a doctor, as a professional person, will try to cure your illness, and in so doing he or she must not be negligent; in other words the doctor must use reasonable skill and care. However, the doctor does not guarantee to cure you, although the doctor will hope to. Any medication must be prescribed professionally and without negligence, so a competent doctor will make reasonable enquiries to ensure that you have no existing medical condition that could make the prescribed medication dangerous. If a doctor did not make sensible and reasonable enquiries then the doctor would be negligent. Importantly however, there is no *guarantee* that the medication will work or that you will get better. A guaranteed cure would be a standard of fitness for purpose, which for the medical profession and others would be unrealistic.

It may seem unfair that these two different standards are applied by the courts to the tort of negligence. The reason may be that many goods and products that we buy are designed and manufactured (built) to accepted standards and after thorough design and development. They will almost certainly have gone through an extensive testing regime. So when you buy a kettle, you expect it to boil water. If it only reaches a temperature of 90 degrees Celsius then your tea will not taste as it should and the kettle will not be 'fit for purpose' even if it complies with normal manufacturing standards.

Unfortunately, applying this standard of fitness for purpose to a design and build construction contract is not quite the same thing, since most construction projects are 'one-off' and do not have the benefit of extensive development and refinement. As a result, construction contracts for the design and build strategy usually convert the fitness for purpose standard into one for reasonable skill and care. Accepting a design standard of fitness for purpose puts the organisation carrying out the design at great risk, particularly since this level of risk cannot usually be insured against. The way that construction contracts, with design as a part of them, approach this important issue is explained in more detail in the chapters on NEC3 ECC and JCT SBC05 (Chapters 21 to 27).

Design contracts

Wherever any significant work takes place it is always wise to have a contract setting out what needs to be done, by when and how payment will be made. The major professional bodies in the construction industry have written forms of contract for construction, such as NEC3 ECC and JCT SBC05. They also have written forms of contract for design work, which is essentially different even though the design and construction both result in the constructed asset.

The design contract is a completely different form of contract from the construction contract; it is written to cover design issues rather than the more complicated construction issues found in contracts such as NEC3 ECC and JCT SBC05. Historically design contracts were written by the professional institutions such as the Association of Consulting Engineers and the Royal Institute of British Architects (RIBA). They are usually intended to be complementary to the forms of construction contract published by these bodies.

There are also specific professional services contracts written for the engagement of structural engineers, building surveyors, quantity surveyors and the many other 'design' professions that make up the construction industry. The earlier forms of design contract have now been added to by contracts published by organisations such as the NEC who publish the NEC3 professional services contract (NEC3 PSC). Whilst RIBA forms of engagement are still very common, the government is encouraging the widespread use of NEC3 PSC through the auspices of the Office of Government Commerce (OGC).

Contracts for design have much similarity, and in order to illustrate their main elements the NEC3 professional services contract (NEC3 PSC) has been selected here. NEC3 PSC follows a similar format to NEC3 ECC, which may at first be unfamiliar to those who use the JCT forms of contract.

Construction contracts (including those for design services) are discussed further in Chapter 20.

NEC3 professional services contract (NEC3 PSC)

NEC3 PSC is written using the same layout and style to the NEC3 ECC contract which is used for construction work.

- It begins with the core clauses one to nine, which are a 'designer' version of the core clauses in the construction contract NEC3 ECC.
- The employer then selects one of four payment options:
  - Option A: Priced contract with activity schedule
  - Option C: Target contract
  - Option E: Time based contract
  - Option G: Term contract.
- Dispute option W1 or W2 is selected.
- There follow fifteen secondary options (X options) which are largely similar to the ones in NEC3 ECC, from which employers can select, if they choose to.
- The contract data part one: data provided by the employer, includes
  - Name of the employer
  - Name of the adjudicator

- **Services**
- **Scope**
- The starting date
- Details of insurance
- The completion date
- Option statements (these largely relate to different methods of paying fees).
- The contract data part two: data provided by the consultant, includes
  - Names, qualifications and experience of key persons
  - Rates of payment for different personnel.

The guidance notes to NEC3 PSC suggest that the 'scope' should describe what the consultant is to provide. So clause 11.2(11) of NEC3 PSC states:

The scope is information which either
- specifies and describes the services or
- states any constraints on how the consultant provides the services.

The scope will vary between appointments depending on the complexity of the consultant's tasks and also the type of project itself. Lists of tasks are published by various organisations, a good one being the *Royal Institute of British Architects (RIBA) Plan of Work Stages* (available from RIBA Bookshops). Employers who are unfamiliar with the professional services associated with construction may ask the architect, consulting engineer or building surveyor to draw up a suitable list for the particular project concerned.

The 'services' would specify what is required of the consultant. Services might simply be design work, but could include feasibility and appraisal, assistance with the tender process, contract administration, advice on dispute procedure, assistance with gaining approvals such as planning approval, services relating to land acquisition and so on.

Contractors as designers

It is usual for contractors to design any temporary works that are required. Temporary works can be very extensive and difficult to design; they are anything that is needed to facilitate construction such as formwork, falsework, scaffolding, settling ponds and lagoons, temporary fences and all temporary supports and shoring. Whilst many contractors employ their own temporary works designers, and some process contractors may employ their own process engineers and mechanical/electrical engineers, very few contractors employ their own architects or designers of civil and structural engineering works. They normally go to a design organisation to do this work.

JCT devotes an entire contract, JCT DB05, to the case where the contractor designs the whole works. Whereas in JCT SBC05, based on the traditional contract strategy, partial contractor design is called the contractor's design portion, and this contract goes to great lengths to define it and the procedures to be used. The standard ICE7 contract assumes no contractor design unless 'expressly stated' (refer to clause 8), and ICE publish a design and build version of ICE7.

NEC3 ECC is completely flexible on the subject of contractor design and it is possible to have a construction contract with:

- No contractor design
- Partial contractor design
- Complete contractor design.

Contractors with responsibility for part of the design

Partial contractor design in particular is a complex subject fraught with difficulty because there is usually an interface between the constructor's design and the designer's design, and managing that interface can be problematic. Where the contractor designs nothing or everything, the matter is much more straightforward because there is no design interface between two design organisations.

Another problem is adequately defining the extent of partial contractor design and the standards to which it must be designed. Take a lecture theatre ceiling for example. If the contract says the contractor shall design the ceiling, what does this mean? Does it just mean the main support beams, and if so for what loading and what fire resistance? Does it include the ceiling panels, the lights, the supports for the AV projector, the brackets for the screen, the sprinkler system for fire and so on?

Unfortunately, NEC3 ECC is very cursory on the subject of contractor design, whether it is partial or complete contractor design. NEC3 ECC covers the issue rather briefly in clause 21.1 which simply says 'the contractor designs the parts of the works which the works information states he is to design.' Where there is any specified contractor design, clause 21 goes on to say the contractor submits the particulars of his design for acceptance by the project manager in whatever way the works information dictates. The contractor must not proceed with the relevant work (relevant to his design), until he has the project manager's acceptance.

Thus in comparison with JCT SBC05 for example, NEC3 ECC is remarkably brief in its treatment of this subject, which can be crucial to a project's success. However the NEC guidance notes do give some extra advice. The other peculiarity is that the NEC3 ECC design contract, the NEC3 PSC (written specifically for design work), is 58 pages long.

Perhaps users of NEC3 ECC who wish to specify contractor design should also look at some of the provisions of NEC3 PSC, or consult the clauses relating to the contractor's design 'portion' in JCT SBC05, which also has a useful design approval procedure in schedule one.

Design and build contracts

The designer role in 'design and build' contracts

The designer role in design and build strategy was discussed in Chapter 9. However, to extend this idea, there are really two designer roles.

The first designer role is again to advise the employer on option selection, and the feasibility of the options. It is particularly important that the designer explores the employer's business needs in detail, so that they can draw up a suitable performance specification and any design or material standards that must be communicated to the contractor, in the works information. What is crucial to the success of design and build contracts is a clear performance specification.

The second design role will be to produce the detailed design, but this time, on a design and build contract, *the designer is employed by the contractor.* Since the contractor carries design liability to the employer, it is crucial that the contractor has a similar design liability in its contract with the designer. This second element of design is usually greater and more time-consuming than the first. However, both are essential.

It is important to remember that design and build does not reduce the total amount of design effort, it simply changes the contractual links, and allows the contractor to have a much greater influence on the design itself.

Chapter summary

- Everything that is built has to be designed by someone. Changing the contract strategy simply moves the responsibility to carry out the design; it does not reduce the *cost* of design.
- The design organisation usually does much more than simply design a project. It usually offers a wide range of essential services to take a project to tender stage.
- In most strategies, the design organisation works for the employer, but in the design and build (D and B) strategy it works for the contractor.
- This means that in D and B the employer must clearly specify what is wanted at a very early stage in the investment process. This is often more difficult than many employers realise.
- Designers have many professional obligations such as to act ethically and for design safety.
- In more modern forms of contract and procurement the design organisation may be *required* to cooperate with the contractor, rather than simply to administer the contract.
- There are forms of contract such as JCT DB05 written specifically for design and build contracts.
- NEC3 ECC, which is a construction contract, can be used for the design and build strategy, but this must be done with caution.
- There are forms of contract written specifically for design services such as NEC3 PSC.
- Where contractors are responsible for partial design their role, services and the interface with other designers must be specified with great care.

11

Contract strategies: partnering and commercial arrangements

Aim

This chapter aims to improve knowledge of the different contract strategies available for construction procurement. It focuses on partnering and commercial arrangements.

Learning outcomes

On completion of this chapter you will be able to:

>> Describe the contract strategies based on partnering, commercial arrangements and funding arrangements.
>> Apply the best contract strategy to different cases.

Partnering

Partnering is a philosophy of working together in a more effective and less adversarial way, with the expectation of cooperation and trust. It is a structured management approach to developing and improving working relationships to the overall benefit of all the participants and the project itself. It usually stresses relationships and culture, and relies on the contract itself to cover the commercial issues. The partnering strategy has developed largely as a result of the Latham and Egan reports, and a desire to reduce the conflict, waste and adversarial attitudes that plagued the construction industry in the United Kingdom particularly in the last two decades of the twentieth century. These government-sponsored reports into the construction industry were Sir John Egan's report *Rethinking Construction* published in 1998, and its forerunner, Sir Michael Latham's report, *Constructing the Team* published in 1994.

The term 'partnering' is applied to a wide variety of relationships; some contractual, many not. So although partnering can have a contractual basis, it is *usually a non-legally binding agreement* to cooperate, share information and work together on the basis of mutual objectives and trust. The binding agreement is the contract itself. This contract would probably be a construction contract chosen from the JCT or NEC suite of contracts. The employer will often be involved in a partnering agreement with the contractor or contractors. In fact this is the main point of partnering agreements, although they can be a supplement to any contractual relationship. This is quite unlike alliances and joint ventures where the employer is rarely involved directly as one of the participants. By having a non-binding partnering agreement, employers can contribute to an open relationship based on trust with their contractors, whilst still having the comfort of a legally binding contract such as those published by NEC and JCT. It is usually hoped that the more onerous terms of the legally binding contract will not need to be used.

One of the reasons for having a non-binding partnering agreement is that it will often include aspirations and behaviours which use words and ideas that have not been tested in the courts because they do not form part of a normal construction contract.

Partnering is not the same as 'partnership', and the words should never be confused. A partnership is a legal business entity with rights and obligations in which the partners share gains and losses. By contrast, partnering is usually a non-legally-enforceable agreement to cooperate and improve towards mutually agreed targets.

Partnering is usually based on a charter which is not contractually binding.

However, there is a middle road between a non-binding partnering agreement, which is usual, and a binding partnering agreement, which is rare. This middle road is to link some of the partnering measures into the main construction contract, possibly as key performance indicators or kpis as they are known. Such kpis will often relate to measurable performance against time, cost, safety or environmental targets. Partnering agreements will frequently include such measures, but in the partnering agreement itself there will rarely be 'money on it'. The 'money' or in other words payment by the employer linked to performance by the contractor will be in the main construction contract such as NEC3 ECC or one of the JCT suites of contracts. NEC3 ECC for example has provision for payment against kpis in option X20. The payments against any given kpis should be indicated in the 'incentive schedule' which will become part of the contract with the inclusion of option X20.

The partnering strategy is sometimes called 'discretionary'. This is because it is usually added to one of the other contract strategies. It is important to be very clear whether a partnering agreement is contractually binding or not, and the standard forms of partnering agreement are always very certain in this respect. As we have said, in most forms of agreement partnering is a philosophy of working together more effectively, rather than creating contractually binding arrangements. The partnering agreement therefore incorporates management concepts and ideas that may be less familiar to many construction professionals, particularly those more used to contractually defined procedures and relationships.

There are legally binding forms of contract for partnering available, such as PPC 2000, which combines the normal, formal contract terms with partnering terms about working together in a more effective way. PPC 2000 (revised in

2008) is the first standard form of project partnering contract, and was launched in September 2000 by Sir John Egan. It is published by the Association of Consultant Architects (ACA). However it is not used frequently, unlike the NEC3 and JCT forms of contract. Whilst the ACA website has an impressive list of users, the 2007 RICS survey of contracts in use *by value* shows ACA as 5.5 per cent, as opposed to 61.5 per cent for JCT forms. The RICS analysis of contracts *by number* shows ACA as 2.2 per cent, against JCT at 79.3 per cent.

Some people may find the concept of partnering difficult to understand when the main aim of construction contracts is to be comprehensive and clear in themselves. Most people would agree that clear and well-defined contract terms are a good idea, and forms of contract seek to be as definite as possible, and try to cover the main eventualities that arise during construction. Unfortunately, the *application* of these contract terms to actual problems and events is not always easy, and there are often disagreements between the contractor and the contract administrator as to whether contract terms apply to particular events and hence who should pay for the problem – the contractor or the employer. These disagreements can take a long time to settle, and occasionally lead to formal dispute procedures and even end up in court. In the process, relationships can become strained and cooperation can diminish; all to the detriment of a successful project. This can lead to the many problems that have dogged UK construction over the years. The worst scenario is for the contractor and the contract administrator to retreat to entrenched positions where they write carefully worded formal letters to each other, rather than meeting to try and sort out the problem, before it escalates to involve the employer.

The New Engineering Contract (NEC) has gone a long way to try to address this state of affairs in all their forms of contract, which have much similarity. For example the NEC3 Engineering and Construction Contract uses phrases such as 'at a risk reduction meeting those who attend cooperate' in clause 16.3, and the well-known first clause 10.1 which says 'the employer, the contractor, the project manager and the supervisor shall act as stated in this contract and in a spirit of mutual trust and co-operation.' NEC also uses words and phrases that are less in the 'command and control' mode of more traditional contracts such as ICE and JCT. Having said this, it must be emphasised that more recent editions of the JCT contracts go some way to follow this change in philosophy. The problem is that ICE and JCT contracts have their origins over sixty years ago, when construction and indeed society were very different places.

So partnering is about team working, better communication and improved relationships. In this way many of the problems of construction can be solved, or their worst effects minimised. In a really successful partnering arrangement, the employer may even make payments to the contractor which are not strictly justifiable contractually, and the contractor may waive many of the smaller claims for financial loss and expense that may be justifiable but could lead to strained relationships. However, for partnering to work well the employer or the staff have to be experienced in the design and construction of projects, so that they can 'talk the same language' as the contractor, and really understand any problems or obstacles and help to resolve them constructively.

The three main features of partnering
Partnering is usually seen as having three main features originally attributed to the Reading Construction Forum, *Trusting the team*. These are shown in Figure 11.1.

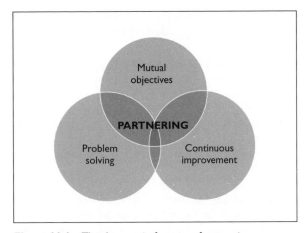

Figure 11.1   The three main features of partnering.

- **Mutual objectives** should be agreed by the team of project participants and set down in the partnering charter and supporting documents, which are described below.
- A **problem-solving** mechanism should be developed and agreed. It should be separate from any contractual stipulations such as arbitration or adjudication. This mechanism will usually be the referral of any problems or disputes to a panel made up of senior representatives of the participating organisations.
- **Continuous improvement**, which should be based on measurable criteria where possible, well-publicised results and a **feedback loop**.

A well known feedback loop is the Deming cycle, named after its inventor Dr W.E. Deming and used frequently in business and quality improvement programmes. This is shown in Figure 11.2.

- **Plan** – establish the objectives and processes necessary to deliver results in accordance with the expected output.
- **Do** – implement the new processes.
- **Check** – measure the new processes and compare the results against the expected results to ascertain any differences.

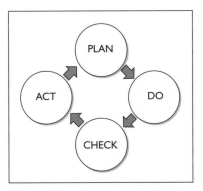

Figure 11.2   Plan – do – check – act.

- **Act** – analyse the differences in these results to determine their causes. Then determine where to apply changes that will bring about improvement.

The important point is to keep going around this cycle in order to obtain continuous improvement.

The two main categories of partnering

There are two main categories of partnering and within these two categories there are nuances depending on the way individual partnering agreements are set up. The two categories of partnering are **strategic partnering** and **project partnering**.

- **Strategic partnering** takes place when two or more firms use partnering on a long-term basis to undertake more than one construction project. It is sometimes called 'multi-project partnering'. It is often seen in 'framework agreements' which last from three to seven years, and are much of the UK construction industry's response to the demands of the EU Procurement Directives (the 'regulations').
- **Project partnering** occurs when two or more firms come together in a partnering arrangement for a **single** project.

For companies who are not subject to the advertising rules of the regulations it is quite possible to extend a successful project partnering relationship into further projects, and here project partnering begins to approach strategic partnering. The response of many 'public' bodies to the regulations has been to use framework agreements lasting a number of years. These are an ideal basis on which to build partnering agreements, which by their long-term nature will be strategic partnering.

The partnering workshop

If the employer wants a non-binding partnering arrangement with the successful tenderer then this should always be made clear in the tender documents. A satisfactory response to this section of the tender could of course be a condition of contract award. It is also a good idea to include questions in the tender documents specifically to determine the experience of partnering and likely attitudes of tenderers to this type of arrangement. After contract award, the successful tenderer will be invited to send representatives to a **partnering workshop**. These workshops are sometimes called 'team builds'. The aims of the partnering workshop are to:

- Introduce the main project participants to each other.
- Allow the participants to interact and get to know each other.
- Explain the intentions of the project and what the employer would see as a successful outcome.
- Explore the business needs and key drivers of the participating organisations.
- Allow key objectives to surface and be recorded.
- Inspire people to think creatively and flexibly.
- Identify any key difficulties or obstacles.
- Write a **charter** (more on this below).

- Start the process of building the team that will deliver the project.
- Agree a programme of future meetings, and set up any sub-groups.

The partnering workshop should have appropriate representatives from the employer, the designer, the contract administrator and possibly any of the employer's support departments that are closely connected with the project, such as operational staff. If there are any crucially important suppliers, then their representatives would also be a valuable addition.

The number of representatives and their level in each organisation are a difficult choice. A workshop with over twenty participants becomes hard to manage, and has an impersonal flavour, which is the opposite of the intention of these events. A senior representative from the employer and the contractor is always a good idea, to show commitment and leadership, and if nothing else an opening statement by such people will get a workshop off to a good start. Other representatives should be the people who will manage and organise the project on a daily basis, such as project managers, the contractor's agent and senior site staff, the contract administrator, representatives of the design organisation and any quantity surveyors who will be closely involved.

It is usual to have an external facilitator to run the workshop. This demonstrates an even-handed approach to the participants. A good facilitator is vital for success. Such people may not be construction professionals, and often have an HR or training background. However they must know something of the construction process as well as being perceptive people with well-developed interpersonal skills and the ability to organise and run group events.

The workshop is likely to take place over at least a day, and an evening session around a meal and a drink will help participants to relax and get to know each other better on a more social level. A venue away from the normal working environment is also a help, and all 'work interruptions' such as mobile telephones and personal computers should be avoided. Good venues are hotels, conference centres or training centres, but the cost will always be a consideration. Cost is part of the balance and with senior people costing upwards of £500 per day with overheads, a day for twenty participants will start at £10,000 before any catering or other arrangements are paid for. Good facilitators can also be expensive, and charges running to a few thousand pounds a day are not unknown.

Of course we should remember that for a project costing millions of pounds, the cost of such a workshop is a relatively small proportion and well worthwhile if it results in a better run, more harmonious project with a good outcome. Smaller, shorter and hence cheaper workshops can be used for smaller projects. Even a morning spent on the right activities can be a big help towards a successful project. The use of sub-groups to work on specific issues, reporting back to the main group, is a good way of reducing future costs. At the end of the workshop, it is important to record key ideas and intentions. This is usually written as a 'charter'.

Yet another advantage of long-term framework agreements is that the parties should already have good working relationships, and understand each others' systems and processes. Hence for successive projects in a framework agreement, team builds and workshops are much less necessary. A short half-day event onsite with the main participants of that particular project may be all that is needed.

The partnering charter

The partnering charter or partnering agreement is the recorded output of the partnering workshop, which lists common goals and objectives, describes how the participants intend to work together and details how they intend to resolve any difficulties. The idea is that the partnering agreement becomes the normal method of working, and the parties attempt to resolve any problems by using the partnering agreement rather than resorting to the terms of the contract itself to do this. There will also be many non-recorded outputs such as better cohesion of the group, a change in attitudes and intentions and the beginning of team bonding.

The partnering charter is sometimes referred to as a 'moral contract' because it largely deals with aspirations and intentions and a commitment to future working practices and behaviours. Having said this, it should also contain more tangible objectives that can be measured so that progress and improvement can be documented and problems can be addressed. The charter will usually contain a short mission statement to encapsulate the group's commitments. Further goals are then listed such as:

- Project completion on time
- Project completion to budget
- A reasonable financial return for the contractor
- Overall employer satisfaction
- A quality target for minimum or zero defects
- A health and safety target such as zero reportable accidents
- An environmental target
- No disputes
- Team working commitment
- Enhanced reputation for all parties.

It is important to follow this charter with supporting documents that say exactly how these goals can be achieved on the particular project in question, and how progress against them can be measured. Selected working groups will usually develop these supporting documents after the main partnering workshop. Measurement is very important. The old saying 'what gets measured gets done' is very true of construction. Any measurement system should be set up to feed results to all important participants on a regular basis, often monthly. Further meetings of selected participants should be set up to review progress, address problems and ensure that successes are publicised. It is all about maintaining the initial momentum of the partnering workshop.

At the end of the project a final review meeting is essential to close the feedback loop, so that all participants learn from the experience and use this to improve future projects. Much of the documentation could also prove useful for future projects. Finally, ways of maximising any gains in reputation and further project opportunities can be discussed by the participants.

Commercial arrangements

Alliances

Alliances are a middle stage between partnering and joint ventures. In partnering, there are often no direct financial gains or losses that are shared between the partners (although there may be). Partnering agreements are usually based on the intention to cooperate based on trust for everyone's benefit. This benefit may be relatively intangible but the participants are likely to hope for efficiency savings and better working relationships, which can help deliver a successful project for the employer and potentially secure future work for the contractor. A joint venture on the other hand is usually a separate company in its own right.

Alliance was a word originally used to describe an agreement between countries to advance common interests, or to support each other particularly in times of war. In construction, alliances tend to be a public declaration by companies intending to work together on a structured basis for a period of time, or for a particular project or series of projects. Hence we often see alliances form in response to invitations to tender for framework agreements. Alliances rarely become an actual company, although this is of course possible. Instead, alliances will often have some formal mechanism for sharing financial gains and losses, probably a contractually binding agreement.

One major advantage of an alliance is that it can bring together disparate skills to produce a whole which has much more appeal to a prospective employer. An example might be the response to an advertised framework agreement for road and bridge maintenance for a county council. If a contractor specialising in concrete repair formed an alliance with a 'blacktop' contractor (a contractor specialising in road construction, particularly resurfacing), they would be much more likely to demonstrate the expertise and experience needed to win the framework. They would be unlikely to succeed with separate bids. Similarly, if this framework included structural design work, then the incorporation of a suitable design organisation into the alliance would make it very attractive to such an employer.

We also see alliances forming when the value of the work is too much for any one company. This may be the case with a major project, but it is more likely to happen with one of the many framework agreements that are advertised by local authorities and utilities. The work in any category of the framework may be worth hundreds of millions of pounds over a period of years. There are not many UK contractors who could or would wish to bid alone for such work.

The management structure of the alliance is normally an executive team, formed from representatives from all the alliance partners, with reporting lines to the employer's board of directors. Below this structure will be reporting lines into each organisation in the alliance. Joint working groups will often be set up to share knowledge on certain aspects of the work. There will almost certainly be a group set up to discuss programme and resourcing issues, to give as much forward visibility as possible to the contractor and to seek to resolve any resource or supply issues before they become a problem.

In alliances the contractors involved may be asked to help develop projects with the employer, often before final financial approval by the employer's directors. They can thus be involved at the appraisal and feasibility stage and in design development alongside the design organisation and probably some of the employer's staff. Hence an alliance is yet another modern way of bringing contractor input and expertise to the design stage, thereby minimising the main drawback with the traditional strategy. The main problem is that after the alliance is formed, it is difficult to obtain competitive prices to a programme of work, such as a framework, and so some pricing mechanism has to be agreed with the employer when the alliance is initially awarded the programme of work.

Sharing office accommodation

Alliance partners will also frequently share offices, often located in the geographical area of the future business opportunities. Thus the alliance partners may each move relevant staff to an office in the employer's operational area. It is not unusual for the employer to offer office space, if the employer has spare space available. There are not only utilities but many city or county councils that offer this type of accommodation. Many water companies accept the tenders of alliances between a contractor and consultant, to design and construct the assets in an area, over a five-year framework period. Kelda and Thames Water are two such examples. The alliance members are separate companies, but share the same offices in the water company area, and receive an input from water company staff acting as both employer and operational advisor. The alliance partners design and build a programme of work, and share systems and processes. The alliance is a 'virtual company' in many respects in that the alliance partners may behave as if they were one company, but legally and financially they are separate.

Design organisations in alliances

Alliances often include design organisations as well as contracting ones. The design organisation effectively forms a long-term 'design and build' team with the contractor or contractors. The difference for employers is that they know who the design organisation will be, whereas in design and build the employer may not know, and the contractor is at liberty to change the designer at will. Hence an alliance of a design organisation and contractor is a little like traditional procurement, or management contracting for the employer. The employer knows who the contractor and designer are, and one assumes will not appoint them unless satisfied with their credentials and likely performance.

Alliances with the employer

Alliances can be formed with the employer (as a corporate body) as a member of the alliance, but this is unusual in the construction industry. It would not be appropriate for a public body such as a local authority to have an alliance agreement with contractors to share profits and risks for example. Equally, many utilities see their capital programme as a *supporter* of their main activity, not *the* main activity. This main activity would be delivering water, gas or electricity to customers. Such employers are not in the 'construction business', they are in the water, gas or electricity business. An alliance that they might consider is with another similar utility, but not a construction company which is simply delivering part of its capital programme. However, to rather confuse the issue, some employers call their arrangements an 'alliance' or a 'virtual alliance', but they rarely have any legally binding terms. Presumably such employers want a title that sounds collaborative.

Another reason why employers do not need to join alliances with contractors and design organisations is that it is easy for employers to set the terms of the main contracts to reflect their needs, which might include sharing gains and losses on projects. Target price contracts incorporating pain/gain share give the employer a stake in efficiency savings (and unexpected cost overruns) as part of the form of contract with the contractor. It is not necessary for the employer to enter into a potentially difficult or risky alliance agreement to achieve the same objective. The employer simply uses an established form of contract such as NEC3 ECC option C or D.

However, the employer will often contribute to an alliance with operational expertise and so on, but not in a legally binding way, as it might be if the employer was actually signed up as part of the alliance. The employer's contribution might be regular programme meetings, release of forward data on likely projects, secondment of operational staff and so on, all of which can bring efficiency gains. Another variant of this is for the employer to formally transfer staff to the members of the alliance, often design staff who previously worked for the employer. These staff then become direct and permanent employees of one of the alliance members. When they transfer employer their terms and conditions are protected under TUPE which is the abbreviation for the Transfer of Undertakings (Protection of Employment) Regulations 2006. The problem, of course, is that for public bodies the Public Contracts Regulations apply to their procurement, and so all alliances must be advertised in the EU. More importantly, they are time limited, indeed to four years in the case of local authorities and government departments. This can produce a real problem for an alliance partner (and its staff) that now has transferred workers, if the alliance partner does not then secure similar work at the end of a framework agreement.

Joint ventures (JVs)

Joint ventures, or JVs as they are usually known, are formal arrangements between contractors or design organisations and contractors to undertake economic activities together. In the case of construction projects this is likely to be design or construction services, together with the provision of plant and equipment by the contractor. If the scale or diversity of the work justifies it there may also be formal agreements between *groups* of contractors (often called a consortium) and one or more design organisations. This formal arrangement is a Joint Venture Company, a separate legal organisation in its own right. The employer signs a contract with the JV. JVs will often set up their own independent offices, often within the employer's operational area. In many ways this is similar to an alliance, but in a JV a new company is formed.

JVs frequently form because the individual companies are not large enough, or do not have all the expertise required to handle a very large project or the programme of work offered by the employer, under a framework agreement. Some large employers use many contractors for their investment programmes. However, the trend is to have very few contractors, with perhaps £100 million of work per contractor per year for example. Few contractors have the staff or expertise to handle this volume of work, particularly for one employer. Even if they do have the resources, many contracting companies do not wish to commit more than about ten per cent or so of their turnover to one employer, because of the commercial risks related to dependency on one large customer. These risks include for example if the employer's work reduces significantly, or the employer cancels a framework agreement, or, at the end of the framework agreement when it is usual for it to be re-advertised, the contractor did not secure another framework agreement.

Then there are the really large projects such as some London Underground projects or projects for the Olympic Games which are much too large for any contractor to handle alone. So instead, contractors may bid as a JV containing perhaps two or three contractors. If design services are also required, the JV may include a design organisation and possibly specialist financial or project management consultants. If they are successful in their bid, these companies then form a true company, the joint venture, which becomes a legal entity as a company with a board of directors and staff taken from the contributing organisations together with new staff, specifically recruited for the JV. The original organisations usually continue in their own right.

Benefits of alliances and joint ventures

Alliances and JVs have many attractions, but they are no panacea. It is very difficult, expensive and time-consuming to bring the culture and systems of two or more companies together. Whilst a five-year framework agreement sounds a long time, it may take the first year to get the alliance or JV running efficiently, and in the fifth year, the companies in the alliance or JV will be considering where to go next if unsuccessful in securing a further framework.

Advantages of alliances and JVs
- Increased opportunity to bid for large or complex projects, or framework agreements.
- Market opportunities that did not exist for any of the constituent companies alone.
- Increase in knowledge base of the companies involved, arising from exposure to the new skills of the 'partners'.
- The opportunity to share specific resources or expertise. Where two contractors are involved, one may have an expert planning department, and another may have geotechnical expertise or proficiency in ground remediation for example. These skills can be made available to all participating companies.
- The possibility of buying specialist plant or equipment because JVs or alliances are usually formed for larger projects or frameworks. Hence there is likely to be better utilisation and the cost is effectively shared among the JV/alliance partners.
- The closer working relationships of the JV/alliance partners should result in reduced administration, less duplicated working and greater efficiency.
- Risks can be shared, which means that a project or framework could be undertaken which contained too much risk for one contractor alone.

Disadvantages of alliances and JVs
- It can be very difficult to effectively combine the culture of different organisations.
- Organisations will have different systems and processes, which may not match easily. These could include payment processes, lists of approved sub-contractors, health and safety procedures, IT systems and so on.
- Matching such processes and procedures into a workable model may take many months, divert a large number of staff and cost a lot of money.
- There may be a 'power struggle', particularly if the participants do not have equal shares.

- There is the related issue of leadership. Any single organisation will have a management structure and a 'figurehead', possibly the managing director or chairman. What happens when two or more organisations combine?
- If the joint venture or alliance is of significant size, it may necessitate some restructuring of the original businesses. Again this could represent considerable risk and cost.
- Existing employees could feel threatened by the joint venture or alliance, either because they are transferred to the JV or because they continue to work in the original company which now has a reduced staffing level and may not be as commercially viable in its own right as it once was.
- Whilst knowledge transfer and sharing are usually a good thing, individual companies may be concerned about confidentiality, intellectual property or patents.

Employers' concerns over partnering, alliances and JVs

Many employers are very supportive of partnering, alliances and joint ventures. However, there is some 'comfort' for employers in traditional tendering – the knowledge that you have the cheapest price out of a number of tenderers (preferably carefully pre-selected). Traditional tendering gives, at least in theory, the current market price for that particular project, bid in a competitive environment. It has its drawbacks of course, as described in Chapter 6. In many ways these long-term arrangements such as strategic partnering, alliances and JVs require a leap of faith on the part of the employer. The employer is giving up the comfort of a competitively bid price for the more intangible benefits of these long-term arrangements. The problems with partnering (strategic), alliances and JVs for employers are:

- Setting and agreeing a realistic price for each project over a long period of time that is not simply driven by the contractor.
- The potential change in market conditions (prices) over the period of the agreement, which can often be four or five years and possibly longer for utilities.
- Maintaining incentives for efficiency (especially on cost-reimbursable contracts).
- Only paying for appropriate work (especially on cost-reimbursable contracts).
- Concern about the 'team' becoming too cosy, and inefficient. The team can often include employer representatives who may also have contract administration duties. This can lead to employer concerns over:
  - Whether the contract terms are being properly applied
  - Whether incentive schemes are being administered rigorously
  - Whether prices are being fixed rigorously and at a suitable level, rather than being 'talked up' by the contractor.
- Worry about becoming 'locked-in' to a contractor, and 'shutting-out' the market for the period of the agreement.
- Concern about what happens at the end of the agreement period:
  - After a stable agreement period, reselection can be very disruptive.
  - Deselected existing contractors are now engaged on a short-term basis (until the end of the current agreement) and may lose commitment or see no harm in pursuing contract claims, when they know that they will probably not work for that particular employer again for four or five years.

Funding arrangements such as PFI

In private finance initiative (PFI), the public and private sectors enter a contract which shares between them the risk of undertaking an investment project and receiving long-term income from it. Such a project would usually be a major capital asset such as a school or a hospital and related support services like repairs and maintenance. These PFI projects, including maintenance and income, are often for 25 years or more.

Hence PFI usually avoids the need for large public sector investment in the capital project at the outset. Instead the *capital cost is financed by the PFI partners*, who then receive payment for the original cost plus 'maintenance or service' over a very long period. At the end of the period, the asset reverts to the public-sector employer. An example is the M6 Toll Road:

> 'The M6 Toll Road is a freestanding concession with no ongoing government financial involvement. The concessionaire generates investment and return from direct user charges. Government has transferred risks, such as planning, design, construction and revenue risk, to the concessionaire. The contract is for 53 years and at expiry the asset will revert to the public sector at no cost.'
> (*HM Treasury Infrastructure procurement: delivering long-term value, 2008, crown copyright*)

The private-sector supplier (usually a consortium) is involved in constructing and maintaining the asset, and is therefore incentivised to have the highest regard to **whole-life costing** as it has the risk of operation and maintenance for a substantial period of time. Incentivising 'normal designers' to consider a whole-life cost has always been difficult, since the fees are often linked to the capital cost (the cost of constructing the project). It is said, for example, that the capital cost of a submersible pump is just two per cent of the total cost over its lifetime (which is mainly power). So, in this example, a designer concentrating on capital may specify a pump at an initial capital saving of a thousand pounds that subsequently costs hundreds of pounds extra in electricity over every year of its twenty-year life. It is rather like the decision to buy a more expensive diesel version of a car that will do so many more miles to the gallon, so that the additional cost will be recovered in a year or two by the savings in fuel costs.

PFI was introduced in 1992, and there has been a significant increase in its use in delivering public infrastructure and services. The Treasury publish figures which indicate that over £7 billion pounds was spent on PFI projects for example. The Treasury also state that by 2003, a total of 451 PFI projects had been completed across a broad range of sectors, delivering over 600 new public facilities including 34 hospitals and 119 other health schemes, and 239 new and refurbished schools.

However there are a number of problems with PFI. Its critics say that the public sector has effectively been mortgaged to the private sector for many decades, and that the total cost of these assets over a lifetime will be much greater than if they had been traditionally funded with public capital.

Another problem is the very high cost of bidding for such projects, which require not just design and construction, but operation, maintenance and overall funding. There are few bidders for many projects, because of this very high bidding cost and hence there are criticisms of a lack of competition. In the credit crunch of 2010 another problem has emerged, and that is the government cancelling PFI projects before the final bid stage. The legal question is whether, and in what circumstances, bidders can recover their costs, if at all. This uncertainty over the government's intentions as to future projects is bound to deter bidders.

As a result of these concerns the coalition government of 2011 is looking seriously at alternatives to PFI. However it is difficult to see any real alternative, unless the government is willing to consider funding the *initial cost* of hospitals, schools and so on, which is most unlikely. Instead it would seem that the government are concentrating on:

- Effective management of existing contract terms such as managing existing pain/gain share mechanisms well.
- Optimising the use of asset capacity possibly by looking at mothballing or subletting of surplus capacity.
- Reviewing the specification of soft services – avoiding 'gold-plating', ensuring the consistency of standards across PFI and non-PFI facilities, reviewing and standardising soft services specifications and terms whilst maintaining operational standards.
- Having discussions on the implementation of a voluntary code of conduct in relation to operational savings matters.

Hence it looks as if PFI will continue to be with us but in a modified form, and possibly with a revised name.

Chapter summary

- Partnering is usually a non-binding agreement that is additional to the legal construction contract and not part of it.
- Partnering has many definitions, but it is generally seen to be a process of working together in a more effective way with improved cooperation, openness and trust.
- The three main elements of partnering are agreed mutual objectives, a problem-solving mechanism and the commitment to continuous improvement.
- Partnering agreements are usually based on a charter, which is a list of objectives, intentions and behaviours that is usually developed at a partnering workshop.
- The partnering workshop will usually be followed by further agreed documents and processes developed by sub-groups.
- It is important for partnering agreements to include measurements which are as objective as possible.
- Alliances are groups of companies who agree to cooperate and to share gains and losses in some way.
- Joint ventures are a development of alliances where the participants form an actual company as a legal entity.
- The private finance initiative (PFI) is a mechanism for funding government projects such as schools and hospitals. The private sector design, construct and maintain the asset (possibly operating it as well) in return for a financial return over a long period, often exceeding 25 years. At the end of the period the asset is transferred to the government.

12

Framework agreements

Aim

This chapter aims to explain the nature and application of framework agreements.

Learning outcomes

On completion of this chapter you will be able to:

>> Discuss the nature and operation of framework agreements.
>> Describe the main factors to take into account when determining suitable numbers and types of frameworks.
>> Relate the concept of phasing work and capital programmes.
>> List the advantages and disadvantages of frameworks.

Framework agreements (frameworks)

A framework agreement is particularly important to the construction professional because it is probably the most frequently used mechanism for establishing construction contracts in the public sector and the 'ex-public sector' affected by the **'regulations'**. The 'ex-public sector' refers to the utilities (such as gas, water, electricity) and other bodies such as rail who were once publicly owned. Most were privatised in the latter part of the previous century. Here the 'regulations' mean the Public Contracts Regulations 2006 and the Utilities Contracts Regulations 2006, which are described in detail in Chapter 13. Of course a framework can also be used by any employer with a programme of work over a period of time.

A framework agreement is both a tendering option and a contract strategy.

A framework agreement is a *tendering option* because it establishes prices for the overall framework (as far as possible) and appoints a service provider. This service provider will usually be a contractor or supplier, but it may also be a design organisation or other provider of professional services. A framework is also a *contract strategy* because it is a 'serial strategy'. In other words it is a way of setting up a number of similar contracts over a period of time (often a number of years). These contracts will be subject to the overall framework conditions, but having their own particular attributes and details.

> An understanding of framework agreements is essential for construction professionals.

Every year in the UK, tens of billions of pounds of design and construction work are carried out under framework agreements. Framework agreements are often simply called 'frameworks'. A framework agreement is defined in the Utilities Contracts Regulations section 2 (interpretation) as follows:

> 'A framework agreement means an agreement or other arrangement, which is not in itself a supply contract, a works contract or a services contract, between one or more utilities and one or more economic operators which establishes the terms (in particular the terms as to price and, where appropriate, quantity) under which the economic operator will enter into one or more contracts with a utility in the period during which the framework agreement applies.'

The definition of a framework agreement in section 2 of the Public Contracts Regulations 2006 is virtually identical except that it refers to a '**contracting authority**' rather than a utility. A contracting authority is the name these regulations use for a local authority, a government department and so on, and a complete list is given in section 3 of these regulations.

The definition in the Utilities Contracts Regulations 2006 is very apt. A framework agreement establishes price terms for further contracts which will be awarded over a period of time under the overall terms of the framework. A framework agreement is thus a 'serial contract', or a 'contract to make further contracts'. It is also sometimes called an 'umbrella agreement'. Thus a framework is never intended to stand alone. It relies on the details provided for each subsequent contract, awarded under the overall framework.

> A framework is an 'umbrella agreement' that relies on the details included with each individual order.

Why have framework agreements?

Historically, procurement has been based on the assumption that the employer (or design organisation acting for the employer) specifies what is required for a particular project with drawings and details; and a number of tenderers bid for the work. The assumption is that this provides the lowest current market price for the contract which the employer then accepts. If the employer only has one project to have constructed, or a few infrequent projects, then this process serves their needs well. When the next project requires a contractor, the employer simply repeats the process; hence tendering is on a serial basis and the employer (and design organisation in some contract strategies) has to work with a succession of different contractors.

Repeating this process for every project theoretically results in the lowest price for any particular project. However there are many drawbacks. Serial tendering will often result in the need to form new relationships for every project, no forward visibility or continuity of work for the contractors, and no opportunity to improve communication and relationships or to learn from previous projects. If the employer and contractor (and design organisation where appropriate) can stay together for a number of projects then a range of benefits are possible. There is the opportunity to improve communication and relationships, for incremental learning and efficiency gains, to invest in processes and people and to develop the supply chain.

In serial contracts the focus tends to be on maximising gain, often by 'becoming contractual' rather than working together. Becoming contractual is a phrase that is often used to denote a contractor that frequently resorts to contract claims and disputes, or even seeks out possible contract claims to maximise financial return. It can equally apply to employer or design organisations which operate the contract in a bureaucratic way and without flexibility or concern for the contractor's problems. However, in frameworks all parties are more inclined to take a longer term view, because the next contract is 'just around the corner' and the relationship is expected to last for many years, not just for one contract.

Another major advantage of frameworks is an *overall* reduction in tendering costs, and the time required to tender for every separate project. Of course, the tendering costs for the initial framework are very high but, once in place, there are no further tendering costs as such, although there will be a necessity to agree the prices of any work in a particular project that is not covered in the framework agreement itself. For all these reasons, frameworks have become the contract strategy of choice for most employers with *programmes* of work. A programme of work is simply a number of projects carried out over a period of time. Many of the larger employers such as local authorities and utilities will have programmes of work comprising many hundreds or even thousands of projects extending over a period of four or more years into the future (see also Chapter 2).

A simple way of viewing a framework is to think of an amateur football team. If there were a large number of available players and the team was chosen from scratch before every match, we can imagine the result. There would be poor communication, poor strategy and cohesion, 'personalities', no team spirit and quite possibly conflict and strained relationships. What we do of course is to choose a team to play a season of matches, and then make small adjustments to personnel as the season progresses. In this way the team should develop and improve and hopefully win. This is very similar to a framework agreement, where we are effectively choosing and then developing a long-term team.

Establishing a framework

The first step is for the employer to decide on the number and extent of frameworks, to cover the work which the employer needs constructing. The employer (such as a utility or local authority or government department) attempts to predict the type and extent of work over the forthcoming framework period, or regulatory period, which for utilities is generally five years. The utility or local authority or government department then groups similar work over that period of years into separate logical categories. They then advertise each of these categories of grouped work in the Official Journal of the European Union (OJEU) as a framework agreement in a similar way as any single contract affected by the regulations. They must then follow a tendering and contract award procedure in compliance with the regulations and appoint a contractor, or contractors, to carry out each category of work (each framework) for the relevant period. Where the process is followed by contracting authorities (organisations affected by the Public Contracts Regulations such as local authorities) then their frameworks are restricted to four years' duration.

There is nothing to stop an organisation that is not subject to the regulations (such as a supermarket chain) but who has a programme of work, from setting up a framework; if this is a sensible course of action. In this case, the employer will not be subject to the regulations, and can use conventional tendering and assessment methods to establish the framework.

Frameworks usually ascertain as many price terms as possible based on technical and other specifications, conditions of contract and as much detail as is available at that time. This is because the initial framework tender is the only one exposed to competition from a number of interested contractors.

Each framework must also have a facility to set up (or 'call off') the individual contracts. In order to ensure clarity, this call-off facility is usually a formal letter called a 'works order' or similar term. The NEC3 Framework Contract uses the term 'package order'. Legally each separate contract within a framework is made when this package order is sent to the appointed contractor. Once each separate contract is made, it cannot be broken without incurring all the legal remedies for breach of contract. However, most framework agreements have a clause allowing the whole framework agreement to be terminated at the end of every year, typically for some default of the contractor. Such termination would not affect contracts already placed under the framework by means of package orders.

It is possible to have few terms in the framework agreement and rely on many terms and details in each individual contract. This is not a good practice from the employer's point of view. The framework agreement itself is usually the only agreement subject to competitive tendering, and hence competition. Once a contractor or supplier is awarded the framework, subsequent changes, details or other matters will be subject to negotiation. There is nothing wrong with this, except that many employers generally prefer competition to negotiation. Hence many employers will put a standard specification and conditions of contract (such as NEC3 ECC) into the framework and as many details and schedules to be priced as possible.

As an example of competition, let us suppose the employer wished to delete the 'weather clause' in NEC3 ECC or JCT SBC05. The 'weather clause' enables the contractor to be compensated for exceptionally adverse weather in clause 2.29.8 of JCT SBC05, and is a compensation event under clause 60.1(13) of NEC3 ECC. If this deletion was built into the original framework, then competitive tenders would be obtained for this modification, which puts a lot more risk on the contractor. However if the weather clause was originally included in the framework, then a subsequent proposal to delete the weather clause in later contracts under the framework would be likely to attract large price rises by the framework contractor, because of the increased risk. A subsequent proposal such as this would then be subject to price negotiation rather than price competition which would not be a good idea for the employer.

It is also possible for a contractor to use a framework agreement with selected sub-contractors. However, this is not very likely unless the contractor has a predictable flow of work of an appropriate type for a long enough period to justify setting up a framework agreement. This is possible if the contractor has a framework agreement with an employer, such that workload type and frequency can be estimated with some accuracy, but the contractor is not in such a good position to do this as the employer. Instead, many contractors will have lists of their approved and competent sub-contractors and suppliers, who they have checked and are satisfied with. It is also quite likely that they will have some pricing agreements or price lists. In many ways this operates like a framework, but with less of a guarantee of work and more informality.

Deciding on numbers and types of framework agreements

Many of the larger employers will have a capital programme of hundreds of millions of pounds per year; some programmes may even approach a billion pounds per year. Such employers will usually be public bodies or 'ex-public bodies' and subject to the 'regulations'. Many such employers will decide to set up framework agreements. The difficult question is how do we split up half a billion pounds' worth of work a year into sensible categories which contractors can bid for, and that can be run efficiently? Many employers will devote considerable time and resource to analysing this question. There are many factors to take into account, and the main ones are discussed below.

The impact of the 'regulations'

This is a primary consideration for employers affected by the 'regulations'. Each framework has to be properly and correctly *described*, advertised and appointments must be made on *objective criteria*.

Another consideration is that the original EU advertisement sets the ultimate scope of the framework, and unconnected work should not be added in later; this is the importance of the word 'described'. Hence, for example, if a county council advertises for road maintenance contractors, it should not later add bridge repair work into the resulting framework. If this was done, the county council could face challenges from bridge repair contractors who may claim that they would have submitted a tender for the framework, had they known that it contained bridge repair work.

The 'regulations' may also specify a maximum duration for frameworks. There is no direct restriction in the Utilities Contracts Regulations, but the Public Contracts Regulations limit frameworks to a maximum of four-years' duration in section 19(10).

The type and scope of the work

It is very important to estimate the type of work in an employer's programme, its value, its geographical distribution, its complexity or simplicity and any attendant risk, particularly for new or novel processes. Another consideration will be the constitution of the work. Are there hundreds of small projects, possibly repetitive, some large projects, and even some massive projects? Even more difficult is that many programmes of work will contain all these variations. This is why an employer with a large programme of work will divide it up into logical categories, each with its own framework, or number of frameworks.

Large frameworks or smaller ones?

The advantage of a large framework is that it gives economies of scale. It also makes the employer's role in communication more straightforward, at least at first sight. Suppose the employer only has to deal with one large framework contractor. The difficulty is that a large framework contractor is likely to sub-contract most of the actual construction work to second-tier contractors, with whom the employer may have no direct communication. Here, the employer will be entirely dependent on the main contractor to pass on any messages. These messages may be about the employer's requirements, standards, ways of working, people and processes, customer relations and organisational culture. The question is whether an employer can rely on a third party (the framework contractor) to perform this service. Messages are often lost in the transmission. As the old story from world war one illustrates, 'send reinforcements we are going to advance' became 'send three and four pence, we are going to a dance' after a number of radio transmissions. Perhaps there is an argument for the employer to have a number of framework contractors so that these messages are correctly communicated to them all, directly by the employer.

Another problem with a large framework is that the employer may become too dependent on the framework contractor, and lose the usual employer recourse of 'going elsewhere'. Many employers will therefore have a number of framework contractors to aid communication, create flexibility and break this dependence, even though this reduces buying gains from economies of scale.

The employer's knowledge and skill base

Where the employer has more knowledge of particular processes than framework contractors, the employer may wish to set up a particular framework with more employer control and intervention. In contrast, in areas where the employer has little skill or knowledge, the definition of the framework is crucial, so that only truly competent contractors are appointed. Another possibility in this scenario is to appoint a design organisation with the required skill and knowledge on a design framework in addition to contractors on a construction framework.

Considerations of the UK contractor base

Since most contractors specialise in only a few sectors of work the employer needs to have an understanding of the UK contracting base. For very large projects, such as Crossrail or the Olympics, it is likely that European contractors will apply but, for most frameworks, it is not worth European contractors trying to set up a UK base or moving into a completely different market. Hence for most normal frameworks, we are looking predominantly at the UK contractor base.

If we take a water company example, sewerage contractors are accustomed to laying pipes in fields and importantly in highways, whereas water treatment contractors are accustomed to process plant and complex control systems and process-driven interfaces on sites entirely owned by the water company. These are fundamentally different sorts of work requiring different expertise, labour, supervision and construction plant and equipment. Equally, there are contractors who specialise in such diverse activities as tunnelling, gas mains, road building, process plants, telemetry and control systems and so on.

Value of the framework

Another very important consideration is the size of any given framework relative to the turnover of contractors in that sector of work who may wish to bid for it. Many contracting companies will have an informal limit on the amount of work that they will undertake for one employer. It may be ten per cent of turnover for example. A contractor placing too much reliance on work from one employer is 'placing all its eggs in one basket'. The contractor becomes very vulnerable to the rate of work from that employer, to ongoing relationships and possibly to payment. If things are going well, the contractor has no problems, but no contractor wants to be reliant on a large proportion of its turnover coming from an employer where relationships or work flow may deteriorate.

Alliances and joint ventures

If an employer decides to make a framework very large relative to the turnover of contractors in that sector, then the employer virtually obliges contractors to bid together as alliances or joint ventures which were discussed in Chapter 11. The same happens if the employer includes significant design work in a construction framework. Most contractors do not possess the design expertise or resource needed and will have to 'buy it in' in some way. Additionally, if the employer mixes work in a framework involving different contractor specialisms, then an alliance or joint venture is virtually assured.

In some county councils for example we see a single framework for design, highway construction and bridge repair and renovation. This is likely to result in perhaps two specialist contractors (one specialising in highways and one in bridges) and a design consultant producing a joint bid as an alliance or JV. There is nothing wrong with this, but it could be argued that the employer gets less direct control than having a number of more targeted frameworks. However alliances and JVs are more likely to work closely and cooperatively together because they also have a more formal relationship as a result of their alliance or JV.

Frameworks decided by geography

Another possible way of splitting an employer's capital programme into frameworks is by geography. Some types of work are very location dependent, such as replacement of small sewers or water mains, small gas mains, or electricity or telecommunication cables. It tends to be much more efficient to run these on an area basis, possibly by city (especially the larger ones like Birmingham or Manchester) or by county. Contractors will probably set up a local office and storage depot somewhere within the area, which improves communication and reduces transport costs. This gives the employer a local contractor base, which can be much more responsive to problems, stakeholder issues or emergencies. Much maintenance work is similar. We see for example that the Highways Agency divides its work on an area basis.

A number of single *large* projects however are unlikely to benefit from being grouped by geography and there may be a great deal of benefit in grouping similar *types* of work, wherever it is located, particularly where there is specialist expertise required. A novel water treatment process requiring installation at ten sites across a water company's area, for example, would warrant grouping this into one framework rather than thinking geographically. Thus the contractor and designer could learn from experience, develop better techniques and form beneficial relationships with specialist suppliers and installers. By definition, of course, local authorities are area based, and so do not have to make the choice between geography, and scale and type of work.

Frameworks for the supply of manufactured plant or materials

A very important consideration for most employers is that of standardisation of plant. As we will see in Chapter 13, employers subject to the 'regulations' cannot simply name a supplier in any contract falling within the 'regulations'. What they can and will do is to advertise supply frameworks in the EU. Once a supplier has satisfied the objective criteria and been appointed to a supply framework, then it can be named.

As an example, if we advertised in the EU for submersible sewage pumps to be supplied to all operational sites under construction frameworks for five years, and 'Super Pumps' were successful, then we could simply specify in all construction framework agreements that Super Pumps were to be the pump supplier. Ideally, we would have determined this supplier before the construction frameworks were set up, so that the provider of the submersible pumps and any price terms were clear to tendering contractors for the main construction framework.

However, this is often not possible. Hence many frameworks contain a term that obliges the successful framework contractor to use all current and future employer framework agreements for manufacturers of goods, equipment and materials. Where full details are not included in the original construction framework there will of course be a period of negotiation with the employer, where the contractor's addition to the supplier's prices will need to be agreed. This addition may include offloading, craneage, storage, installation, supervision or one of the many other services that contractors provide. This is not ideal from an employer point of view, since all employers usually want as many items as possible to be bid competitively in the original framework. However, because of the complexity and range of most employers' work, particularly the larger ones, incorporation of the *detail* of all supply frameworks in the main construction framework is rarely possible.

Framework agreements for professional services

Professional services will usually be design services, but they could also be architectural services, quantity surveying, cost consultants or even non-construction-related services such as legal and accountancy services. Such professional services are called simply 'services' in the regulations.

The price threshold (the contract value above which the 'regulations' apply) is much lower for services contracts than works contracts such as civil engineering. The services threshold in the Utilities Contracts Regulations is about £300,000. The Public Contracts Regulations define different categories of 'services' with limits set at about half this level. It is not possible to be precise because thresholds are reset every two years and largely depend on the relative values of the pound and the euro.

Employers subject to the 'regulations' generally follow a similar process for the establishment of frameworks for professional services as that for construction frameworks. Frameworks for professional services are very common because of the reduced number of employers with their own design staff. Many years ago county councils would employ hundreds of staff in their roads and bridges departments, and most utilities would have similar numbers of engineers. This is much less common today owing to progressive outsourcing and consequently there is now a greater reliance by employers on design organisations such as consulting engineers and architects.

Hence we see frameworks for consulting engineers, structural designers, architecture, quantity surveying, cost consultancy and all the many services that large employers require. Similar benefits accrue from these frameworks in terms of improved communication and team work, streamlined administration, and the development of synergies and knowledge transfer both with the employer and with framework contractors.

Possibly the main difference between these professional services frameworks and construction frameworks is that the employer will see the organisation on the services framework as a professional advisor, replacing the in-house expertise that has been lost. In some ways this makes the establishment of a services framework more important to an employer than applies in construction frameworks. Employers will be expecting the services framework organisation to explore and define their needs. There is really nothing more important to an employer than this.

The alternative to establishing a framework for design services is to use the design and build contract strategy. Here, the contractor procures the design service directly via the contractor. There are many advantages with design and build as a contract strategy, but perhaps the most important feature for an employer to consider is the much reduced control which the design organisation has, and the continued employment by the contractor. This may not be a major issue for a one-off project but, for a framework lasting four or five years, many employers wish to retain control of the appointment of the design organisation, and over how long they are appointed for, and hence they use a design framework agreement. The result of this is actually similar to the idea of the management contracting strategy, since both the designer and contractor are appointed early, and can cooperate and improve communication. However they have separate framework agreements with the employer.

Phasing work using a framework

The issue of phasing is very important. Suppose a large employer has a framework for £500 million of design and construction work covering hundreds of projects over five years. The total programme is thus £2,500 million. This value of work will require hundreds of designers every year to design, specify, produce drawings and carry out any contract administration.

In theory the employer could carry out the framework as in Figure 12.1. However they are most unlikely to do this, since no construction is begun until year three and a massive number of designers are needed in years one and two, and then have little work for the next three years.

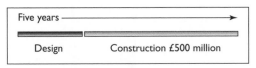

Figure 12.1 An unlikely framework agreement.

What the employer will almost certainly do is to phase the work, so that as soon as a project is designed, construction can commence, and the designers move onto the next project. This ensures continuity of work for designers and contractors, and a smooth flow of work for the parts of the employer's organisation connected with the capital programme, such as payments and customer services. This is shown in Figure 12.2, which only shows eight projects of the many dozens or even hundreds that may be involved. Quite frequently the remains of some projects will be constructed outside the five-year framework period, and this can also be seen in the figure.

The other advantage of phasing work in this way is to achieve a relatively flat expenditure profile, thereby avoiding undesirable large peaks and troughs. Such peaks and troughs produce financing difficulties for the employer and also potentially upset customers, who make payments and see no early benefit in terms of construction. A local authority for example will receive income from local taxes on an annual (or possibly monthly) basis, and will probably wish to balance its budget on an annual basis. Imagine the reaction of payers of local taxes, who make payments for three years before seeing any improved roads, or other services.

The regulators of most utilities wish to see a fairly flat profile, because they argue that income is being received on an annual basis, but expenditure is being deferred if a programme is 'back-end loaded'. Hence many employers with five-year programmes will seek a profile like the one shown in Figure 12.3, which is the result of phasing construction to follow design work. The numbers on the *y*-axis are in millions of pounds, and give a total of £2,500 million (which is £500 million a year on average, for five years).

Much of the expenditure in year one on this graph is design and site investigation work, with construction commencing towards the end of year one, when details of the first batch of projects is available. Years two to four will see a mix of design and construction expenditure, whilst year five will be largely construction.

Another consideration is that a fairly flat profile gives continuity of work to suppliers who have a factory base. All suppliers wish to have continuous production and full utilisation of their assets. Suppliers cannot usually suddenly increase

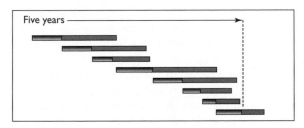

Figure 12.2 The phasing of a framework of eight projects.

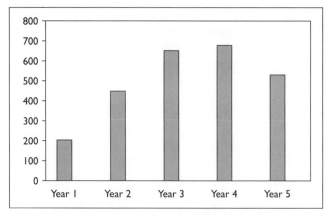

Figure 12.3   An example expenditure profile of a five-year investment programme.

or decrease production without large attendant costs. Hence a fluctuating or peaking profile will increase their costs and hence their framework prices.

A final issue for many utilities is the question of other utilities' financial profiles. Let us take ductile iron pipes for example. There are few UK or European suppliers of such items. If a number of utilities 'back-end' their programmes then it will almost certainly lead to shortages and price increases. A utility with a flatter expenditure profile will almost certainly benefit in such a scenario.

Applying framework principles to a water company capital programme

Let us look in more detail at the example of a framework agreement for a water company with a total capital programme, over five years, of two billion pounds. Water companies are regulated in five-year cycles of work with the consequent charges to customers. This enormous sum of money is typical for one of the larger UK water companies. It is a tremendous amount of work for construction professionals, which is why knowledge of these regulations and the operation of frameworks are so important.

About three years before the start of the forthcoming five-year programme, the water company will estimate the extent, type and value of work that is likely to be determined by the **regulator**. All the privatised utilities have a regulator, appointed by the government to agree the extent of work in the capital programme to meet the company's obligations, and also to agree appropriate charges to customers. For the water industry, the regulator is the Water Services Regulation Authority known as Ofwat. This is the economic regulator of the water and sewerage sectors in England and Wales. Its role is to ensure that water companies provide household and business customers with a good-quality service and value for money.

The work will comprise improvements and new construction to water treatment, sewage treatment, water mains and sewers but the balance, value and type of work involved will be different in every five-year regulatory determination period. In the 1990s for example there was a great emphasis on water treatment to meet enhanced standards of drinking water quality. The early part of this century has been characterised by concerns over flooding of properties from sewers overloaded by new weather patterns together with the separate issue of leakage from water mains. Each of these categories of work is unique and will have different contractors who are competent to carry it out.

If the sewers category is worth one billion pounds over the five-year period, then a number of framework contractors may be beneficial, because few UK sewerage contractors could cope with this turnover on their own. Alternatively, if the value of the sewers category was just £100 million over five years, then one or possibly two framework contractors could be selected. If the sewers category comprised many small contracts spread over the region then a decision could be made as to whether to have one framework, where only large contractors could bid, or a number of regional frameworks, where local contractors could also apply and be considered with possible efficiency gains arising from lower overheads and transport costs.

If part of the water treatment work included new and technically advanced treatment (as it often does), then very specialist contractors would be needed. Increasingly, water treatment is expected to remove small traces of chemicals that may have no established removal process, such that one has to be developed. Water treatment also frequently involves a lot of mechanical and electrical plant that many normal 'civil' contractors may not have the expertise to coordinate well.

Suppose that in this particular five-year period there was a need to replace the whole telemetry and control system for the water company at a cost of £100 million. This is very specific high-risk work, and would justify a framework agreement in its own right, again with contractors with very specialist expertise. A design and build based framework may be advisable here, since many such specialist contractors carry out their own design work.

A major decision is how to procure the design resource needed for this programme. Programmes such as this require many hundreds of designers every year to carry out the design work. The water company will need to consider its strategy for procuring this design expertise if it does not have the capacity in-house. It may decide to have frameworks with contractors only on a design and build strategy. Or it may follow the traditional contract strategy with one or more separate design organisations with framework contracts directly with the employer for design work together with separate construction-only frameworks. Alternatively the water company might define a strategy to encourage alliances of consultants and contractors to form in order to bid for the work.

Finally, the water company will need to consider the benefits of standardisation which can be achieved by setting up separate frameworks for electrical and mechanical plant and equipment. There would be justification for dozens of such frameworks covering submersible pumps, dry well pumps, screens, steel pipes, concrete pipes, ductile iron pipes, valves, actuators, control equipment, electrical panels and so on.

After taking into account all these considerations the water company finally determines how many framework agreements to set up, and what each will comprise. It then advertises for suitable tenderers for each framework in the OJEU, specifying as far as possible the type and extent of the work in each framework, its value and its approximate distribution over the five years. At this stage, none of the work has been *finally* detailed or valued, since the 'final determination'

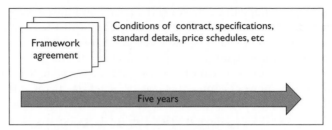

Figure 12.4 The overall framework agreement.

(the final agreed extent and cost of the five-year programme) has not been determined by the regulator.

However, with care, framework agreements can still be set up, and a typical one is shown in Figure 12.4 which is for improvements to water treatment works. This framework agreement is based on a selected form of contract (such as NEC3 ECC), standard specifications, standard details and as much relevant information as is available at that time, so that as many rates and prices as possible can be agreed at the start of the framework agreement. These rates and prices are what the regulations call 'terms as to price'.

Once the final determination has been made by the regulator, the water company can begin designing projects in detail and producing drawings and information relevant to the particular projects within each framework agreement. Frequently this work will be carried out by a design organisation appointed under a design framework. Defining *all* the work in the five-year programme in the detail needed for construction is another long process and may take a further three or four years. However, with a framework agreement, construction of each individual project can begin, as soon as the details for it are finalised.

This speed of implementation is a major advantage over more traditional methods of procurement, with the added benefit that further advertising in the OJEU is not necessary, since it was done at the outset of the framework. It is really only by means of this successive phasing of projects that the massive programmes of most utilities and similar bodies can be achieved.

Each specific contract is set in place with a package order containing *project-specific* information and drawings. Many of the rates and prices will be based on the original framework agreement, but inevitably there will be specific prices to be agreed, that relate to that particular project. So we may have a works order for improvements to Nottingham water treatment works (WTW), six months later another works order for Leicester WTW and nine months later another one for Derby WTW. Once each works order is issued, a legally binding contract for that particular work is put in place. This is shown diagrammatically in Figure 12.5.

Advantages of framework agreements for the employer

Framework agreements have many advantages for the employer such as:

1. The costly and time-consuming process of advertising in compliance with the regulations is only done once (usually every five years).

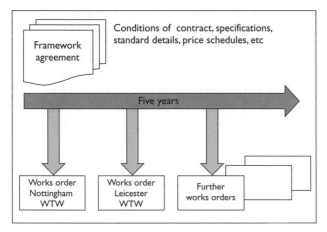

Figure 12.5 The framework showing package orders.

2. A great deal of care and effort can be put into the process to ensure best selection and full compliance with the regulations, thereby minimising the risk of a challenge and court action, and giving more likelihood of securing the most appropriate contractors.
3. The selection and award process can be completed before a full detail of the actual work in each contract is known.
4. Individual contracts can be put in place very quickly once the framework is set up, allowing a quick response time to planned contract work and any work of an emergency nature.

Advantages of framework agreements for the contractor

Frameworks also have many advantages for the contractor:

1. A long-term relationship with the employer (the contractor's client) is established.
2. Work flow can be predicted with some certainty, allowing better planning and use of the contractor's people, equipment and other resources.
3. Joint problem-solving can take place between the contractor, employer (and design organisation, if one is used) at an early stage of a project.
4. The contractor can set up its own longer term relationships with suppliers and sub-contractors on the basis of a larger and more predictable work flow. This will often give efficiencies and buying gains.
5. It becomes more worthwhile to train and develop the contractor's staff who work with the particular employer on the framework.
6. Where a separate design organisation is used, selected contractors can develop long-term relationships with the designers, thereby improving the exchange of ideas, reducing misunderstandings and simplifying administration.
7. The long-time span makes it financially viable for selected contractors to set up a specific local area office where appropriate, thereby minimising travel and improving team work and communication.

8. The selected contractors can also
   (a) Develop long-term understanding of the employer's business needs and challenges, so they can give a better service.
   (b) Develop a much better understanding of the employer's specifications, contract arrangements and required processes.
   (c) Develop long-term relationships with the employer's operational and investment staff, thereby minimising misunderstandings.
   (d) Be involved in early planning, phasing and development of projects in conjunction with the employer's staff and any external design organisations.
   (e) Be given an early view of likely up-coming work, to aid their business planning and staff deployment, thereby producing greater efficiencies.
   (f) Be encouraged to work together, sharing ideas, equipment and even staff, for everyone's benefit.
   (g) Become familiar with any specialist suppliers of plant and equipment.

Additional advantages of framework agreements for all parties

Framework agreements have many advantages for the employer, the selected contractors, suppliers and design organisation as follows:

1. A long-term arrangement is set up which facilitates team work, fosters good relationships and encourages a partnering philosophy.
2. Because of the high work value and long time frame, it is economically viable to set up specific joint IT and administrative systems and payment mechanisms between the employer and the selected contractors and suppliers, thereby improving efficiency.
3. Economies of scale are available for all participants.
4. Joint purchasing arrangements can be put in place, thereby producing buying economies and reduced administration.
5. Long-term incentive arrangements can be put in place. These are often based on key performance indicators such as performance against time, cost, quality and safety criteria.
6. Contract conditions can be simplified because they involve repeat order work, and such items as retention and liquidated damages (these are described in Chapter 18) may be dispensed with by the employer. This simplifies contract administration.
7. Shared office accommodation becomes a viable possibility. It is now common for the design organisation, selected contractors and relevant employer staff to share the same office. The potential for improved team work, efficiency and better relationships is immense.
8. Finally, because this is a long-term arrangement rather than a one-off contract, both the employer and contractor are more likely to adopt a 'give and take' approach to contract issues, thereby reducing confrontation and the very high cost of determining and settling individual contract claims.

Disadvantages of framework agreements

Framework agreements do have some disadvantages however, such as:

1. Tendering for frameworks can be very costly for contractors. Tendering costs of many hundreds of thousands of pounds are not unusual. Because so much work and money are at stake, employers usually have very rigorous selection procedures for frameworks which may take many months and involve interviews and presentations, site visits and detailed requirements for document submission. However, the successful tenderers are virtually guaranteed a secure, stable and long-term workload.

2. Unsuccessful tenderers may be forced out of that particular sector of the market. So for example a contractor who specialises in water treatment and fails to secure a framework with *any* UK water company has no direct UK business for possibly five years or so. In this instance, contractors will either seek to work as a sub-contractor, diversify or even shut down areas of their business.

3. In a similar way, frameworks 'shut out the market', so unsuccessful tenderers do not work for a particular employer for five years, thereby losing touch with personnel and that employer's business needs and processes.

4. Frameworks are set up with as many price terms as possible, but inevitably many prices have to be agreed once the detail of each project is known. This will suit contractors, who are now in a relatively secure position without competition. However, the employer no longer has the comfort of a market price based on a competitive tender process.

5. Whilst frameworks can contain price fluctuation clauses, unexpected changes in market prices, or even shortages, are hard to plan for over a forthcoming five-year period. This can be a risk for either the employer or the contractor, depending on who takes the risk of price fluctuations and shortages.

6. If, despite everyone's best efforts, a framework starts to turn sour, it is hard to break and takes another long period of tendering and selection to set up a new framework contractor. This is why many employers will have more than one framework contractor for a certain category of work. This reduces the benefit of economies of scale, but brings greater flexibility for the employer. It can also allow limited competition between framework contractors on the basis of financial incentives or the allocation of work linked to key performance indicators.

The difficulty of defining a framework with any precision

In addition to the drawbacks listed above there is a more general difficulty for employers and contractors, and that is the rather uncertain nature of the work when a framework is set up. This is because the whole process of advertising and setting up framework agreements can take between one and two years, and it is beneficial to have them in place for the start of any regulated programme (which is often of five-years' duration) since the regulator usually requires a prompt start to the actual construction work in that programme. There is therefore very little time between having the programme agreed with the regulator and a start on-site. This time lag to set up frameworks is a real dilemma facing most utilities and contracting authorities (the term for local authorities in the regulations).

Even where a programme does not have to be agreed with a regulator, such as with a local authority, a similar problem arises. The problem is how to define work up to four years in the future. This is no easy task. However the *probable* scope, value and extent of the work are known from the submission made to the regulator by a utility or from deduction by a local authority based on previous practice. Employers will therefore set up frameworks that are *realistic* (the EU advertisement has to be fairly specific) but also *flexible* in order to respond to the detail of the capital programme, once it is known. Hence with care, this difficulty of foreseeing the future can be overcome.

The costs of setting up and bidding for a framework

Finally there is the cost of setting up a framework, both to the employer and the tendering contractors. It is not unknown for a contractor to spend over a million pounds bidding for a large framework. This expense is rarely *directly* recoverable. It is normal for contractors to allow for all tendering costs, including those costs for framework tendering, in their overhead. This overhead is applied to all contracts where they have been successful in tendering. However, many contractors take the view that one big tender expense for a framework is better than twenty smaller ones. The problem of course is that a framework is 'all or nothing', probably for five years. If the contractor is unsuccessful then future contracts must carry an increased overhead based on unsuccessful framework bids, and here the expense of an unsuccessful bid can be very significant whereas bidding for a number of tender packages should at least realise some work. However, the prize for the successful is so great that it is not difficult to find many contractors willing to bid.

Duration of framework agreements and limits on numbers

In the utilities contracts regulations there is no specific maximum duration stated for a framework agreement. However five-year agreements are probably the most common. There is some debate over whether very long-term agreements, possibly exceeding ten years, would themselves be anti-competitive and illegal. Further, some of the disadvantages listed above would become very apparent in agreements lasting longer than five years. Perhaps the main reason for five-year agreements is that this is frequently the regulatory period of price determination for utilities. As described above, frameworks need to mirror the type, scope, location and value of the work. This is only known with any clarity for any given regulatory period. In the Utilities Contracts Regulations there is also no restriction on the number of suppliers/contractors in any framework agreement.

However the **Public Contracts Regulations** are much more prescriptive in respect of framework agreements. Here framework agreements are restricted to a maximum duration of **four years** in section 19(10) which states '*the contracting authority shall not conclude a framework agreement for a period which exceeds four years except in exceptional circumstances.*' The Public Contracts Regulations also have a stipulation on the number of contractors or suppliers (economic operators) in a specific framework in section 19(6) which says:

'Where the contracting authority concludes a framework agreement with more than one economic operator, the minimum number of economic operators shall be three, insofar as there is a sufficient number of –
(a) economic operators to satisfy the selection criteria; or
(b) admissible tenders which meet the award criteria.'

Framework agreements as a means to standardise

A potential problem with the regulations for many of the large employer organisations affected by them is that they can seem to stand in the way of good business sense. Large organisations usually wish to standardise on plant and equipment for the following reasons:

- Reduced spares holding and stores requirement.
- Interchangeability of equipment.
- Simplified training for maintenance and operational staff.
- Maintenance staff becomes more efficient because of the repetition of basic tasks, and operational staff becomes familiar with the characteristics of equipment.
- Specific employer requirements to standard plant become economically viable ('we want ours with blue stripes').
- Real potential for long-term improvement opportunities with suppliers.
- Simplified and standardised control equipment.

In order to standardise the employer must be able to specify by name. At first sight this is not possible for contracts falling under the 'regulations', because specifying by name contravenes regulation 12, unless the named supplier has been advertised in accordance with the 'regulations'. This then is the way to achieve standardisation in compliance with the regulations. Firstly the employer decides which types of plant and equipment require standardisation. This will determine a number of categories of plant and equipment which will become the subjects of framework agreements. For a water company such frameworks will almost certainly relate to pumps, electrical panels, pipes of various materials and so on. Each of these frameworks is advertised in compliance with the 'regulations'. This follows a proper assessment of the total likely value of the particular plant and equipment in each framework over the selected time period. The employer selects objective criteria for tender assessment very carefully to reflect all the 'whole-life' needs, not just initial capital cost.

A framework is placed with one or more suitable suppliers with 'terms as to price' after advertising and appointing the supplier in accordance with the 'regulations'. These suppliers can now be named, since the framework has been properly advertised and set up. The framework supplier is named in the specification documents for future contracts and a price schedule is normally included. Future tenderers for the construction frameworks therefore bid for a contract in full knowledge of the suppliers of plant and equipment to be used under their own construction framework.

Let us take an electricity company as an example. Suppose the electricity company wishes to standardise on switchgear and floor tiles in its sub-stations. These items are advertised in accordance with the 'regulations' and frameworks are set up with SW Switches and Smith Tiles. Construction contractors tendering for a five-year framework programme of sub-station construction receive a framework specification in which SW and Smith are named as specific framework suppliers. The successful construction contractor now has a framework contract with the electricity company in which Smith and SW must be used. Price terms for SW and Smith should be built into the construction framework as far as possible.

Inevitably some project-specific prices may require later agreement but hopefully the bulk of the pricing has been written into the original construction framework. A further major advantage of this method is that the construction contractor can also strike up a long-term relationship with SW and Smith, to everyone's benefit.

Frameworks reduce employer risk under the 'regulations'

A final major driver for the extensive use of frameworks is to minimise employer risk. If an employer affected by the regulations does not comply with them, then that employer faces the prospect of massive fines and loss of reputation. Hence wise employers follow the regulations diligently. However they are complex and compliance is costly and time-consuming. Moreover the whole process can be risky to an employer, because a successful challenge that the regulations have not been followed properly can mean that even a conscientious employer faces court action. Hence an employer advertising say fifty separate contracts under the regulations, rather than one large framework, faces much greater risk of non-compliance. The simple answer is to combine contracts for similar work, advertise a framework and put the time and effort into doing it really thoroughly. Overall for the employer this minimises risk, cost and the time taken. All the additional benefits such as improved team work and efficiency, as described above, follow.

Forms of contract for framework agreements

The NEC contract

A framework is an 'umbrella agreement', which initially includes conditions of contract, specifications and price schedules. The framework is a 'contract to make further contracts'. Later the specifics of each project will be issued with each package order. NEC3 now publishes a *Framework Contract* (NEC3 FC) which is consistent with their other contract forms such as the Engineering and Construction Contract (NEC3 ECC). This framework contract is a very short document of just five pages. The parties are described as the employer and the supplier. Here 'supplier' could be a supplier of design services, a contractor or a supplier of equipment or materials; so the drafting of the NEC3 FC makes its use very flexible. A 'work package' is the description NEC3 FC uses for each individual contract and this is put in place with a 'package order'.

The contract data in the NEC3 FC is very short, as befits an 'umbrella agreement'. It defines the scope, selection procedure, quotation procedure and the end date of the framework. It also sets out data which will apply to all package orders and time charge orders.

In NEC3 FC there is also provision for advice on a proposed work package, by the use of a 'time charge order'. Time charges are frequently used to pay for design services rather than the total fee, where work cannot easily be defined. A time charge is simply a rate per hour, day or week that includes not only staff salaries but also overheads and profit. A typical time charge rate per hour for a graduate engineer is about £30. A time charge order could also be used to engage one of the framework contractors in providing construction or financial advice for example.

The NEC3 FC has a 'break clause' which says *'either party may terminate their obligations under this contract at any time by notifying the other party'* (clause

90.1). The NEC3 FC reminds the parties of the legal principle that all contracts that have already been set in place by a package order must be completed (clause 90.2).

The JCT contract

The Joint Contracts Tribunal (JCT) also publishes a Framework Contract, but approach it in a more comprehensive way. In the JCT version the work to be done by the 'provider' (this will be the contractor or supplier) is called the 'task'. The employer sends out an 'enquiry' requesting a price for the task which the provider prices in accordance with the 'pricing documents' (these will have been agreed when the framework agreement was set up), and if satisfied places an 'order' using the form in Annex 2. Again, clause 22 of the JCT framework contract allows cancellation by either party, but not of any 'orders' already placed. An interesting feature of the JCT framework contract is that it includes many features of partnering and collaborative working such as sharing information (clause 11), collaborative risk analysis (clause 14), sustainable development (clause 16), value engineering (clause 17) and team approach to problem solving (clause 20). This JCT contract also includes an early warning procedure in clause 19: a very brief clause, but one that follows the principle first set out in NEC3 ECC. Early warning is described in Chapter 22.

It is very important in any framework agreement to be very clear when a specific contract under the framework is set in place, and this is done in all forms of contract for framework agreements. Hence we have the words 'package order' (NEC3 FC), 'order' (JCT) or 'works order' in the ICE7 version.

Chapter summary

- The Utilities Contract Regulations and the Public Contracts Regulations are the UK's legal response to the EU Procurement Directives.
- The result of the 'regulations' has been the extensive use of framework agreements.
- A framework agreement is a 'serial contract', or a 'contract to make further contracts'.
- Frameworks usually last for five years in the utilities sector, but are limited to four-years' duration by employers falling under the Public Contracts Regulations.
- Suitable forms of contract for frameworks are now published by the NEC, ICE and JCT.
- Framework agreements have many advantages and few drawbacks.
- Some advantages are the potential for standardised plant and equipment, reduced administration, better communication and teamwork, and the potential for the development of synergies and knowledge sharing.
- Framework agreements have many advantages for contractors and suppliers, particularly a more predictable work flow and improved efficiency and relationships.
- Finally, the use of frameworks reduces employer risk, principally the risk of inadvertently breaking the 'regulations'.

13

The EU Procurement Directives

Aim

This chapter aims to explain the application of the EU Procurement Directives to the construction industry and to describe the major impact of these directives, particularly in respect of specifying and tendering.

Learning outcomes

On completion of this chapter you will be able to:

>> Explain the main features of the Utilities Contracts Regulations 2006 and the Public Contracts Regulations 2006.
>> Apply these regulations to the specification of design and construction work affected by them.
>> Describe the main routes for obtaining tenders under the regulations and the need to set acceptable selection criteria.

The EU Procurement Directives

The **EU** (European Union) **Procurement Directive**s were enacted to encourage competition and opportunity across Europe. They are based on the principles of transparency, non-discrimination and competitive procurement, and are intended to facilitate the achievement of value for money for the taxpayer as well as promoting the single European market. In other words, the intention is to make it easier for companies and suppliers to sell their products and services across the whole of Europe, rather than being restricted by national standards, specifications or procedures. A further intention is to enable a united Europe to compete on a world stage. The construction industry across Europe spends immense amounts of money every year, and hence is a prime target for legislation to open up competition and markets.

The explanatory memorandum to the Utilities Contracts Regulations 2006, states in section 7.1:

'The total public procurement expenditure across Europe represents over €1500 billion, which is over 16 per cent of the EU Gross Domestic Product. The procurement directives, which are based on the principles of transparency, non-discrimination and competitive procurement, are of great significance in promoting the single European market and in facilitating the achievement of value for money for the taxpayer.'

It is important to distinguish between **European regulations** and **European directives**. Article 249 of the Consolidated Version of the Treaty Establishing the European Community states:

- A **regulation** shall have general application. It shall be binding in its entirety and directly applicable to all member states.
- A **directive** shall be binding, as to the result to be achieved, upon each member state to which it is addressed, but shall leave to the national authorities the choice of form and methods.

In other words, *European* directives are transferred into the particular law of each member state, by that particular member state. However such laws in any member state have to achieve the objectives of the European directive. If a member state fails to pass the required national legislation, or if the national legislation does not adequately comply with the requirements of the European directive, the European Commission may initiate legal action against the member state in the European Court of Justice.

Once it forms part of UK law, the legislation is also frequently called a 'regulation', which can be a little confusing. The difference is that here it is a regulation made by the UK government. A major set of procurement regulations was issued by the UK in the 1990s, so these ideas have been with us for some time. The earlier UK regulations have been refined and extended into the 2006 regulations, which are:

- **The Public Contracts Regulations 2006** which implement directive 2004/18/EC and apply to Local Authorities, Government Departments and so on (the '**public-sector**').
- **The Utilities Contracts Regulations 2006** which implement Directive 2004/17/EC and apply to companies operating in the water, energy, transport and postal services sectors (the '**ex-public-sector**').

This chapter will simply refer to both sets of regulations as the 'regulations'.

The 'public sector' refers to central government departments, local government and state-controlled enterprises. Here we find the civil service, government departments, county councils, district councils, London borough councils, parish councils and so on.

The 'ex-public sector' is a term sometimes used for bodies that were once state owned and that were privatised in the latter part of the twentieth century. It may come as a surprise to many people to learn that the private water companies in the UK were state-owned water *authorities* until 1974. Similar considerations apply to gas, electricity, rail and many transport services. Ex-public sector bodies are not subject to competition in quite the same way as truly private companies, since they often control extensive underground assets in geographic areas, and are hence subject to government-sponsored regulation. The relevant regulator agrees outputs, targets, levels of service and importantly the prices that can be charged.

This previous state ownership and 'semi-monopoly' position may explain why the EU Directives apply to such 'ex-public sector bodies', but not to independent private companies such as Sainsbury or Tesco or house builders and developers whose construction work is not subject to these two sets of regulations. What it does mean, however, is that the two sets of regulations apply to the majority of civil engineering work in the UK (and much 'building' work as well).

These two sets of regulations are often known in the construction industry as the '*EU Regs*'. Another common term for such work is '*OJ work*'. OJ stands for the Official Journal of the European Union. This is the journal in which all relevant work must be advertised.

The Public Contracts Regulations and the Utilities Contracts Regulations are very important, and they apply to most construction work in the UK, particularly in civil engineering.

Non-compliance with the regulations can lead to enormous fines, sometimes running to tens of millions of pounds, and so a potential breach of the regulations is a major risk for employers affected by them.

The regulations mainly impact on tendering and specifications and so are of great importance to construction professionals.

This chapter will describe the Utilities Contracts Regulations 2006 and then highlight some of the differences to be found in the Public Contracts Regulations 2006.

Who do the regulations apply to?

The Utilities Contracts Regulations

The Utilities Contracts Regulations list affected organisations in schedule one. This is a fairly complex list of definitions but, in simplified terms, they apply to:

- Water or sewerage undertakers
- Electricity
- Gas
- The coal industry
- An airport operator within the meaning of the Airports Act 1986
- A harbour authority
- Network Rail
- Eurotunnel plc

- London underground
- Transport for London.

It takes little imagination to see the tremendous amount of construction work generated by these bodies in the UK. It will amount to tens of billions of pounds of work per annum.

The Public Contracts Regulations

The Public Contracts Regulations apply to a wide array of 'public entities', known in the regulations as **'contracting authorities'**. A slightly simplified list of contracting authorities, abstracted from section 3 is:

- A Minister of the Crown
- A government department
- The House of Commons
- The House of Lords
- The Northern Ireland Assembly Commission
- The Scottish Ministers
- The Scottish Parliamentary Corporate Body
- The National Assembly for Wales
- A local authority
- A fire authority
- A police authority
- A National Park authority.

It may be slightly confusing that such public bodies are called 'contracting authorities', but that is simply a term used in the regulations. Of course such public bodies are not contractors as we generally use the word; they are the 'employer'. The two bodies affected by the Public Contracts Regulations, of most importance to the construction professional in terms of the enormous amount of construction that they generate, are government departments and local authorities.

Definitions

To understand the Utilities Contracts Regulations 2006 we need to start with some definitions (the terminology used in the Public Contracts Regulations is generally similar):

- A *'works contract'* means a contract for the carrying out of a work or works for a utility. Works are listed in schedule 2 and are very extensive. They include virtually all varieties of civil engineering and building work. So in everyday speech the body carrying out the work would be the contractor.
- A *'supply contract'* means a contract for the purchase or hire of goods by a utility and for any siting and installation of those goods. This would usually apply to electrical and mechanical plant and equipment. The relevant body is called a 'supplier'.
- A *'services contract'* means a contract under which a utility engages someone to provide services. This would usually apply to design services obtained from a consulting engineer or architect. The relevant body is called a 'services provider'.

- '*Economic operators*' is a generic term used in the regulations to include a contractor, a supplier or a services provider.
- The *threshold* is the **estimated value** of the contract in euros (converted to pounds) above which the regulations apply.
- The *estimated value* is the total consideration (a legal word meaning payment) payable, net of value added tax.

When do the Utilities Contracts Regulations apply?

There are a number of special areas where the regulations do not apply, but in simple terms they apply to any utility (employer) defined in the regulations (a company supplying water, gas or electricity for example). They then only apply to categories of contracts above a threshold 'estimated value'. Section 11(2) sets these estimated values as

> (a) 422,000 euros for a supply contract or a services contract
> (b) 5,278,000 euros for a works contract.

The value in pounds sterling of any amount expressed in euros is calculated by reference to the rate for the time being applying for the purposes of the Utilities Directive as published from time to time in the Official Journal (section 11(3)).

How does this threshold apply to a framework agreement, or a series of contracts? A framework agreement is straightforward in that the threshold applies to the estimated *total value* of the framework. Where we have a series of contracts which have similar characteristics or are for the same type of goods or services, or a contract which is renewable, then the appropriate value shall be *the total cost* of this work over the previous financial year of the utility. Utilities have to be particularly careful about this rule. It is easy to think that a series of contracts below the threshold is being awarded, whereas if they are considered to 'have similar characteristics and are for the same type of goods or services' then their value over a financial year must be aggregated. Hence a series of small but similar contracts could actually fall under the regulations and require advertising and assessing in accordance with the regulations.

A contract which exceeds the threshold should not be deliberately split into smaller contracts to avoid the regulations. This is often called 'disaggregation'. Clause 11(18) applies, and states that 'a utility shall not enter into separate contracts nor exercise a choice under a valuation method with the intention of avoiding the application of these regulations to those contracts.'

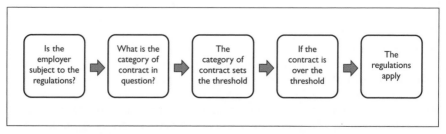

Figure 13.1 Flow chart to determine whether the regulations apply.

Figure 13.1 shows a simple flow chart for determining whether the regulations apply to a particular contract. The first thing to do is to check whether the employer is a utility as defined in schedule one of the regulations. If this is so then the 'category of contract' needs to be determined. In other words is it a works, services or supplies contract? Each has different thresholds. Once the threshold is checked we can see whether the regulations apply or not.

What processes do the Utilities Contracts Regulations apply to?

The utilities regulations apply mainly to tendering procedures, contract award procedures and specifications. They have far-reaching effects in these areas.

Tendering

The intention of the regulations is to make tendering as inclusive as possible: for tendering procedures and timescales to be clear and fair; and for selection of the successful tenderer to be transparent and based on objective criteria.

> Many of the provisions of the regulations override traditional tendering and contractor selection procedures.

In order to give prospective tenderers due warning of likely future contracts a utility must send a notice in the form of the periodic indicative notice (PIN). There are complex rules around this, but basically the PIN specifies for any given utility the likely types of contract and approximate values that are expected to be awarded over the coming year. Tendering is controlled by clause 14, which uses the term 'seeking offers'.

Clause 14 says *'for the purposes of seeking offers in relation to a proposed contract a utility shall use the negotiated procedure, open procedure or the restricted procedure.'* These are defined in clause 2 as follows:

'Negotiated procedure means a procedure leading to the award of a contract whereby the utility negotiates the terms of the contract with one or more economic operators selected by it.
Open procedure means a procedure leading to the award of a contract whereby all interested economic operators may tender for the contract.
Restricted procedure means a procedure leading to the award of a contract whereby only economic operators selected by the utility may submit tenders for the contract.'

The economic operator is the generic term for contractor, supplier or services provider, as we saw above.

In normal tendering language these procedures would be called negotiated, open tendering and select list respectively. It should not be imagined that the negotiated or restricted procedures give a utility great freedom. They do not, and are subject to strict rules and controls.

A utility must make a 'call for competition', which is the phrase used in the regulations for 'inviting tenders'. There are a number of exceptions to the requirement to make a call for competition (abbreviated here to C for C) and they are listed in clause 17. However, for all normal purposes, it is easier to simply make a C for C unless the utility is absolutely certain that the requirements of clause 17 can be met and, if necessary, defended in court. The open procedure satisfies the C for C in itself. The negotiated and restricted procedures are satisfied by the use of a PIN or if a notice indicating the existence of a qualification system for economic operators has been sent to the Official Journal.

There are other alternatives to the negotiated procedure, open procedure or the restricted procedure. These are:

- Dynamic purchasing systems
- Electronic auctions
- Central purchasing bodies
- Framework agreements.

Dynamic purchasing systems are electronic systems of limited duration which are used to purchase commonly used goods, work, works or services. Electronic auctions are electronic processes which take place after the initial evaluation of tenders. They allow tender prices to be revised downwards in a dynamic and open way. A central purchasing body is basically a utility which is a contracting authority and which acquires goods or services intended for one or more utilities or awards contracts intended for one or more utilities. Framework agreements are long-term arrangements to award a number of serial contracts. They are so important to construction professionals that they are covered in Chapter 12.

Very importantly, utilities are obliged to tell prospective tenderers *how their tender will be assessed*. Again, sensible good practice showing openness and transparency is required by clause 22(17):

The utility shall include the following information in the invitation

(a) the final date for making requests for further information and the amount and method of payment of any fee which may be charged for supplying that information;

(b) the final date for receipt by it of tenders, the address to which they must be sent and the one or more languages in which they must be drawn up;

(c) a reference to any contract notice;

(d) an indication of the information to be included with the tender;

(e) the criteria for the award of the contract if this information was not specified in the contract notice, the notice on the existence of a qualification system used as a means of calling for competition published in accordance with regulation 16(2)(a)(ii) or the contract documents; and

(f) the relative weighting of the contract award criteria or, where appropriate, the descending order of importance of such criteria if this information was not specified in the contract notice, the notice on the existence of a qualification system or the contract documents.

Contract award procedures

Whilst they may sometimes seem onerous, the contract award procedures under the regulations actually represent good practice. They force employers to think in advance about what constitutes the best potential tender, to be transparent in their dealings and to use previously constructed objective criteria for their choice. The regulations also include feedback mechanisms, which facilitate improvement for unsuccessful tenderers, should they choose to use them.

Contract award should be based on either the most 'economically advantageous offer' or the lowest price. In assessing economically advantageous offers, utilities might consider delivery dates, running costs, sales service and so on. For most purchases, whole life costs are more important than initial purchase costs. Construction professionals often concentrate on initial capital costs, but this can be a mistake. When buying a car, most people will also factor in fuel consumption, servicing costs, other running costs and depreciation. Whole life costing is no more than an extension of this idea.

Clause 30 states:

(1) Subject to regulation 31 (dealing with offers of third country origin) and paragraphs (6) and (9) (abnormally low offers) of this regulation, a utility shall award a contract on the basis of the offer which –
(a) is the most economically advantageous from the point of view of the utility;
or
(b) offers the lowest price.

(2) A utility shall use criteria linked to the subject matter of the contract to determine that an offer is the most economically advantageous including delivery date or period for completion, running costs, cost-effectiveness, quality, aesthetic and functional characteristics, environmental characteristics, technical merit, after sales service and technical assistance, commitments with regard to parts, security of supply and price or otherwise.

(3) Where a utility intends to award a contract on the basis of the offer which is the most economically advantageous, it shall state the weighting which it gives to each of the criteria chosen in the contract notice or in the contract documents.

(4) When stating the weightings referred to in paragraph (3), a utility may give the weighting a range and specify a minimum and maximum weighting where it considers it appropriate in view of the subject matter of the contract.

(5) Where, in the opinion of the utility, it is not possible to provide weightings for the criteria referred to in paragraph (3) on objective grounds, the utility shall indicate the criteria in descending order of importance in the contract notice or contract documents.

Clause 30 goes on to deal with offers that are abnormally low.

Informing unsuccessful tenderers

Informing unsuccessful tenderers of their performance is very important, if they are to understand where their bid lies relative to the 'current market'. It is good procurement practice, and is covered in detail in clause 33 of the regulations. This clause basically says that the utility shall, as soon as possible after the decision has been made, inform unsuccessful tenderers of:

- The criteria for awarding the contract
- The score of the successful tenderer and the score of that particular unsuccessful tenderer.

There is then a ten-day 'standstill period' before the utility can actually award the contract. This period is basically for unsuccessful tenderers to seek more information as to why they were not selected. This is the area that really 'bites' on employers. They have to set objective criteria, they have to use them to select the successful tenderer, and then they have to give unsuccessful tenderers the criteria, their own score and the score of the successful tenderer. Employers have to be extremely careful in all these areas or they could face a challenge from an unsuccessful tenderer, a delay to their tender process and hence projects and potentially a massive fine.

Time limits

Clause 22 is a long clause specifying time limits for various processes and tendering options. However, it begins with this very sensible advice for any construction professional considering a suitable tender period:

(1) Subject to the minimum time limits specified in this regulation, a utility shall take account of all the circumstances, in particular, the complexity of the contract and the time required for drawing up tenders when fixing time limits for receipt by it of requests to be selected to tender for or to negotiate the contract and for the receipt by it of tenders.

Subject to various provisos, the normal 'tender period' is stipulated as 52 days. Where the 'open procedure' is used (a tender open to anyone to submit a tender) the regulations say 'a shorter period of generally not less than 36 days and in any event not less than 22 days' may be used. In reading these time periods, it is worth remembering that the 'industry norm' for tender periods for projects up to about five million pounds was four to six weeks. The regulations impose much longer tender periods. This aids tenderers but makes planning of timescales much more important for employers, if their projects are not to be unduly delayed.

Keeping records

Finally a utility must keep records for four years of the qualification and selection of tenderers, and the use of a procedure without a call for competition. Additionally when a utility decides not to apply the regulations it shall keep appropriate information to justify that decision. Statistical reports are also required on an annual basis for certain categories of work.

Technical specifications

Construction professionals who work on contracts for employers not affected by these regulations are accustomed to specifying items with reference to standards (often British Standards), or by referring to them *by a supplier's name*. The latter is not acceptable under the regulations. Hence specifying valves to BS EN 1074-2:2000 is acceptable. However specifying valves supplied by Saint Gobain (one of Europe's largest manufacturers of pipes and valves) is not acceptable. In fact such a reference to a manufacturer *would be illegal* in a contract subject to the regulations. In simple terms, the regulations require technical specifications to state characteristics and quality and performance levels rather than manufacturers' names. The full text of the definition of **technical specification** in clause 12 of the Utilities Contracts Regulations 2006 is:

'Technical specifications' means –
(a) in the case of a services contract or a supply contract, a specification in a document defining the required characteristics of materials, goods or services, such as quality levels, environmental performance levels, design for all requirements (including accessibility for disabled persons) and conformity assessment, performance, use of a product, safety or dimensions, including requirements relevant to the product as regards the name under which the product is sold, terminology, symbols, testing and test methods, packaging, marking and labelling, user instructions, production processes and methods and conformity assessment procedures
(b) this section goes on to describe specifying in works contracts.

The restrictions on what can be specified occur in clause 12 sections (4) to (19). Sub-section (4) states:

A utility shall ensure that technical specifications afford equal access to economic operators and do not have the effect of creating unjustified obstacles to the opening up of public procurement to competition.

Sub-section (16) states:

Subject to paragraph (17), a utility *shall not lay down technical specifications* in the contract documents which refer to:
(a) materials or goods of *a specific make* or source or to a particular process
(b) or trademarks, patents, types, origin or means of production which have the effect of favouring or eliminating particular economic operators.

Specifying by name breaks the regulations

So under sub-section (16) it would be illegal to specify 'Coca-Cola' by name. So if we wanted to specify Coca-Cola we could not use the trade name. We would have to use a performance specification describing typical constituents, colour and possibly the requirement for a glass bottle rather than plastic. Consequently Pepsi-Cola and many other brands could then tender for the work. Under sub-section (4) it would also be illegal to describe a product in such a way that only one manufacturer can comply. So if we specified a requirement for the particular 'hour-glass' shape of a Coca-Cola bottle, we would probably be breaking the regulations, because only Coca-Cola could comply, and we would therefore be 'creating unjustified obstacles'.

The use of 'or equivalent'

Many people think that specifying by name ('Coca-Cola' for example) is acceptable provided the words 'or equivalent' are added. *This is a risky practice.* It is only acceptable under sub-section (17) which says:

> Notwithstanding paragraph (16), exceptionally, a utility may incorporate the references referred to in paragraph (16) into the technical specifications in the contract documents, provided that the references are accompanied by the words 'or equivalent', where
> (a) the subject of the contract makes the use of such references indispensable
> (b) or the subject of the contract cannot otherwise be described by reference to technical specifications which are sufficiently precise and intelligible to all economic operators.

'Translating' sub-section (17) then, it means that if you specify by name with the addition of the words 'or equivalent', you can only do so if you really *cannot describe the item in any other way.* So if you specified 'Coca-Cola or equivalent', you may have to convince a court that the reference to Coca-Cola was indispensable or that a specification stating constituents and colour would have been unintelligible. This would not be an easy task!

> Specifying by a manufacturer's name is a very risky practice in contracts that fall under the regulations. The use of the words 'or equivalent' is often insufficient to avoid breaking the regulations.

Performance and functional specifications

Many items can be defined by *performance and functional specifications.* These are acceptable provided they comply with sub-section (7) which states that:

> 'A utility may define the technical specifications in terms of performance or functional requirements (which may include environmental characteristics) provided that the requirements are sufficiently precise to allow an economic operator to determine the subject of the contract and a utility to award the contract.'

A normal specification defines items by description or reference to standards as described above under 'technical specifications'. By contrast, a performance or functional specification defines what an item should do. A specification defining the details of a car engine would be a technical specification, whereas one that required an engine to achieve an output of 50 brake horse power at 5000 rpm would be a performance specification.

Chapter summary

- The Utilities Contracts Regulations and the Public Contracts Regulations are the UK's legal response to the EU Procurement Directives.
- They apply to employers in the public sector such as local authorities, government departments and the 'ex-public sector' such as utilities.
- These two sets of regulations affect the majority of civil engineering work in the UK and much building work in addition.
- The regulations control tendering, contract award and specifications in all contracts exceeding the threshold values.
- The result of the regulations has been the extensive use of framework agreements.

14

Pricing and payment mechanisms

Aim

This chapter explains the different pricing mechanisms and payment mechanisms commonly used in construction contracts.

Learning outcomes

On completion of this chapter you will be able to:

>> Differentiate between pricing mechanisms and payment mechanisms and explain how they work in practice.
>> Describe the range of pricing/payment mechanisms used in common forms of contract.
>> Select a suitable pricing/payment mechanism for different circumstances.
>> List some of the difficulties in assessing contractor costs.
>> Calculate payments in pain/gain situations.
>> Describe concepts such as target price moves and disallowed costs.

Pricing mechanisms and payment mechanisms

A **pricing mechanism** basically represents the 'consideration' or fee that must be paid to the contractor, in return for a completed project. The **payment mechanism** describes how and when the contractor will be paid.

The employer will want to have a pricing mechanism which indicates as closely as possible what the final cost of the project will be, at the time that tender prices are received. It is also an advantage to be able to compare the *detail* behind the tender prices submitted by different tenderers, in addition to a statement of the total tender price. This detail is often called the 'tender make-up'.

Once the contract is awarded to the successful tenderer, this tenderer now becomes the contractor under the construction contract. Clearly contractors want to understand how and when they will be paid. The way this is done is to specify a payment mechanism in the tender documents. The payment mechanism specified will affect the tender price. Regular and even early payment will tend to produce lower tender prices for example.

Frequently the pricing mechanism and payment mechanism are the same, but this is not always the case. It is possible to have a pricing mechanism to determine and explain the tender price, and a completely different payment mechanism to make payment. This occurs in some NEC3 ECC options

Example of payment mechanisms and pricing mechanisms

Let us suppose I am fortunate enough to be able to order 144 bottles of wine, at £10 per bottle, from a wine merchant, to be delivered, when available, in cases of twelve, every month for a year for a quotation of £1,440. If I paid in advance as a lump sum then £1,440 would be a pricing mechanism and a payment mechanism. If I paid monthly on the basis of wine delivered, then £1,440 would only be a pricing mechanism, the payment mechanism would be monthly payment on the basis of wine delivered. If in some months ten bottles were delivered, and in other months, fourteen bottles, and I paid per bottle delivered every month, then some months I would pay £100, and some months I would pay £140. This is exactly the way a remeasured payment mechanism works such as a bill of quantities contract (more on this shortly); monthly payments represent the work actually done with a final corrected price on completion.

The main functions of pricing/payment mechanisms

As we have outlined, the main point of a pricing mechanism is to represent the 'consideration' that must be paid, in return for the completed project. The payment mechanism describes how and when and how much the contractor will be paid. Paradoxically in many forms of contract, the final payment determined by the payment mechanism is different from the price predicted by the pricing mechanism at tender. There are many reasons for this, which will be explained in this chapter. It is a very important concept for the construction professional to understand clearly.

> The final price paid under the payment mechanism is often different from that predicted by the pricing mechanism at tender.

The five main functions of pricing/payment mechanisms are shown in the box below. It should be noted that some functions are common to pricing and payment mechanisms, and some are particular to one or the other. We shall come back to this later in the chapter.

> 1. To allow the employer to compare tenders.
> 2. To give the employer a likely contract price, so that money can be raised, or budgeted for.
> 3. To make accurate regular payments to the contractor (often monthly).
> 4. To provide the designer with a database of rates for estimating the likely cost of future projects.
> 5. To provide a sound base for pricing variations (depending on the form of contract used).

Making the contract

A valid contract in the UK must have **offer**, **acceptance** and **consideration**. Consideration is usually the payment of money in return for goods or services. The employer sends out tender documents (an 'invitation to treat' in contract law), and the tenderers respond with their offers. By the use of a particular form of contract (or option in NEC3 ECC), the *employer determines* the pricing mechanism and the payment mechanism that will be used. They are often the same, but may not be, particularly in NEC3 ECC contracts.

The legal offer is made by the tenderer when the tender is returned to the employer. One of the main elements of this offer is the completed pricing mechanism, with prices inserted by the tenderer. This establishes the price of the contractor's offer. Once the tender is accepted, a legal contract is formed and the payment mechanism defined in the contract becomes the basis of payment to the contractor.

Whichever payment mechanism is chosen, it becomes incorporated in the contract itself, and hence becomes a term of the contract.

Types of pricing/payment mechanisms

The *employer* decides on the pricing/payment mechanism, either by choosing a specific form of contract such as ICE7 or JCT SBC05 which incorporate a prescribed payment mechanism, or by choosing an option *within* a form of contract. The employer does this particularly in NEC3 ECC, which contains options A to F. NEC3 ECC options A to E are a mixture of pricing mechanism and payment mechanism. Option F is a management form of contract and is a contract strategy (as described in Chapter 9) rather than a pricing/payment mechanism.

There are five main types of pricing/payment mechanisms
1. Lump sum
2. Remeasurement
   (a) Bill of quantities
   (b) Schedule of rates
3. Activity schedules
4. Cost reimbursable
5. Target price.

Lump sum

Lump sum is a pricing and payment mechanism in which the contractor is paid a pre-determined sum for completing the whole contract works or in some contracts for completing particular stages as well as the whole contract works. This lump sum is the price inserted by the contractor in the tender. The sum is not adjusted to take into account any change in the *extent* of work from that which was assumed *by the contractor* in the tender. The contractor therefore carries the risk of correctly estimating the *extent* of work required to be carried out. The lump sum will usually be adjusted for changes (called 'variations' in JCT SBC05) made by the contract administrator.

So for example in a lump sum contract to build a domestic garage, if the contractor forgot to include the cost of the door lintel in the tender, then there will be no additional payment for the lintel, just the original lump sum. It would be a different matter, of course, if the employer wanted a window, when *none was shown* on the tender drawings. This would be an addition to the contract (a variation) and, under most forms of contract, would allow the contractor to increase the original lump sum price by the cost of the extra work needed.

Many contractors produce their lump sum bid by using outputs and costs, or by producing their own bill of quantities (see the next section). So each tenderer incurs the cost of breaking the contract down into work items of one kind or another. Some lump sum contracts require the contractor to supply their own bill of quantities, even though it is not used for payment purposes. Where this is done it provides a good means of checking the build-up of the lump sums provided by tenderers. It can be a useful 'reality check'.

Use

A lump sum pricing/payment mechanism is the most appropriate for small and simple contracts, design and build contracts or for other contracts where the contractor's method of construction is a major factor in determining the work to be done. These contracts can be of very high value, often running to millions or tens of millions of pounds. At the other extreme lump sum is used for well-defined, simple contracts of perhaps a few thousand pounds. Many 'domestic building projects', for example house extensions, roof renovation or window replacements, are lump sum contracts. On a larger scale, a new office block might be a lump sum design and build contract, especially if the method of construction such as the use of travelling formwork were a major decision factor for the contractor.

Advantages

1. It is easy to administer during construction, provided the contractor administrator does not order changes to the works.
2. The final account is simply the original lump sum, if there are no 'claims' or changes.
3. It is easy to compare tenders for quoted price.

Disadvantages

1. The contractor may have to wait until the project is completed before receiving payment thereby increasing the contractor's financing costs which are usually reflected in a higher tender price. However many lump sum contracts, particularly the larger or more complex ones, provide for monthly payments as well, which avoids this problem.
2. Any changes on lump sum contracts can give rise to difficulties in reaching agreement over cost and so the resolution of the final amount to be paid to the contractor may take some time.

3. If a bill of quantities or other cost breakdown is not requested from tenderers, it is difficult to see whether a very cheap tender offer is genuine, or a mistake. This is because all the employer has to go on is the lump sum. Further enquiries can of course be made of the tenderer.

Remeasurement (bill of quantities)

The bill of quantities (BoQ) is described in more detail in Chapter 15 but is basically a document listing all the items of work in a project against which the contractor inserts its rate for doing the work when completing its tender. So there might be an item for the provision of 15 cubic metres of concrete, and the contractor might price it at £100 per cubic metre. In contracts using remeasurement (bill of quantities) as a pricing mechanism, the quantities are **taken off** (as accurately as is reasonable, they should not be guessed) by the design organisation or quantity surveyor. Taking off is a term used to describe the action of abstracting the quantities from the drawings and summarising them in a bill of quantities. The item descriptions for each item in the bill of quantities are normally prepared in accordance with a standard method of measurement.

During construction, the BoQ is used as a payment mechanism and the contractor is paid the item price for the *quantities actually constructed*, usually on a monthly basis based on a valuation. At the final account stage a reconciliation of all the constructed items is carried out, and this is the total amount that the contractor is paid for what is called the **measured work**. The value of the measured work is simply the actual quantity of each completed item of work multiplied by the rate for that item in the bill of quantities. Any contract claims for financial loss and expense (compensation events in NEC3) are paid in addition to this amount together with any allowance for inflation. Provisions to allow for inflation have to be written into the contract and are fairly rare in these times of low inflation.

> Bill of quantities contracts were once the norm and still have much to recommend them.

Use

Remeasurement (bill of quantities) is used on the whole range of contract types and values. However, very small and simple contracts are better treated as lump sum, simply to avoid the time involved in taking off quantities.

Traditional contracts (such as JCT SBC05 and ICE7) use the BoQ as the payment mechanism almost exclusively. The design has to be complete or almost complete, in order to take off reasonably accurate quantities, so this mechanism is not suitable for the design and build strategy, where the contractor produces the design after the contract is awarded. This mechanism also gives the employer a fairly accurate estimate of the final price to be paid (subject to any changed quantities or claims).

Taking off quantities also provides a very useful check on the completeness and consistency of the drawings. In fact it was once usual to mark up a print of each drawing to show that the quantities had been taken off for every item and added to the bill of quantities. This was an attention to detail that is lacking in much modern design office practice particularly in civil engineering, where BoQs are used less often than other mechanisms such as target price. It had the real benefit of highlighting errors and omissions so that they could be corrected before the project started on-site, where delay is very expensive.

Advantages

1. The drawings have to be complete (or almost) with all details shown so misunderstandings are less likely because:
   (a) issues have been thought out early
   (b) designers' intentions are clear
   (c) taking off quantities is a useful cross-check on drawing accuracy and consistency with the specification.
2. All tenderers price the documents on the same information.
3. Clear rules in the standard method of measurement mean tenderers and later the contract administrator should know exactly what the pricing covers.
4. It is relatively easy to compare tenders.
5. By comparing tenders, 'tendermanship' such as front-loading and back-loading are very evident (see Chapter 6 on tendering and contractor selection).
6. Although the contractor should order materials from the drawings, the bill of quantities gives an approximate check.
7. Valuations for monthly payments are straightforward. The amount of work done against each item that month is measured on-site and multiplied by the rate to give the amount to be paid per item.
8. Variations are relatively easy to price, using existing rates, or using existing rates as a basis (where the conditions of contract permit this).
9. The bill of quantities is only produced once, by the *designer*. In many other pricing mechanisms, each tenderer has to individually produce some sort of pricing or estimating mechanism, which is inefficient. It takes time and adds to tendering costs which simply increase the 'overhead' in the construction industry and are hence ultimately passed onto all employers.
10. The employer has a fairly good idea of likely outturn cost, which is the contract sum produced from the bill of quantities. To this must be added any contract risks that the employer pays for, such as unexpected ground conditions, and of course these are harder to estimate.
11. The contractor bears less risk because measurement errors or deviations from the standard method of measurement are usually corrected at the employer's cost, so tender prices can be lower. Of course, the employer will expect a competent designer (or quantity surveyor) to keep such errors and deviations to a minimum.
12. The designer (or quantity surveyor) can use previous contracts using bill of quantities to estimate prices on future projects.

Disadvantages

1. Since the drawings must be virtually complete, a later (but more defined) start on-site may occur. This later start is often seen as a disadvantage. However, whilst an earlier start theoretically leads to an earlier completion, this is often not the case if there are errors, omissions or misunderstandings. Because of

the better definition of contract documents using bills of quantities, claims due to incorrect or misleading information are less likely.
2. Disagreements can occur over whether the measurement rules given in the standard method of measurement have been applied properly. Disagreements can also arise over the contractual treatment of incorrect quantities, and omissions.
3. This mechanism is less likely to encourage team work and innovation as other mechanisms which we shall come to shortly, such as target price (especially with pain/gain share), tend to do.
4. It cannot be used for cost-reimbursable type contracts (again we shall come to this shortly).
5. It cannot be used for the 'design and build' contract strategy, since the design has not been done at tender stage so there is insufficient detail to take off quantities.

Remeasurement (schedule of rates)

This method is similar to remeasurement (bill of quantities), except that all items are listed as 'per metre', 'per square metre', 'per item' or similar terms in a document called the **schedule of rates**. In other words there is usually *no indication of the quantity* involved. The tenderer inserts a price without any real idea of the final quantity (because the employer does not know either at this stage). An item such as 'per beech tree planted' may be used without knowing clearly whether there will be one, five or fifty beech trees. Hence the tenderer has to make some assumptions over likely quantities. In order to assist with this, the employer may give approximate ranges of quantities. Exact quantities cannot be given to tenderers since they have not been determined at tender stage.

> Remeasurement is a very flexible mechanism but can easily lead to disagreements over the quantities assumed by contractors in building up their rates at tender stage.

Remeasurement is clearly a payment mechanism because the contractor is paid for the amount of work done multiplied by the rate. It is also a pricing mechanism, but a rather more indeterminate one than a bill of quantities. In order to compare tenders, the employer will usually have a list of estimated quantities. Hence the pricing (by the employer) of different tenders is dependent on this assumption of likely quantities.

Use

The schedule of rates mechanism is often used where the design responsibility lies with the employer and the design is to be completed during construction. This is typically for work where the design is straightforward but completed in phases, so that construction can commence whilst subsequent phases are designed.

This mechanism is also useful for simple contracts where the amount of work is not known at tender, or for contracts where work is repeated (in a slightly different form) or for contracts for maintenance or emergency work. Typical examples would be an annual contract for erecting new fencing around different sewage works, or an annual contract with a county council for planting trees or maintaining lighting columns. Another example might be the replacement of small-diameter water supply pipework in a region, over a period of two years, where an early start is required, but where the detailed design in any given road depends on locating existing services and obtaining permissions under the Traffic Management Act or similar legislation.

Advantages

1. Construction can commence early by overlapping phases of design and construction.
2. This mechanism is flexible and can deal with changing quantities.
3. It can provide a good basis for paying for maintenance or emergency work.
4. In theory changes (variations) are easy to value, since if we change from five trees to fifty, we might expect to pay ten times as much. In practice is does not work as simply as this – see disadvantages below.

Disadvantages

1. It is difficult to compare tenders, since there will be an assortment of differently priced items, and no accurate quantities. Employers will usually have some estimated quantities to use for the purpose of assessing tenders. However, depending on the quality of the estimates, the 'wrong' contractor can easily be chosen.
2. The flow of information for construction must be carefully managed if claims for delay by the contractor are to be avoided.
3. Since a standard method of measurement is rarely used, item descriptions in the schedule of rates have to state very clearly what is included and what is not. So if descriptions are not clear and comprehensive, then the contractor may claim an entitlement to a more appropriate rate for the correct item description.
4. A number of measurement claims peculiar to this mechanism can easily arise unless the schedule of rates is written very carefully. Buying or constructing in bulk is usually cheaper. Hence the contractor may argue that the tendered rate was based on an assumption of quantity at tender, and this assumption has changed significantly with the actual quantities constructed. The argument is that the balance of plant, labour and materials may change so that this *of itself* makes the unit prices inapplicable or inappropriate, so that they need to be adjusted (usually upwards). The term 'of itself' is used in ICE7 clause 56(2) for exactly this issue.

   As the actual quantities become apparent another argument is that the **overhead element** is different from that assumed at tender. An overhead element could be the cost of supervision or small plant and tools for example. To try and avoid these problems, approximate ranges of quantities are often given. Take the example of a contract to plant trees at various sites, but where the employer is unsure exactly how many trees will be needed at each site – sometimes a few, sometimes dozens. Then one way around the problem is to obtain tender rates in ranges such as a unit rate for planting 5 to 10 trees, a unit rate for 11 to 50 trees, a unit rate for 51 and over trees, and so on. To overcome the problem of 'overhead element', the travelling cost to sites at different distances could be stated separately. There is clearly a difference in cost between planting one tree 10 miles from base and one tree 100 miles from base. So constructing a suitable schedule of rates to avoid claims is actually quite difficult, and the schedule of rates mechanism is not as simple as it first appears.

Activity schedules

An activity schedule is a list of the *activities* which the tenderer expects to undertake to complete all the obligations under the contract. So for example a tenderer may choose to denote an activity as 'construct base slab' and price it at £10,000. Compare this with a bill of quantities contract where this activity would be broken down into dozens of measured parts such as excavation, formwork, reinforcement and concrete. For straightforward items such as this, the simplicity both for tendering and payment purposes is an attraction of activity schedules over a bill of quantities. However, the contractor carries the risk of ensuring that all the work required in the item is included in the activity price. Because contractors carry this additional risk, they may increase the overall tender price somewhat over a bill of quantities contract, where the employer usually carries the risk if items are missed.

An additional cost to the contractor in NEC3 ECC contracts using option A is that the contractor is only paid for *completed* activities as stated in clause 11.2(27). In our example, this would be the completion of the base slab in accordance with the contract and without defects. In a bill of quantities contract, the contractor would be paid for the work done in any given payment period (often monthly), so if the construction of the slab took two months, then at least some payment would be made at the end of month one.

The total of the prices for all the activities is the tenderer's price for providing the whole of the works. Activity schedules are used increasingly in more modern forms of contract such as NEC3 ECC options A and C. These NEC3 ECC options use activity schedules to show the build-up of the total contract price for comparison of tenders. The activity schedules are used to determine the (monthly) payments in option A. However the activity schedules are not used for payment in option C which is a cost-reimbursable option, with a target cost. Hence in option A the activity schedule is a pricing and a payment mechanism, whereas in option C it is only a pricing mechanism.

> NEC3 ECC option C is probably the option used the most in UK construction of civil engineering works.

Unless the employer prescribes the make-up or number of activities (which some do), the tenderer decides what activities to include and what price to put against each. This means that activity schedules can reflect the tenderer's view of the work. However, this makes tender comparison difficult. In a five-storey tower block for example, one tenderer may have an activity for completion of each floor, whereas another tenderer may break this down into dozens of activities. Because the contractor is only paid for completed activities in NEC3 ECC option A, tenderers may be tempted to break the work down into hundreds of activities to ensure a reasonably steady cash flow, unless the contract prescribes otherwise. A tender with a large number of minor activities can make the administration of the contract inefficient and expensive.

Activity schedules also have the attraction of saving the designer a great deal of work in the preparation of a bill of quantities. However there are significant drawbacks to this and designers used to newer forms of contract have arguably lost many of the estimating skills possessed by their predecessors who used ICE7 and JCT contracts with a bill of quantities.

Use

Activity schedules, like a bill of quantities, can be used on the whole range of contract types and values. However, very small and simple contracts are better treated as lump sum to avoid the cost of preparing an activity schedule and administering the subsequent payments.

Advantages

1. Activity schedules save the designer a lot of work in the preparation of a bill of quantities.
2. They reduce the employer's risk, but put a lot of risk on the tenderer, to ensure each activity is identified and priced fully.
3. They can reflect the contractor's planned method and sequence of work, since the contractor prepares them. There is therefore a link between the programme of work and payment for it.
4. This link to the programme with an activity schedule is easier to make than in a contract using a bill of quantities, so better prediction of cash flow is possible.

Disadvantages

1. When using activity schedules there is the temptation to give the contractor incomplete drawings and other documents. Preparing a bill of quantities to a standard method of measurement demands that all the detail is properly completed.
2. It is difficult to compare tenders, since each tenderer will break the work down into different activities.
3. Each tenderer has to do a significant amount of work to price each activity. Many contractors produce their own bill of quantities in addition, in order to price the activity schedules, which rather defeats the object.
4. If activity schedules are used for payment purposes, there can be arguments as to exactly what each activity covers.
5. It is difficult to use activity schedules as the basis for valuing changes (variations).
6. Importantly, designers lose the ability to prepare estimates (see below).

In traditional contracts, designers or quantity surveyors would prepare a bill of quantities. They would also compare tenders. In time, they would build up a mental picture of how much things cost, for example concrete at £80 per cubic metre, and £10 to place in foundations, normal fair-finish formwork at about £30 per square metre and so on. There were also proprietary databases of such rates. They could therefore estimate the likely cost of a contract to within about five or ten per cent. They could also identify unusual rates in tenders, which often indicated a mistake by the contractor, or deliberate loading of items with small quantities, in the expectation that these quantities would increase. In the author's experience these valuable skills have been lost for many younger designers, particularly in the civil engineering side of the industry, where they tend to be brought up on activity schedules. They (and their employers) are now at a commercial disadvantage compared to the contractor's estimator who will certainly understand the likely cost of construction activities.

Cost-reimbursable

There are a number of varieties of this payment mechanism, which is increasingly becoming the 'norm' for civil engineering contracts in its **target price** form, which will be covered in detail in the next section. Cost-reimbursable means the payment of actual direct cost plus fee. It is often called 'cost-plus'. Payment is usually made on an 'open-book' approach. In open-book, all contractor costs should be open, and can be audited by the employer. The idea of cost-reimbursable contracts is to reduce much of the work and time involved in preparing a bill of quantities or an activity schedule; the contractor is simply paid costs plus a pre-agreed addition for 'fee'.

The fee represents the contractor's overheads (such items as the costs of the contractor's headquarters, directors and other staff) and profit. But of course 'simply paying costs' is not as simple as it sounds, and because it is a potential issue with this payment mechanism it is described in detail in the section below. Because of these payment issues, administration of cost-reimbursable contracts can be expensive and time-consuming. It is very important to allocate sufficient competent people to undertake this task, especially for larger contracts, and have them in place before construction payments begin.

> Cost-reimbursable contracts usually result in higher costs for the administration of payments than those using a bill of quantities or activity schedules.

Another problem is tender comparison. Employers should also question whether the winning tender with say nine per cent fee addition will deliver **a whole project** more cheaply than a tender with say ten per cent fee addition. This is because the real cost is not in the percentage fee addition, but in the time, people and plant used and reimbursed for the cost of the actual construction work.

On the positive side, there are few contract disagreements in this method, since the contractor is paid costs plus fee, and hence has no real issues with payment or profit. Fixing the fee as a lump sum at tender (rather than as a percentage) is a small help here to the employer, since the *cost* of a lump sum fee does not rise with increased contract costs, as the *cost* of a percentage fee does. However it must always be remembered that the only competitive element in these contracts, at tender, is the fee. This competitive fee is often only about ten per cent of the total contract cost.

The competitive element at tender is much lower in cost-reimbursable contracts.

Then there is the issue of incentive for contractor efficiency. In theory with cost-reimbursable contracts, simpler documentation, flexibility, improved working relationships bringing potential value engineering and better solutions lead to lower costs, from which the employer benefits. Unfortunately, many employers have doubts as to the balance of these benefits against the possibility of lower contractor efficiency, for which the employer pays. The contractor does not for example have the incentive to complete an aspect of a job in one day when he knows he could be paid for two-days' work, plus say ten per cent of those higher costs as the fee.

In a bill of quantities contract, contractors are only paid the bill rate whether they work efficiently or not. However in a cost-reimbursable contract, contractors are paid their costs. There are two main ways to reintroduce incentives, and reassure employers over paying for inefficiency. One is the concept of 'disallowed costs' and the other is to use a target price contract, with 'pain/gain' share incentives. Both are described in the relevant section below.

NEC3 ECC option E is a cost-reimbursable option.

Employers tend to see cost-reimbursable contracts as a 'blank cheque'.

Use

The cost-reimbursable payment mechanism is often used for poorly defined contracts, 'emergency' contracts or contracts with little apparent risk, and can be a best-choice option in these circumstances. It is sensible to use a trusted and efficient contractor.

Cost-reimbursable can also be used where the employer controls the distribution of the work, and informs the contractor as to what needs doing next. However, within this overall employer distribution of work, the contractor will usually determine the *method of work*. The employer may issue an order to '*plant 50 beech trees in east embankment to the A46 junction 3*', *but* the contractor decides *how* to do this, and what plant and labour to use. If the employer actually 'directs' the work and determines the contractor's 'method of working', then the employer takes legal responsibility for the contractor's health and safety. Most employers do not wish to do this, and so distributing the work must be done with care, so as never to prescribe the contractor's method of working.

Advantages

1. Work can commence on-site before all the details and drawings are completed, which is
   (a) Ideal for emergency work
   (b) Good for some types of investigative work that cannot be clearly defined
   (c) Good for a quick start. It must be remembered that a quick start does not always guarantee a quicker finish.

2. The employer can use the cost-reimbursable payment mechanism for very ill-defined work, and can simply direct the contractor to undertake specific items of work as and when required.
3. Contract documents are simple since there is no need for a bill of quantities or activity schedules.
4. Cost-reimbursable contracts tend to improve teamwork because of the reduction in potential disagreements over 'measurement issues'.
5. Only risks which actually materialise are paid for, so the employer does not pay the tenderer's price allowances for risks that may not occur.
6. Disagreements over unforeseen events and other 'claims' are minimised.
7. Any problems of agreeing the price of variations or extensions of time are removed since the actual cost is paid.

Disadvantages
1. The employer carries almost all the risk.
2. There is no incentive for the contractor to be efficient; the contractor is paid all costs, plus a fee.
3. It is impossible to compare the total price of tenders, except for 'fee', which is often only about ten per cent of the total cost.
4. Agreeing *actual* costs, as construction work proceeds, can be difficult and time-consuming.
5. Disagreements over what constitute **disallowed costs** can occur. Disallowed costs are explained in the section below.
6. It is relatively easy for contractors to overcharge costs or charge twice; frequently by mistake as a result of accounting difficulties.
7. The final cost of the contract is not known, so the employer has difficulty in budgeting.
8. There is no 'feedback to the designer', to enable the designer to estimate the cost of future contracts.
9. If the work is ill-defined, it is very difficult for the contractor to know what resources to allocate to it. This in itself introduces inefficiency for which the employer ultimately pays.
10. If the contractor has resources 'available and waiting for deployment' (possibly to undertake emergency work) then there may be no payment for them until they arrive at the site to do actual work. In reality this means that the employer may not have the 'rapid-deployment-force' that the employer may be expecting (for emergency work for example) since the contractor cannot afford to have resources tied up on 'stand-by' unless a specific payment will be made for this.
11. Finally, it may be difficult on larger contracts of this type to know whether they exceed the price thresholds and hence fall within the Public/Utilities Contracts Regulations.

Target price

At first sight, target price contracts are a pricing mechanism – the target price mechanism. However they are frequently used in conjunction with a different payment mechanism, such as cost-reimbursement. Most target price contracts also incorporate **pain/gain share** which is explained below.

Target price contracts are often called target cost contracts. NEC3 ECC avoids the problem of semantics by simply calling them 'target contracts'. There is a subtle difference between cost and price of course, but the industry seems to use the two terms as though they are interchangeable. This chapter will use the term 'target price'.

Target price contracts are a version of cost-reimbursable contracts with additional refinements to overcome some of the drawbacks of a straight cost-reimbursable contract. That is why they are used so frequently in modern construction, particularly civil engineering contracts. Target price contracts have three major benefits over pure cost-reimbursable contracts:

- They provide a means of comparing tenders.
- They give the employer a good idea of the likely contract outturn cost.
- They provide an incentive for contractor efficiency that is lacking in pure cost-reimbursable contracts.

However target price contracts still have difficulties over the definition and assessment of cost which are described in more detail in the section below.

> Target price contracts, like cost-reimbursable contracts, usually result in higher costs for the administration of payments than those using a bill of quantities or activity schedules.

The way target price contracts work is as follows. Each tenderer bids a target price and a fee, so both these items generate tender competition and a good way of comparing tenders. The target price is the amount that the employer should expect to pay. Payments are made based on allowable contractor costs plus the fee until the target price is reached. The fee usually consists of the contractor's **overhead costs** plus profit. The contractor's allowable costs plus fee are usually assessed as in a cost-reimbursable contract and are paid as the work proceeds. A contractor's overheads include the cost of running the headquarters office and associated staff and are usually about six to ten per cent. These costs have to be recovered as part of the payment that the contractor receives on ongoing construction contracts.

In theory, when the contract is completed, the outturn (or final) cost is the target price although in practice this rarely happens owing to unforeseen events and the real difficulty of predicting an accurate final price at the tender stage. The big advantage in the target price mechanism is that there is a process for sharing an outturn contract cost which is either above or below the target price.

Pain/gain

Outturn costs above the target price are usually called 'pain', whereas those below are usually called 'gain'. This 'pain' or 'gain' is shared between the contractor and employer in some stated proportion. This share proportion is set in the tender documents. It is often 50/50, up to a limit of say ten per cent over the target price, after which the contractor often suffers all the loss. This limit is often called the **ceiling price**. Once payment reaches this ceiling price, no further contractor costs are paid.

An added sophistication is to set pain/gain percentages at different rates such as 50/50 and 60/40, applied to different bands above and below the target price.

Figure 14.1 shows an example of the operation of pain/gain on two contracts. The hollow arrows indicate how much the contractor is actually paid. Each contract has a target price of £1,000,000 and a pain/gain share proportion of 50/50.

In the left-hand example, which is gain, the contractor's costs and fee are £900,000, so the contractor receives a final payment of £900,000 plus 50 per cent of the gain, which comes to £950,000. The hollow arrow halfway up the Gain share box shows the amount that the contractor is paid. Whereas in the right hand example, which is pain, the contractor's costs and fee are £1,100,000, so the contractor receives a final payment of £1,000,000 (the target price) plus 50 per cent of the pain, which comes to a final payment of £1,050,000 (again shown by a hollow arrow).

The employer's position is that, in the gain example, a contract that was expected to outturn at £1,000,000 is delivered for £50,000 less. In the pain example, the employer has paid £50,000 more than target price, but not the whole excess of £100,000 which the employer would have paid if this had been a normal cost-reimbursable contract.

The ceiling price

Many target price contracts have a ceiling price. This is the price above which no further payments are made to the contractor. The ceiling price is often set at ten per cent above the target price. Such methods of capping payments at a ceiling price give a contract with the illusion of a 'guaranteed maximum price' (GMP). It is an illusion because under most contracts, employer risks will normally increase the target price and with it the ceiling price. The next section explains this.

Target price moves

In NEC3 ECC employer risks are known as compensation events and there are over nineteen of them, depending on the option chosen. When a compensation event is agreed, its cost moves the target (and ceiling price) by that amount. Such increases in target price are usually known as **target price moves**. A target price move (or TP move as they are often known) increases the target price and ceiling price by the appropriate cost of the risk provided that the contract recognises this risk as an employer risk, such as a compensation event in NEC3 ECC. The bands of pain and gain share also move with the new target price. Let us take an example on a contract with a target price of £10,000 and a ceiling price of £11,000. An agreed compensation event occurs with a cost of £300. The target price moves to £10,300 and the ceiling price moves to £11,300.

Figure 14.2 shows a contract before and after a target price move. The target price move is the difference between the target price and the new target price. The target price move sets the new target price, and carries the pain/gain bands with it, since they are set at a percentage of the target price, whatever that might be. If there is a ceiling price, then this too moves in proportion.

Depending on the contract used, employer risks which produce these target price moves could be changes (variations), unforeseeable ground conditions, late access or late information and so on. Unfortunately, arguments over the validity and value of target price moves can become the same as the 'claims industry' of

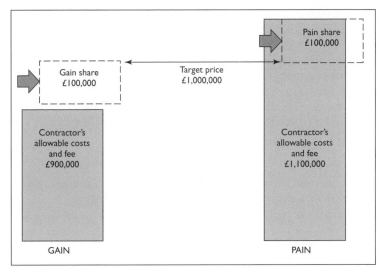

Figure 14.1 The operation of pain/gain.

the 1990s, thereby diminishing cooperation and teamwork and adding greatly to the contract administration costs. In many ways they are similar to claims for financial loss and expense in more traditional contracts such as JCT SBC05.

The first decision for the contract administrator is whether the event in question is an employer risk at all (called a compensation event in NEC3 ECC). If the contract administrator accepts that it is an employer risk then the cost and time implications have to be determined and agreed with the contractor. This process can take days or even weeks for more contentious or high-cost target price moves. It is not unknown for larger contracts to have hundreds of target

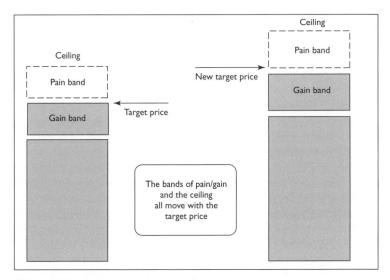

Figure 14.2 A contract before and after a target price move.

price moves, many of which have still not been agreed long after the work itself is completed. This is not the intention of such contracts but it often happens in practice.

If the target price move cannot be agreed at all then it potentially has to be taken to a formal dispute resolution procedure such as adjudication. This is likely to undermine the team work principles of target price contracts. So unfortunately the agreement of what constitutes allowable cost for payment and hence the determination of target price moves can negate most of the benefits of target price contracts.

> The determination and agreement of target price moves in target price contracts can create a major administrative burden and cost.

Other issues with target price contracts

In most target price contracts, the pain or gain is paid on completion of the contract. However, it is even possible to calculate (and pay) it on a monthly basis. This does of course add considerably to the work involved in agreeing monthly payments.

A further problem with this target price payment mechanism is if the final account calculation shows that the operation of the pain/gain mechanism means that the contractor owes the employer money. Securing payment of such debts can be problematic and it is best if a watch is kept on the likely final cost and final target price as later payments are made to the contractor, so that they can be adjusted to keep the total payment within the ceiling price.

Target price contracts – summary

So in summary, the employer sets the ranges of pain/gain percentages in the original contract, and a ceiling price at a percentage of the target price (often 110 per cent) and the contractor tenders a target price. In some contracts, this target price is simply stated as a sum of money. However it is usual also to show the make-up of the target price in some way. Three common methods of doing this are bills of quantities, lists of major elements of work or activity schedules. It is normal for the employer to prepare the first two of these, and in the third for the contractor to decide on the number and type of activities in activity schedules. The make-up is used to compare tenders, to compare prices with previous similar contracts and to give the employer some idea of the cost of various parts of the work, and hence estimate the payment profile.

NEC3 ECC options C and D are target price contracts with provision for pain/gain, called the 'contractor's share' under clause 53 which can be found in the additional clauses relating to options C and D. Option C is target price with activity schedules, and option D is target price with a bill of quantities. In options C and D these schedules or bills of quantities are not used for payment purposes, but are pricing mechanisms intended to facilitate the comparison and assessment of tenders.

JCT SBC05 allows for the employer to require activity schedules *in addition* to a bill of quantities (see recital two). Payment against the activity schedules is explained in clause 4.16.1.1. Unlike NEC3 ECC option A which only pays on completed activities, JCT SBC05 makes payment on the basis of the amount of each activity that is completed satisfactorily.

Use

The target price payment mechanism is used more and more on larger contracts. It is not really suitable for small contracts because of the increased administration of payment and the agreement of target price moves. The great benefit is the increased sharing of ideas and risks between the employer and the contractor, and the more cooperative atmosphere it engenders. It is possible to see excessive and repeated gain shares in serial contracts with the same contractor (such as in a framework agreement). If an excessive 'gain' is made without evident reason, then it is likely that the initial target price was tendered at too high a level. The employer may want a full explanation for this which does put some constraint on contractors who are pricing repeat order work without competition, such as in framework agreements.

Another major use of target price is for contracts where the work is not fully defined at tender, or where some risks are hard to estimate. This is because of the facility for target price moves and the operation of the pain/gain mechanism which in effect share costs of unallocated risks between the employer and contractor, thereby reducing the overall risk to the contractor. Unallocated risks could be such matters as price increases of materials or labour.

Advantages

1. There is a good basis for comparing tenders.
2. The employer has a reasonable idea of the likely outturn cost of the project.
3. Employer's risk is capped to a certain extent (apart from target price moves).
4. There is an incentive for contractor efficiency (gain share).
5. There is an incentive for employer cooperation (gain share).
6. There is an incentive for contractor ideas to reduce costs and create value engineering solutions (gain share).
7. They tend to improve teamwork because both parties are working to minimise outturn cost, and hence gain share.
8. They produce simpler contract documents without a bill of quantities.

Disadvantages

1. Agreeing *actual* costs as construction work proceeds can be difficult and time-consuming.
2. It is relatively easy for unscrupulous contractors to overcharge costs or charge twice. Equally, genuine mistakes can lead to the same result.
3. There is no 'feedback to the designer', to enable the designer to estimate the cost of future contracts.
4. Disagreements over what constitute disallowed costs can occur (see below).
5. Agreeing target price moves can be time-consuming, expensive and difficult, and reduces the cooperative nature of this form of payment mechanism. This can be a major disadvantage.
6. Determining final payment is complicated because of the difficulty of calculating pain/gain, disagreement about target price moves and disallowed costs.

Difficulties over agreeing actual costs in cost-reimbursable and target price contracts

The cost-reimbursable and target price mechanisms rely on paying the contractor's *costs*, rather than using a pre-priced schedule of items like a bill of quantities or schedule of rates. Measuring work done and paying the rate for it from a bill of quantities are relatively simple in a more traditional payment mechanism. However it can represent a few days' work every month for the contract administrator on a larger contract.

'Paying costs' in a target price or cost-reimbursable contract sounds very simple but it is not straightforward in practice. What a contractor's 'costs' are and what they include can be a difficult accounting exercise. For example, company cars may be a separate item in the contractor's accounts, or they may be shown as an on-cost on the daily rates for staff time. The contract administrator must ensure that the employer is not charged twice. Another example is where a contractor has a subsidiary company that hires plant. If the subsidiary hires plant to the contractor at double the normal industry rate, is that a true cost to the contract? This is a very obvious example but such pricing can be difficult to identify.

A further problem can be to do with personnel. How does the contract administrator know that the three quantity surveyors in the site office and charged to Contract A are working full time on Contract A, and that two of them are not actually preparing tenders for Contract B or even another project for a different employer altogether? The author has had instances of the same contractor's agent who is actually supervising two separate projects, being charged in full to both projects. If deliberate, this case would be clearly fraudulent, but more subtle overcharging can be difficult to detect. These types of overcharging by the contractor may be quite innocent mistakes arising from some of the accounting difficulties inherent in these payment mechanisms.

Assessing the cost for payment has become an increasingly expert job and target price contracts in particular frequently utilise cost assessors, quantity surveyors or even specialist firms of cost consultants to carry out this work. This usually results in increased administration costs and time when compared with more traditional payment mechanisms such as bill of quantities. So what at first seems temptingly simple can turn out to be quite the opposite.

NEC3 ECC target contracts

NEC3 ECC has some clever processes for minimising some of the drawbacks with target price contracts. One is the 'working areas' concept. These are defined in the contract and payment is only made for contractor's staff and equipment working within them. All other staff and equipment costs should be allowed for in the contractor's fee percentage. The idea behind the working areas concept is that the contract administrator can actually see (and assess) the staff and equipment in the working area, because they are local to the contract work and in a pre-agreed defined area. They are not being claimed at some remote location, to which the contract administrator does not have easy access.

Let us take for example an extension to Coventry University. Here a local disused car park might be rented by the contractor and designated as a working area for use by a team of quantity surveyors. Personnel working in offices in this disused car park would be allowable costs and paid for. Staff working in the

London head office would not be paid for in this way because they are assumed to be included in the fee percentage.

The other NEC3 ECC process is the use of pre-determined rates for reimbursing contractor costs other than sub-contractor costs. These pre-determined rates are inserted in the schedule of cost components at tender. In some ways this acts rather like a schedule of rates contract, and again produces some controls on the extent of costs that can be claimed. The detail of these NEC3 ECC mechanisms can be found in the NEC3 chapters (Chapters 21 to 26).

Disallowed costs

As we have seen, another difficulty with paying costs is agreement over exactly what costs are appropriate for the contractor to charge. Appropriate costs are usually called 'allowable' costs, and those which are not appropriate are usually called **disallowed costs**. One major source of contention is whether the contractor should be able to charge for the cost of correcting contractor defects such as bad workmanship, failure to meet specified standards or tests and so on. Under a traditional contract the contractor is only paid the bill of quantities rate for completed work that is satisfactory. In fact, many traditional contracts have the provision for making a deduction on previous payments for work which is *later* found to be defective. So in an ICE7 contract, a contractor may have received payment at bill of quantities rates for constructing a pipeline which subsequently fails a specified pressure test. Here the contract administrator would be entitled to make a correction (deduction) to a future payment to reflect the work needed to pass the test.

At first sight, a target price or cost-reimbursable contract would pay the contractor for all work needed to pass this pressure test, including relaying or rejointing previously constructed pipe. In some contracts such as NEC3 ECC options C to E, this is actually the case. If we wish to avoid this situation we could use the concept of disallowed costs. Disallowed costs are a list of areas of cost which are not considered appropriate for payment by the employer. They generally attempt to ensure that the contractor is not paid for inefficiency or failure to follow the procedures incorporated in the contract. NEC3 ECC defines disallowed costs in additional clause 11.2(25) which can be found in each of the options C, D and E. Typical disallowed costs might include:

- Costs which cannot be verified from invoices or records.
- Costs paid by the contractor which are outside the provisions of the contract.
- Costs where the contractor does not follow a prescribed procedure.
- Use of inappropriate plant or resources (this is hard to prove).
- Inefficiency (this is very hard to prove).
- Correcting defects after completion.

Many employers consider that payment for correcting defects *before* completion is also unwarranted, and amend the contract accordingly. Unfortunately there may also be the argument as to what exactly constitutes a defect, as a result of interpretation of the definition of a defect.

As an example of one of these difficulties, consider a contractor who owns its own excavation plant. Suppose a three-metre deep excavation is needed. A normal small excavator would be appropriate. However, if the contractor has a very large and expensive excavator standing idle in its yard, there is the temptation to use it for this three-metre deep excavation. It would be an inappropriate use of plant, and the employer may pay twice as much as the employer should pay, but the contractor is getting better utilisation of the expensive plant. Imagine the arguments on-site between the contractor and the contract administrator as to what represents 'appropriate plant' for this work.

Another example is where the contract administrator considers that the contractor is working inefficiently and disallows some costs. First of all it is slightly insulting to accuse a contractor of being inefficient and is hardly conducive to good site relations, and secondly it is very hard to prove.

More on the main functions of pricing/payment mechanisms

Now that we have discussed in detail the different types of pricing and payment mechanisms, we can look again at their main functions in more depth. Here is a reminder of the five main functions. Remember that some functions are common to pricing and payment mechanisms, and some are particular to one or the other. This distinction is explained in the notes below the box.

---

1. To allow the employer to compare tenders.
2. To give the employer a likely contract price, so that money can be raised, or budgeted for.
3. To make accurate regular payments to the contractor (often monthly).
4. To provide the designer with a database of rates for estimating the likely cost of future projects.
5. To provide a sound base for pricing variations (depending on the form of contract used).

---

1. All pricing mechanisms allow the employer to compare tenders; this is the first function. However some pricing mechanisms make tender comparison easier than others. Bill of quantities tenders are easier to compare than those based on activity schedules for example.

2. Some pricing/payment mechanisms such as lump sum give a definite contract price whilst, at the other extreme, a contract which uses the cost-reimbursable payment mechanism gives no contract price at all, so the employer has to estimate the likely final cost. Pricing/payment mechanisms such as target price and remeasurement give a *likely* contract price, which could easily vary up or down by five to ten per cent simply because of the operation of the payment mechanism in making regular payments and adjusting the final account for pain/gain. Claims for delay, loss and expense can further increase final contract prices but this increase depends on the form of contract used.

3. Once a contract is awarded, payment has to be made to the contractor on successful completion of the contract. This is done by using the prescribed payment mechanism. In many contracts payments are regular, usually monthly. Where regular payments are made there has to be a sound and auditable method of calculating the amount payable; this is the main function of payment mechanisms. Monthly payments are used to reduce the financial risk and burden on the

contractor, which usually results in a lower tender price, because of reduced financing costs. Most employers have better and cheaper access to funding than contractors. Hence it makes sense for employers to make regular payments to contractors from their income or profits, or even to borrow the money on a loan of one sort or another. Many large employers finance their capital programmes by taking out long-term bonds.

4. The further function of pricing/payment mechanisms, particularly those using bills of quantities, is to provide the designer with a database of rates from which the likely cost of future projects can be estimated. Competent designers with experience of traditional contracts are able to estimate likely tender prices to within five or ten per cent. Unfortunately, modern methods of payment encouraged by NEC3 EEC, on 'civil engineering' contracts such as target price and cost-reimbursable, do not provide any such database. Today, civil engineering designers have much less real data to produce estimates from, and hence face a disadvantage compared with the estimators who work in the tendering departments of contractors.

5. Another function of pricing mechanisms, again on traditional contracts such as ICE7 and most of the JCT suite of contracts, is to provide a sound base for pricing changes (called *'variations'* in these contracts) ordered by the contract administrator. NEC3 ECC does not normally use this method, but relies instead on contractor quotations for pricing changes (which is the NEC3 ECC term for variations). That having been said, NEC3 ECC option B does allow the use of the bill of quantities if the project manager and contractor agree.

Cost, price and value and their importance

Cost, price and value have different meanings and they are important because of their effect on pricing and payment mechanisms, particularly bill of quantities and target price contracts. This effect is not immediately obvious but needs careful attention.

Price is the amount of money charged for something. So a cup of coffee may have a *price* of two pounds, but *cost* a café one pound to produce. As we know, the price of a cup of coffee is usually reflective of where you drink it; perhaps fifty pence outside a roadside van and five pounds in a hotel in London's Park Lane (perhaps with a biscuit!). If you are willing to pay higher prices, then presumably the *value* to you is greater, perhaps because of nicer surroundings or better service.

In construction contracts price is typically represented by the rates and lump sums in a bill of quantities (the pricing mechanism). Cost is what the contractor has to pay to provide something. In this context cost will be the cost of labour, plant and materials plus the contractor's overhead costs and profit. So when the tenderer completes a tender, a rate in the bill of quantities for laying pipes *costing* the contractor £50 per metre (made up of labour, plant, materials and overheads) may be inserted at a *price* of £55 per metre, thereby producing a profit of £5, if the contractor wins the tender. Presumably the *value* of this item to the employer is £55, since this is what the employer is willing to pay in accepting the tender.

Or let us take another example. Suppose you can afford to buy a new Mercedes SLK sports car. Its list price is £35,000 and the waiting list stretches for over a year. So the dealer is charging a *price* of £40,000 for one in the showroom. It may *cost* Mercedes £25,000 to manufacture but they can charge a £10,000 premium because of its desirability. Its *value* to you may be £40,000 if you can get it now for the summer so you might be willing to pay this extra £5,000 for one that is actually on the showroom floor. But the *value* of this Mercedes to your mother might only be £10,000; she does not particularly want a Mercedes sports car and prefers her small hatchback which has four seats and is easier to get into.

Bill of quantities contracts

Monthly payments on bill of quantities contracts are usually correctly called *valuations*. They measure and pay the **price** of work done, by multiplying quantities of work achieved against rates inserted in the bill of quantities. In theory, this is their *value* to the employer, since the employer has accepted the tender with these prices as being reasonable and satisfactory.

However the problem with bill of quantities contracts is that the contractor also needs to keep a watch on actual *costs*. So if for example 500 cubic metres of excavation has a billed rate of £5 a cubic metre, the valuation will produce a payment of £2,500 when the work is completed. The rate of £5 will have seemed sensible to the estimator when the tender was produced based on normal outputs. Now suppose the excavator kept breaking down and it cost £3,000 to do the work, then the price of £2,500 paid to the contractor is less than the cost to the contractor, who now has a loss to contend with.

Target price contracts

The opposite effect can be seen in more modern payment mechanisms such as target price which pay for the **cost** of the work done. All the contract work has to be completed, and if the total costs exceed the target price then the contractor has a problem if a pain/gain share is in operation. This final outcome can be very risky for a contractor unless an **earned value** check is done regularly. An earned value analysis compares both cost and value. If costs exceed value then there is a problem in the making. A contractor really ought to know the *value* achieved on target price contracts *as well as the costs*.

An example of the issue of cost and price on a target price contract

A simple example would be a target price contract for the construction of four tanks on a water treatment works at £100,000 each, with a target price for the whole contract of £400,000. After two tanks are completed the value is £200,000. If the contractor's costs were say £250,000 then the contractor would be paid this because we are at an interim stage and final costs are not known precisely. However at this rate of efficiency, when the contract is completed the ultimate cost of four tanks would be £500,000. This would exceed the target price by £100,000 and the contractor would be in the 'pain' zone. If the share percentage was 50/50 with no cap on ceiling price, then the contractor would only be paid £450,000 and would make a £50,000 loss. This is an uncomplicated example. In practice on a normal contract with many work items it is actually very difficult to determine 'value'. So in this example of a contract for four tanks, the contractor is heading towards a 'pain' situation but may not know it, because the *costs* so far are being paid.

This issue really can lull contractors into a false sense of security on target price contracts, only to get an expensive surprise near the end of the contract as work still needs to be done, but there is little headroom below the target price to pay for it. Unfortunately, what may happen in these circumstances is that the contractor's quantity surveyors will be obliged to try and recover this loss and they may well start to look back for compensation events that produce target price moves to increase the headroom. This is not the intention behind target price contracts. NEC3 ECC tries to reduce the effect of this unfortunate practice by means of stipulated time periods for notifying compensation events. A very late notification can be 'time-barred'.

What is the pricing/payment mechanism in JCT SBC05, ICE7 and NEC contracts?

JCT SBC05

- **JCT SBC05 with quantities** is basically a lump sum contract (pricing mechanism), with monthly payment provisions against completed work (payment mechanism). The idea is that the quantities are an accurate representation of the work and the bill of quantities (called the contract bills in JCT) is listed as a contract document in the definitions section of the contract. There is a provision for variations and a means of agreeing their cost in section five of the contract.

  Since this contract is assumed to be a lump sum contract, the *original lump sum* is adjusted for variations and claims for loss and expense. Interim (monthly) payments are made by multiplying the amount of work done by the rate in the contract bills. From the total due, a deduction is made that represents previous payments, to give the net amount representing that month's work. If an activity schedule is required then it must be specified in the second recital. If an activity schedule is used, then payment is made against the proportion of each activity completed as clause 4.16.
- **JCT SBC05 'without quantities'** is a pure lump sum contract and the tenderer has to bear the risk of the method of arriving at the contract price including any omissions or mistakes that the estimator may make.
- **JCT SBC05 'with approximate quantities'** is a remeasurement contract, rather like ICE7, where all quantities need to be remeasured as the construction work is done. Tenderers insert rates and prices against items in the bill of quantities on the understanding that all quantities are approximate and will be remeasured.

ICE7

- The normal ICE7 form is also a remeasurement contract, and the quantities are described as the 'estimated quantities'. They are all remeasured as the construction work is done.

NEC3 ECC depends on the option chosen

Options A to F of NEC3 ECC select different pricing/payment mechanisms and are described in detail in the chapters on NEC3 ECC contracts (Chapters 21 to 26).

Chapter summary

- The pricing and payment mechanisms are chosen by the employer, usually on the advice of the design organisation. It is a very important decision.
- Sometimes the pricing mechanism and payment mechanism are the same.
- However in many NEC3 ECC options the pricing and payment mechanisms are different.
- The final payment determined by the payment mechanism is often different from the price predicted by the pricing mechanism at tender.
- There are five main *functions* of pricing/payment mechanisms, although some mechanisms do not fulfil all five functions. The five functions are to allow the employer to compare tenders; to give the employer a likely contract price; to make regular payments to the contractor; to provide the designer with a database of rates for estimating the likely cost of future projects; and in some forms of contract to provide a basis for pricing variations.
- There are five main *types* of pricing/payment mechanism. They are lump sum; remeasurement (bill of quantities or schedule of rates); activity schedules; cost-reimbursable; and target price. Traditional contracts generally use bills of quantities, whereas contracts such as NEC3 ECC can use bills of quantities but are generally target price.
- The cost-reimbursable and target price payment mechanisms rely on paying the contractor's *costs*, rather than using a pre-priced schedule of items like a bill of quantities or schedule of rates. Agreeing what costs are valid can be difficult and administratively expensive.
- There are advantages and disadvantages with all pricing and payment mechanisms, and selecting the most appropriate one for any specific contract must be done with great care.

15

Methods of measurement and BoQs

Aim

This chapter explains the use of standard methods of measurement and bills of quantities (BoQ) as commonly used in construction contracts.

Learning outcomes

On completion of this chapter you will be able to:

>> Describe the production and operation of bills of quantities.
>> Define a standard method of measurement.
>> Differentiate between pricing and payment mechanisms.
>> Explain the concept of 'measurement claims'.
>> Describe why the outturn price of a project is rarely the tendered price.

The bill of quantities (BoQ)

Until the advent of activity schedules in the previous two decades, bills of quantities were probably the most used pricing/payment mechanism in construction contracts. They are still used extensively in building contracts, whereas civil engineering contracts tend to favour target price contracts of one form or another. The bill of quantities is often abbreviated to BoQ or 'the bill'. Bills of quantities were touched on in Chapter 14 and are explained in more detail here.

A bill of quantities is a list of items taken from a **standard method of measurement**. Against each item the total quantity is abstracted from the tender drawings (usually by the design organisation). The bill of quantities is rather like a shopping list, which the tenderer then completes by inserting the prices that will be charged against each item. Most of these items will be measured as either 'rates' or 'prices per unit'. So we might have rates against metres of pipes, square metres of formwork and cubic metres of concrete, but valves and pipe fittings would be measured as prices per unit against 'number' (nr). The total of all the items is the tender price; this is the use of the BoQ as a pricing mechanism. During the course of construction, regular payments are made by measuring the actual quantities used on-site and multiplying them by the rate in the BoQ; this is their use as a payment mechanism.

In traditional contracts, such as ICE7 and JCT SBC05, the bill of quantities is a pricing mechanism and a payment mechanism. It is used in NEC3 ECC option B in a similar way. Option D also uses bills of quantities, but here only as a pricing mechanism. In JCT contracts the bill of quantities is usually called the contract bill and is combined with the specification items relevant to each item in the BoQ.

The normal assumption of contracts using a bill of quantities is that it will be prepared by the design organisation or a quantity surveyor and issued in the same format to all companies that are tendering. The major advantages of a bill of quantities are consistency and a straightforward process for checking and comparing tenders. Another advantage for the industry as a whole is that the bill of quantities is only prepared once, rather than by each tenderer, who must of course price it. So there is no abortive work effort by unsuccessful tenderers.

The profession that concentrates on the billing and measurement function is the quantity surveyor, or 'QS'. Quantity surveyors are usually employed in the building industry to **take off** quantities, whereas those civil engineers who still have the skill base generally take off their own quantities. 'Taking off' is the term used for abstracting quantities and inserting them against the relevant item in the bill of quantities. Tender documents including the bill of quantities are sent out to the contractors who will be bidding for the contract. The contractor's tendering department normally includes people called 'estimators' who 'price up the bill' by inserting an appropriate rate (or price) against each of the items.

The bill of quantities can also be used by the design organisation to estimate quite accurately the likely cost of the project. This is done by inserting rates against the BoQ items. These rates may be obtained from previous similar contracts, published databases or the designer's own experience. An experienced designer should be able to estimate the likely tender price within five or ten per cent using these methods.

Tender comparison is much easier with contracts using a BoQ, because each BoQ contains the same items; only the tenderers' prices differ and these can be compared, examined in detail and compared with prices for previous projects.

Finally, BoQs are often used in traditional contracts for determining the cost of changes (variations), where BoQ items are reasonably applicable to the varied work.

Standard methods of measurement

Contracts based on bills of quantities are usually measured to standard methods of measurement. The purpose of a standard method of measurement is to ensure that everyone concerned with the contract understands clearly what every item of work includes and hence provides a price that is not open to later disagreement as to what work it actually covers. There are two main standard methods of measurement in common use today.

- **CESMM3 – The Civil Engineering Standard Method of Measurement**
- **SMM7– The Standard Method of Measurement of Building Works**

The numbers 3 and 7 denote the latest version of these methods of measurement which were initiated many decades ago and are still the norm for traditional contracts. Whilst the JCT suite of contracts still makes extensive use of SMM7, the use of CESMM3 has declined radically with the demise of ICE7 and the almost universal application of NEC3 ECC to civil engineering contracts. NEC3 ECC options B and D are designed for the use of a method of measurement such as CESSM3 but the attractions of activity schedules in options A and C has led to a marked decline in the number of contracts using a bill of quantities.

RICS published a new method of measurement called the 'new rules of measurement' (NRM1) in 2009. RICS say that NRM2 will be published in December 2011 and NRM3 will be published in April 2012. NRM2 will be an alternative to SMM7 from 2011. Its use will no doubt increase but, as we see with new editions of forms of contract, extensive use often takes some years.

- CESMM3 is normally used for civil engineering contracts such as ICE7 and NEC3 ECC options B and D, which use a bill of quantities. It should be emphasised that civil engineers today are rarely trained in using standard methods of measurement to produce bills of quantities since option C (target with activity schedule) is the most prevalent option of NEC3 ECC. This does not mean that options B and D are poorer relatives; they actually have many advantages, for those who understand the use of a bill of quantities. The main advantages of options B and D are the easier comparison of tenders and much reduced site administration (in option B) to process monthly payments to the contractor.
- SMM7 is widely used in the JCT suite of contracts for 'building' contracts which make general use of a bill of quantities. SMM7 is more detailed than CESMM3, which reflects the more complicated nature of building work. SMM7 has to provide items for the diverse range of such work from structural items to cladding, windows, roofs and heating systems; whereas civil engineering is more often concerned with structures of steel and concrete.

Standard methods of measurement usually have an introductory section with general directions for measurement and the layout of the bill of quantities. The measurement sections which follow are then separately designated A, B, C and so on. Each section covers one of the main construction materials or activities, such as earthworks, concrete, formwork, reinforcement, brickwork and pipes. Measurement sections are laid out by class and type of work, and have a logical numbering system for the standard items within them. Each measurement section is divided into *classification, measurement rules, definition rules and coverage rules*. Measurement is done in prescribed categories of types of work to a defined set of rules. These rules aim to make measurement consistent and readily understandable for the three purposes of taking off quantities, pricing quantities and paying for completed work.

Example using CESSM3

CESMM3 Class G Concrete ancillaries

Formwork

Number	Item description	unit	quantity	rate	amount
G214	Fair finish, plane horizontal, width 0.4–1.22 m	m²	25	*£20*	*£500*

The bill of quantities is usually produced using a computer program which stores all the standard numbers and item descriptions. The program will also perform addition and other numerical functions. So the computer program will multiply the rate by the quantity to give the total price for that item in the column headed 'amount'.

In the formwork example G214 given above, the 'designer' or quantity surveyor will measure and take off the quantity and insert it in the tender bill of quantities. Here it is 25 square metres. From the standard number G214 the computer program produces the item description and unit (m^2). Where items within a code differ, they are given additional description and coded G214.1, G214.2 and so on.

Deviating from the standard method of measurement

It is possible to deviate from the standard method of measurement provided it is done in a clear and acceptable way, so that it is completely obvious to tenderers. There are two ways of doing this:

1. The preamble is a very important descriptive section at the front of the bill of quantities. 'Preamble' is a word that is often used for a relatively unimportant front section of a book. However in the context of the bill of quantities, the preamble has great contractual significance. It is necessary to state in the preamble where general or specific departures from the stated method of measurement have been made. If this is not done then the contractor will have a legitimate 'measurement claim'.
2. An individual non-standard item can be used. It should be clearly numbered to show that it is non-standard, and have a full description of the work that it covers. In CESMM3 the number 9 is inserted in the code number where a feature of an item is not listed in the work classification.

Special terms used in a bill of quantities

There are a number of special terms used in a bill of quantities, as follows:

- **Preliminaries** (SMM7) or **general items** (CESSM3) are items in a separate first section of the bill which are not related to quantity. Preliminaries/general items will include such things as setting up the site, welfare facilities, security and temporary works. This section of the bill is usually comprehensive and can easily add a cost of thirty per cent or so to the measured items in the bill. In CESSM3 contracts the first section of the bill is made of similar 'preliminary' items as SMM7 together with items called **method-related charges**. Method-related charges are inserted by the contractor and are separated into fixed and time-related charges. An example might be 'set up and remove tower crane, £10,000' as a fixed charge and 'operation of tower crane 20 weeks at £5,000 per week', as a time-related charge.
- **Daywork** is minor work instructed by the contract administrator to be carried out on a time-charged **basis at the cost** of labour, plant and materials used. An allowance as a **provisional sum** for daywork is inserted in the first section of the bill. On a project costing say a million pounds, an allowance of £25,000

would be reasonable, and this represents the **total estimated costs** of all daywork. This sum actually works a little as a **contingency sum**, but is not intended as such. Daywork should be used for minor items of incidental work. It is relatively easy to administer, but there is no incentive for contractor efficiency since the contractor is paid for the cost of doing the work, whatever that cost might be. Daywork is effectively a small version of the cost-reimbursable payment mechanism described in Chapter 14.

- **Provisional sums** are inserted in the first part of the bill by the design organisation or quantity surveyor to represent items that cannot as yet be described fully, and will be described later, after the contract has been awarded, on revised drawings, specifications or other documents. So, for example, if the full scope and extent of the final landscaping work on a project has not yet been determined at the tender stage, a provisional sum of perhaps £20,000 could be inserted in the bill by the design organisation. Provisional sums are used or not used at the discretion of the contract administrator. Of course, before a provisional sum can be used the work has to be properly defined, and sometimes this simply cannot be done at tender stage.

  A major drawback with provisional sums is that they are inserted by the design organisation in the tender BoQ. Therefore tenderers do not price them, so there is no competitive element, and an actual price for the work has to be agreed when the details are issued to the contractor. The other problem is that the contractor cannot realistically allow for the work in the contract programme until details are issued. This can then lead to disagreements over whether the work to carry out the provisional sum can be achieved within the contract period. Finally, if there are extensive provisional sums that are not used, the contractor may make a claim for loss of profit on these items.

- **Prime cost sums** (sometimes called PC sums) were inserted in the first part of the bill by the design organisation to represent items that would be carried out by nominated contractors (these were other contractors or suppliers named by the employer *after* tender acceptance). Nominated sub-contractors are no longer a feature of many modern contracts, but may still appear where the JCT 98 series of contracts is used.

- **Contingency sums** are sometimes inserted by the design organisation or quantity surveyor. Their use is generally discouraged by employers since they really represent a guess by the designer at the possible cost of all the items that have not been properly finalised and whose cost might increase, possibly with a general additional allowance for risk. The idea is to have a total tender price that represents the tendered price for the work itself, with an allowance for risk, so that the employer does not get a shock later if the outturn price proves to be more expensive than the tender price and can hence budget more effectively. Today much more sophisticated methods are used, such as pre-tender risk registers. Here the estimated cost and likelihood of events which are employer risk (such as exceptional adverse weather) are assessed so that the employer can make a financial provision for them but outside the contract itself.

Determining the actual price to be paid

In the rare instances where no employer risk events occur and the tender quantities are absolutely correct, the employer may actually pay the tendered price for the contract. Unfortunately such instances are quite rare, and there is often an increase in contract cost from 'tender to outturn' as it is known. This increase is often five per cent or so, and a prudent design organisation will advise the employer of this issue, and suggest a contingency budget, for a possible 'overspend' as the employer will see it. On contracts that go seriously wrong, an increase of over 30 per cent is not unknown. Good site investigation, a thorough and competent design, good communication and comprehensive contract documents all reduce this risk of overspend. These are all actions that this book advocates.

The final price paid by the employer is typically made up of a number of regular payments. It is possible to have a lump sum contract, when the total price is paid on completion, but this payment mechanism is generally only applied to smaller and less complex projects, because it increases the contractor's financing costs which will probably be reflected in an increased tender price.

Regular payments to the contractor

Under most forms of contract with a regular payment system, the contractor is paid for work actually completed in the agreed payment period. This period is almost always one month. So if the length of pipe of a specified diameter and depth is 50 metres at a rate of £30 per metre, and the contractor has laid 20 metres that month, the payment is £600 (20 metres multiplied by the rate of £30). This regular payment process is generally called a 'valuation'. The contractor submits a list of the quantities that are claimed to have been completed, and these are multiplied by the rates to give a total price to date. The previous payment is then deducted to give the net payment due to the contractor. It is prudent for the contractor and contract administrator to actually agree these quantities before formal submission of the monthly valuation. There may also be provisions for 'goods off-site' and retentions, which are outside the scope of this chapter.

NEC3 ECC calls this valuation the 'price for work done to date'. Unlike traditional contracts such as ICE7 and JCT SBC05, the NEC3 ECC contract has no rules or procedures relating to how this valuation is actually carried out. JCT SBC05 for example has procedures in clause 4.11 for the quantity surveyor (representing the employer) to carry out 'interim valuations' and the architect/contract administrator to issue 'interim certificates' setting out what is to be paid by the employer. There is also a procedure in clause 2.12 for the contractor to submit an application for an interim certificate. The contractor's application describes what the contractor considers is due.

Loss and expense claims add to the final price

JCT also has extensive requirements for a **final certificate** which are absent from NEC3 ECC. The final certificate sets out the final payment to be made to the contractor after completion of both the contract work itself and the rectification period, which is often a further year for the correction of any defects. The final certificate is issued when all measurement errors have been resolved and all claims for financial **loss and expense** have been determined.

The price paid by the employer will usually include additional items for 'loss and expense' claims (compensation events in NEC3 ECC).

The price to be paid at the end of any payment period will also include any agreed claims for loss and expense. These costs should be paid on a regular basis as soon as they are agreed, and not at the end of the contract work. Loss and expense is the term used in clause 4.23 of JCT SBC05 and relates to agreed 'relevant matters'. Relevant matters are events listed in clause 4.24 for which the employer pays an additional cost. They include items such as changes (variations), instructions by the contract administrator and acts or omissions of the employer. In JCT SBC05 the extra time for completion that may arise for these events is covered separately in clause 2.29 for what are called 'relevant events'. NEC3 ECC uses the term 'compensation events' for such items. There are 19 compensation events listed in clause 60.1 of NEC3 ECC, and the options B and D add more which are specifically related to contracts using a bill of quantities, which are discussed below.

'Measurement claims' add to the final price

'Measurement claims' is the name usually given to claims (justified under the contract) for various types of measurement error (by the employer or design organisation) in contracts using bills of quantities. These measurement claims are included in the relevant matters (JCT SBC05) or compensation events (NEC3 ECC in options B and D).

Measurement claims fall into two main categories:

1. Items in the bill of quantities whose descriptions or measurement do not conform to the standard method of measurement (unless this non-conformity is described in the preamble).
2. Items where the *quantity* of work actually carried out is different from the quantity shown in the bill of quantities.

Category 1 claims are covered in NEC3 ECC clause 60.6 which refers to corrections to the BoQ which are departures from the standard method of measurement specified in the contract. This is a compensation event which can entitle the contractor to extra cost and more time to complete the project. Clause 60.6 points out that such correction may lead to reduced prices. Typically, NEC3 ECC is very brief on the matter.

JCT SBC05 'with quantities' covers the issue in clause 2.14, which deals with errors and discrepancies in general, and clause 2.15, which deals with the procedure for contractor notification. Notification under clause 2.15 generates a relevant matter, which is discussed above. The JCT clause 2.14 is quite long because it has to cover errors in the contractor's design portion as well. JCT SBC05 is also available in a 'without quantities' and 'with approximate' quantities form.

Category 2 claims exist because it is often cheaper to buy materials or carry out work in bulk. The contractor will make such assumptions when pricing the tender bill of quantities. Most priced quantities also include an element for overheads such as supervision or possibly specialist plant. Let us take the example of an item for repairing defective areas of the soffit of a service reservoir roof slab. It is difficult to be precise about the real quantities until the reservoir is taken out of service and drained. So an estimated quantity may be inserted in the BoQ by the designer. The work requires scaffolding access. A quantity of 200 square metres is inserted in the BoQ by the designer and the tenderer calculates a rate of £20 per square metre for labour, materials and profit, and also includes the cost of the scaffolding on the assumption of five-days' work. Now suppose the actual quantity is 80 square metres, which is two-days' work. The proportionate cost of the scaffolding is much higher, because the setup/take down cost is constant, but is now spread over just 80 square metres rather than 200.

A very similar 'overhead' argument applies to much excavation or lifting work, because of the cost of the time taken to track a machine or crane to the area of work, to set it up safely and correctly, and then to remove it. These costs are the same regardless of the quantity of work involved and will normally be added into the rate on the assumption that the quantities in the tender BoQ are correct.

JCT SBC05 'with quantities' uses clause 2.14 in the same way as a category 1 claim for loss and expense. However NEC3 ECC is quite specific on incorrect quantities in clause 60.4 which is found in options B and D. This clause only makes incorrect quantities a compensation event if 'the rate in the bill of quantities for the item multiplied by the final quantity of work done is more than 0.5 per cent of the total of the prices at the contract date.' In other words, the correction to the quantity has to be significant. NEC3 ECC also makes increased quantities a compensation event if this delays completion. This is a sensible provision. An increase in the quantity of an item on the critical path is very likely to delay completion unless extra resources can be brought in by the contractor and if they are available in the time frame required.

The different treatment of category 2 claims in JCT SBC05 and civil engineering forms of contract such as NEC3 ECC arises from the basic contractual assumption of these contracts. JCT SBC05 is a lump sum contract and 'assumes' that the drawings are complete and the quantities shown in the tender documents are correct. Civil engineering by its nature is more difficult to define absolutely. This is because most civil engineering contracts have a significant element of work in the ground, which is variable despite the best efforts at site investigation. Civil engineering contracts are also more susceptible to the effects of weather and external events.

The final price to be paid

In summary, the final price that the employer pays will usually be different from the tendered price. A good design organisation will advise an inexperienced employer of this. The final price is often called the 'outturn price' and many experienced employers monitor their 'tender to outturn'. This is often plus five per cent, but in bad examples could be much more. As we said earlier in this chapter, thorough site investigation, good design, accurate drawings and bills of quantities, and effective contract administration all reduce the tender to outturn percentage, which will please the employer. Some of the issues that contribute to this tender to outturn shift are:

- Inaccurate initial quantities, corrected on remeasurement
- Changes and additions
- Measurement claims
- Claims for loss and expense
- Compensation events (NEC3 ECC).

The status of the bill of quantities

Although JCT SBC05 and NEC3 EEC treat the bill of quantities in a similar way there are differences as explained below.

JCT contracts such as JCT SBC05

JCT contracts such as JCT SBC05 are actually lump sum contracts, but with the provision for regular payments (usually monthly). This lump sum is called the 'contract sum' and is inserted in Article 2 by the tenderer. Article 2 states that 'the employer shall pay the contractor at the times and in the manner specified in the conditions the sum of … *here the tenderer inserts an amount.*' There is another sentence in Article 2 which adds 'or such other sum as shall become payable under the contract' to allow for the contract sum to be changed. Later in the contract are provisions to add various items to the contract sum, such as the cost of variations or claims for loss and expense.

Hence whilst JCT SBC05 is technically a lump sum contract it actually operates with a lump sum that can be changed. It also makes provision for regular payments as the work proceeds. In JCT SBC05 the quantities are assumed to be correct and comprehensive. Whereas in the nearest civil engineering version, ICE7, they are only expected to be 'estimated' and both parties to the contract know that they are likely to change.

JCT SBC05 assumes that a standard method of measurement is used and here of course it is SMM7. The relevant clause in JCT SBC05 is 2.13.1 which requires the use of a standard method of measurement which is defined as SMM7 in the definitions section. Again, the contractor will have 'measurement claims' for unstated departures from SMM7 and also for differences between tendered and actual quantities. Clause 2.14 requires that all errors in quantity or omissions are corrected. Such corrections generate a potential claim for loss and expense by the contractor.

NEC3 ECC

NEC3 ECC options B and D use a bill of quantities as the pricing/payment mechanism in option B and as a pricing mechanism in option D. Again a standard method of measurement is assumed and it has to be named specifically in the contract data provided by the employer. It is likely to be CESSM3. Again, inconsistencies or unstated departures from the standard method of measurement are 'measurement claims'. NEC3 ECC uses the term 'compensation event' of course. Differences between tendered quantities and actual quantities are treated differently in NEC3 ECC in comparison with the treatment in JCT SBC05. In NEC3 ECC, such a difference is a compensation event if it delays completion. The other way of generating a compensation event is the '0.5 per cent rule' described above.

Chapter summary

- Most contracts have an outturn cost higher than the tender price. This is due to the impact of compensation events and relevant matters.
- Where a pricing/payment mechanism employs a bill of quantities (BoQ), this BoQ is almost always produced using a standard method of measurement (SMM).

- The purpose of a standard method of measurement is to ensure that everyone understands what every item of work includes and hence the price inserted is not open to later disagreement as to what work it covers.
- The two main standard methods of measurement are CESMM3 and SMM7.
- Unspecified departures from the standard method of measurement and incorrect quantities are usually grounds for additional payment to the contractor.
- The BoQ can be used for an initial estimate of the contract price as a pricing mechanism.
- The main purpose of the BoQ is for tenderers to complete it to produce their priced bids.
- The BoQ is then normally used as a payment mechanism for determining regular payments to the contractor. Such payments are usually monthly.
- When the contract is completed and the rectification period is completed the BoQ with final quantities forms the basis of the final account.

16

Time and programmes

Aim

This chapter aims to develop an understanding of the importance of time in construction contracts, and how it is related to programmes.

Learning outcomes

On completion of this chapter you will be able to:

>> Describe the uses and types of programmes in the construction industry.
>> Define some common programme terms.
>> Describe the role of the project manager.
>> Determine items that a good programme should cover.
>> Check a construction programme to ensure it is realistic.
>> Differentiate between delay, disruption and prolongation and explain their contractual effect.
>> Have an appreciation of the drawbacks with programmes.

> **Important note**
> This chapter is an introduction to time and programme principles. Contract clauses dealing with the *contractual implications* of these issues can be found in the chapters on NEC3 ECC and JCT SBC05 contracts (Chapters 21 to 27).

Time and programmes

Time is of fundamental importance in all contracts, during their preparation, design and construction. Construction professionals plan, manage, monitor and coordinate activities by means of programmes, and thus bring time under control.

> 'But at my back I always hear
> Time's wingèd Chariot hurrying near'

So wrote Andrew Marvell almost 400 years ago in the poem 'To his coy mistress'. It sums up the attitude of the employer, the designer and the contractor in construction contracts, as if it had been written today.

The employer will be waiting for delivery (completion of construction and handover) of the new asset that the employer is paying for. There may be tenants, operators or employees who are waiting to move in. Or there may be legislation to meet, new competitive pressures to address, plans to increase efficiency, reduce risks or any other of the many needs that the construction project is intended to meet. Hence prompt delivery of the asset (contract completion to us) is usually very important to the employer. For all sorts of reasons, which this chapter will explain, few construction projects are completed on time. This is an issue that many employers find hard to understand.

For the contractor, the faster the work can be completed, the smaller will be the *total* of the weekly site overhead cost. Site overheads occur even when little or no work is taking place. They will include such items as fencing and security, lighting, office accommodation and welfare facilities, small tools and equipment, telephones, services and supervision. At contract completion, the contractor will be able to deploy labour, plant and supervision to other contracts (to make more profit from them). Early completion will mean that there is no likelihood of the contractor paying legal damages (the usual contractual form being liquidated damages), and a happy employer can lead to the possibility of further work. Contractors will have another project to add to their reference list and marketing literature. Finally, of course, the sooner the contractor's contractual obligations are over, the sooner the construction risk is over.

Unlike the employer and the contractor, the design organisation has less significant commercial interest in prompt completion, since its main role will probably be as contract administrator, which is of course a contribution to its fee income, albeit fairly small. However, design organisations should always act in a professional manner in the interests of their client, which is usually the employer. Of course, the design organisation will also benefit from a 'job well done', another project for its marketing team, and an increase in its expertise and reputation, and of course a happy employer.

Everyone likes projects that are completed on time and for most projects early completion is even better. Projects which are late cost someone money, and they often lead to poorer team relationships, less job and project satisfaction, and a loss of reputation. They may also lead to a formal dispute. The old expression 'time is money' is very true in construction.

> **TIME** is at the root of almost all construction problems, claims and disputes; being late usually costs the employer or contractor, or both, a lot of money.

It is essential that all contracting parties understand the time available, any constraints on work such as restricted access or working hours, and interfaces and dependencies between work activities. All contracting parties need to understand what they must do to ensure that the contract is completed as expeditiously and efficiently as possible.

Depending on the contract strategy used, contracting parties will include the employer, designer, contractor and suppliers. It is also likely to include numerous sub-contractors, manufacturers, specialists and other professionals. Coordinating this diverse range of companies and individuals and keeping them to schedule is a major and challenging task. This task is the role of the project manager (and during construction the contractor's staff) and the main tool will be the Master Project Programme, described below. It is through careful production and control of these programmes that successful projects will keep to time, and so avoid the costs and frustrations that delays will inevitably cause.

How contracts deal with time issues

Most contracts have lists of events that entitle the contractor to more time or more money or both. In JCT contracts these events are called relevant matters (money related) and relevant events (time related). Most events are listed as relevant matters *and* relevant events, but some are not (these are often called 'neutral events'). NEC3 ECC contracts use the term 'compensation events'. In NEC3 ECC all compensation events entitle the contractor to 'time and money' if the contractor can justify entitlement. What all contracts are trying to do is to predict the main events that may occur, and allocate the risk of these events to the contractor or the employer (or sometimes to both). So in most contracts the contractor is reimbursed for any changes or delays to access (employer risk). Contractors are not usually reimbursed if they have miscalculated activity durations or if they are inefficient or have problems with personnel or labour (these items are contractor risk). Many traditional contracts (but not NEC3 ECC) treat exceptional weather conditions as a 'neutral risk'. Here the contractor may get additional time, but the employer does not pay any additional costs, so in some ways this risk is shared because both parties suffer in different ways. All major forms of contract treat matters relating to time with great care. A full explanation of this treatment in JCT SBC05 and NEC3 ECC contracts will be found in later chapters.

Commencement, progress and completion

Commencement, progress and completion are at the heart of all projects and all contracts. However, different forms of contract treat them differently. Fundamentally, the contractor must know when the contract work has to be completed and when work can start on-site. Hence many contracts set a completion date and a date when the contractor has access to site, to commence construction work. The JCT forms of contract and NEC3 ECC work in this way. The access date (called 'possession' in JCT) is not the same as the date when the contract is signed, and may be many weeks later depending on arrangements for land access or purchase. The ICE7 contract works by setting a date for commencement of the contract (possession of site may be a later date) and a contract period. Adding the two together gives the completion date in ICE7.

If a contract does not set a completion date, then legally 'time is at large' and a court would have to set a date that was reasonable in all the circumstances. It is usual for the employer (on the advice of the designer or project manager) to set the access and completion dates in JCT and NEC3 EEC by inserting them in the contract particulars. Rarely, the contractor is asked to insert the completion date in the contract particulars which they provide. The advantage here is that the completion date can really suit the contractor's resources, workload and programme. The big drawbacks are that the contractor's proposed completion date may not suit the employer, and different dates inserted by tenderers can make tender comparison difficult.

Completion dates have to be considered with great care and must not result in too short a construction period. This could be inefficient owing to the possibility of having more people and plant working than is productive. More importantly it could be unsafe. CDM regulation 9 makes it a requirement for the employer to allocate sufficient time for the construction work to be carried out, as far as reasonably practicable, without risk to the health and safety of any person.

> The contract completion date must be set with great care taking into account the characteristics of the project and with due regard for health and safety.

Some contracts make progress a requirement by the use of such phrases as 'proceed with due expedition', which is used in ICE7, with a similar provision in JCT SBC05. However more modern forms of contract such as NEC3 ECC are silent on this matter, and presumably assume that proceeding at a suitable rate to meet the completion date is what the contractor will do anyway.

Project managers and contract administrators

The construction industry uses the terms project manager and contract administrator to mean a number of different things. Because of this rather confusing terminology they may sometimes be the same person fulfilling two *different roles.*

So this chapter uses the expression **project manager** in its more usual sense as the person responsible for the overall direction, coordination and delivery of the project from an early stage after concept to completion. The project manager should draw up the Master Project Programme, and coordinate all activities in the 'pre-design', design and construction stages. The project manager's prime concern is the effective delivery of the project, on time, to budget, to quality criteria and satisfying the employer's needs.

We shall use the expression **contract administrator** to denote the person who administers the contract (between the employer and the contractor), according to its contractual terms. Most construction contracts are between the employer and the contractor; they are the parties to the contract. Because the employer is usually a company or business, often with little understanding of construction, most contracts allocate an administrative and payment role to a *named person,* the contract administrator (CA). The role of the contract administrator is very demanding. The easier part of the CA role is the routine administration of contract terms, such as payment. However, the contract administrator is usually empowered to order changes and, when unexpected events occur, the CA has to make a *judgment* about the contractor's entitlement to payment or extra time *under the terms of the contract*. This judgment can be over events costing thousands to millions of pounds. The chapters on specific forms of contract will cover this issue in more detail. The CA's judgment determines who pays for these eventualities: the contractor or the employer.

The contract administrator is usually a qualified engineer, architect or building surveyor and is usually named personally in the contract. Depending on the form of contract used, he or she may be called the 'Project Manager', 'Engineer' or 'Architect/Contract Administrator'. Occasionally a company will be named in the contract (such as Jones, Smith and Partners) but where this is done, a post-contract award letter should really be sent to the contractor naming a specific person from this company to act as contract administrator. The CA is usually a member of the organisation that designed the project and for larger or more complex projects would often be a partner or director. For smaller projects the CA would often be a less senior person and that is why it is essential that construction professionals understand how construction contracts work, and how they are administered. The choice of title 'Project Manager' in NEC3 ECC is unfortunate, because the NEC3 ECC role is primarily that of a contract administrator although the title suggests otherwise. The contract administrator's prime concern is to ensure that the contract terms are adhered to, that proper payment is made and that all provisions of the contract are met in full. Of course, the contract administrator would also like a contract delivered on time and to budget, but if events entitle the contractor to more time and money, under the contract, then the CA must certify both properly, even though this action will cause cost and time overruns for the employer.

It is possible for one person to fulfil both roles and this is quite often the case, and of course does save time and effort in liaison between the project manager and the contract administrator! What is particularly important where the contract administrator and project manager is the same person is for that person to separate mentally the two roles: or to 'wear two hats'. So the contract administrator role must be carried out fairly and impartially *according to the contract* and without favouring the employer. All decisions related to the contract itself should be made clearly as contract administrator. At other times the person may be reporting progress and liaising with the employer and advising the employer in the project manager role.

The project manager's coordination role

The project manager has a very important overall coordination role, which should include everyone involved in the investment process. The project manager should therefore ensure that the **Master Project Programme** (described below) is comprehensive, and should allocate enough time to ensure good coordination and communication with everyone with a role to play in the process. Much of the early work of the project manager is with the employer, who is not usually familiar with construction, and the employer's role in a successful project. At this early stage, the project manager must truly understand the needs, constraints and requirements of the employer and ensure that this essential interface is fully integrated into the Master Project Programme.

Ultimately, the project manager should ensure the following are shown on the Master Project Programme:

- Health and safety considerations and communication are planned in to all stages of the investment process.
- There is a comprehensive design programme.
- That design is progressing to this programme and that the employer is properly consulted on all issues which affect the employer.
- There is a suitable tendering process and tendering option, taking account of all relevant legislation, particularly the Public Contracts Regulations or Utilities Contracts Regulations where they apply.
- A *realistic* date for construction completion is set. This date should allow for the *time taken to invite and accept the tender,* and the time taken to give access to the site. This time period for tender invitation and acceptance is usually a number of weeks. On large or complex projects it can be a number of months. A further time addition must be made for arranging access to the site and all the necessary statutory approvals. This time addition can be anything from a week to many weeks, depending on project circumstances, particularly the ownership of land.
- The employer understands the implications of imposing too short a date for completion. These implications are usually additional costs arising from inefficient working. However, more importantly, the employer has a duty under safety regulations to ensure that sufficient time is allowed for the construction work to be carried out safely (CDM 2007, regulation 9).
- The employer provides the land, access and information in his or her possession in good time.
- All service diversions, power and water supplies are properly arranged and programmed.
- All interfaces and constraints on the contractor's freedom to programme are properly defined – usually by the designer (in NEC3 ECC contracts, these are detailed in the works information).
- The contractor's programme is agreed and any implications for the employer and designer are understood by them.
- Sufficient design information is available for the contractor to procure materials, start work and carry out construction to the contractor's programme.
- All key dates are shown on the contractor's programme.
- The employer's staff (these could be operational staff or employees) understand their role and impact on the programme and do not delay it.
- The contractor's programme is updated as necessary and assessed fully for any new implications for the employer.
- The contract is administered properly in accordance with its terms, by the contract administrator. This will particularly include:
  - prompt and correct payment
  - regular programme updates
  - correct assessment of claims (called compensation events in NEC3 ECC)
  - proper extensions of time are given when necessary
  - all certificates are given according to the contract
  - defects are corrected.

- The project manager will need to ensure that a programme for training, handover and occupation is agreed and that the employer understands the part that the employer's staff plays in this.
- Finally a programme for the correction of any defects should be agreed.

Programmes

A programme is the result of the planning process. A programme shows the sequence of significant activities, their durations and interrelationship required to plan, design or construct a project.

1. The **Master Project Programme** (MPP) is used to plan and coordinate the whole project from concept to completion. It encompasses the design and construction programmes in 2 below.
2. The design programme and the construction programme are used for planning, controlling, coordinating, resourcing and monitoring either design or construction activities. This is primarily the concern of the designer or the contractor, but the project manager will have an interest in this aspect as well because it affects the whole project.
3. The construction programme is also used to determine contractual entitlement to delay, extra time, extra costs and expenses. This second use of the construction programme is the main focus of this chapter. The people primarily concerned in this respect are the contractor and the contract administrator.

So the MPP not only shows design and construction but all other essential activities from 'concept to completion'. The designer will concentrate on the design programme, and the contractor will concentrate on the construction programme, but it is the MPP that pulls the whole project together. The project manager should prepare the MPP although, until a construction contract is awarded, the part representing the construction programme can only be an estimate.

The **Master Project Programme** includes the design and construction programmes together with all other activities which are essential for the successful delivery of the project.

Comprehensive programmes are essential to all construction projects. However, they should be seen as living documents for they are never 'set in stone'. Initial programmes at the concept stage may only contain a few major activities since, at this early stage, little detail is known. So there may be one activity line for design as six months, and one for construction as twelve months. As the project progresses, more and more detail will become known and can be shown on the programme, until it eventually becomes the comprehensive programme that is so vital for success. The programme should be reviewed at regular meetings of the project participants, and updated to show actual progress against activities. It should be revised when the activities or logic change sufficiently to warrant a programme reissue.

Programmes come in a variety of types, from simple bar charts of activities, to complex critical path networks showing hundreds of activities, with their logical interaction. These logical interactions are usually called 'dependencies'. A further refinement is to allocate resources to each activity, so that resource estimates can be produced and resources can be levelled or optimised. Scenario planning can also be carried out on a good programme, to test the effect of possible changes or events.

The Master Project Programme in detail

It is important to remember that the Master Project Programme allows the *whole project* to be planned and coordinated. It should include not just the design and construction programmes, but also all other relevant stages of the investment process. It is imperative that all links from and to these stages are shown on the design programme and, where affected, the construction programme. One of the most frequent causes of delay to projects is failure to show these links and allow for the time taken for statutory approvals such as planning consent or permissions under the Traffic Management Act. These approvals often require extensive work such as the preparation of reports or drawings. Finally, the MPP must allow for any time periods for employer gateway approval, which may take days or sometimes weeks before the project can proceed to the next stage.

In addition to incorporating the design programme and the construction programme, the Master Project Programme should show:

- The feasibility and option selection stages
- All 'employer approval gateways', allowing sufficient time for these
- Site investigation in all its forms
- Any statutory or regulatory approvals such as planning consent with adequate time allowed
- All external constraints (such as diversions, power supplies, land acquisition, way leaves and consents), and the programme for resolving them
- A time allowance for public consultation if appropriate
- Stakeholder consultation, communication and management
- Sufficient time for the employer to consider tenders and carry out the legal steps to formally accept a contract
- A programme for the manufacture of any items supplied to the contractor
- Links showing information transmission between the various parties involved in the project
- All key dates or 'milestones'
- A handover and training programme
- Plans for the resolution of any defects after completion
- Adjustment and optimisation of the asset if required by the employer.

At an early stage of a project, the activities shown on the MPP will be approximate estimates. At this point, some major sections of the programme such as construction may only be shown as a single activity line on the MPP. As the design of the project progresses, more detail can be included and, finally, after contract award, the contractor's actual construction programme can be included.

The design programme

The designer should produce a design programme, so that design work can be planned, monitored and carried out efficiently within the required timescales. In the author's experience it is apparent that many designers are less assiduous in producing programmes than their contracting colleagues. Lack of proper design programmes can easily lead to cost and time overruns for the *whole* project, not just the design elements.

In larger design organisations it is necessary to distribute staff and other resources across many projects. Without a comprehensive suite of design programmes, showing critical activities and resource needs, effective allocation of staff is impossible. Managers of design organisations will also use **resourced** design programmes (with resources shown against each activity) as a way of monitoring design costs against fees earned, by comparing design progress with fee allocation. Modern programming software can analyse changes to these resources and, if required, can smooth or level resources across projects to reflect the total staffing available.

Resources for a design organisation would be categories of staff such as architects, designers, CAD technicians, quantity surveyors, specialist staff and so on, together with supervisory or management staff. Without a comprehensive suite of interrelated design programmes it is virtually impossible for a design organisation to allocate staff resources across different projects, in response to the project with the greatest need.

A good design programme will include:

- Any preliminary design or option evaluation
- All important design activities with durations
- People and resources allocated to those activities
- Adequate allowance for risk assessments and compliance with CDM
- Any specialist inputs such as structural design, hydraulic or process design
- Time allowance for the preparation of any reports or other documents needed for approvals
- Any preliminary design or drawings required for land purchase or way leaves
- Layout and elevation drawings which are required for planning consents
- Environmental assessments which have an impact on design
- Production of all necessary tender drawings (this is a major and time-consuming activity)
- Specification preparation and quality management processes
- Discussions with any specialist suppliers or manufacturers
- Bills of quantities or other measurement activities
- Cost estimating
- Amendments (if any) to standard conditions of contract
- Tender document preparation
- Tender period
- Tender assessment and recommendation
- Proper regard to health and safety for all the above activities
- *Finally, the design programme should have links into the MPP, wherever these links affect design activities.*

As can be seen from the above list, design itself forms quite a small part of the time and resources needed to complete the activities in the design programme.

The construction programme

All construction projects, however small, should have a construction programme. This is essential to plan and sequence the work effectively, and ensure completion by the contract completion date. Because construction programmes are essential and often very complex, many contractors employ professionals to produce them. These professionals are usually called 'planners' or 'planning engineers'. They will usually produce any updated programmes as well, and will often give advice on the delay consequences of changes or other proposals. Thanks to the power of programming software, sophisticated 'what-ifs?' (scenario planning) can be undertaken, to test the time and cost implications of alternative courses of action.

A construction programme is in fact *required* by most standard forms of contract. The contractor will usually produce a preliminary construction programme while preparing the tender, in order to really understand the activities, links and constraints and so be able to price them properly and evaluate the risks involved. The employer sometimes requires a preliminary construction programme with returned tenders, especially if the employer has to provide items during construction to match the contractor's programme requirements. Such items might include manufactured goods or special designs or complex access and interface arrangements.

Whether or not a preliminary construction programme is prepared for the tender, one is always produced once the contract is awarded. The construction programme is prepared by the contractor, and submitted to the contract administrator for acceptance, usually a short time after contract award. It should be updated as necessary as the construction work proceeds to reflect changes and events. Most forms of contract have clauses requiring initial construction programmes, and updated programmes at intervals as necessary.

Whilst the construction programme is prepared by the contractor for its own purposes, it is also very useful to the contract administrator and the employer as the list below demonstrates.

Some of the purposes of a construction programme are:

- To ensure that the contractor has planned and programmed the works thoroughly.
- To allow the contract administrator to assess whether the contractor has fully understood all obligations, and any constraints.
- To enable the contract administrator to prompt the employer with any requirements for land, access and consents.
- To enable the employer to plan funding and likely payment flow.
- To ensure that the designer and contractor carry out CDM risk assessments on time.
- To enable the CDM coordinator to coordinate and facilitate communication on safety matters.
- To assist the contract administrator in programming design requirements so as not to delay the contractor.
- To assist the contract administrator in programming site supervisory staff and any specialist supervision that may be needed.

- To enable the contractor to manage and coordinate the work.
- To enable the contractor and contract administrator to monitor progress.
- To enable the contractor to predict and monitor cash flow and finance requirements.
- To enable materials and manufactured items to be produced and delivered when required.
- To enable plant requirements to be predicted and organised.
- To coordinate the work of sub-contractors.
- To enable assessments to be made of the effect of delays or additional works.
- To know when to inform the public of any significant events.
- To programme essential training and handover at the end of a project.

Some programming terms and concepts

This is not a chapter on programming techniques. Some professionals specialise in project planning, and spend their careers in this field. However, all construction professionals need to have some understanding of how programmes work, and in particular the use of programmes in evaluating contract issues. All major forms of contract have clauses which require construction programmes to be produced by the contractor, and many forms expect them to be approved by the contract administrator. The construction programme is primarily used by the contractor to plan, control and monitor construction activities. However, for our purposes, the construction programme, updated as necessary, is the main tool used in the evaluation of delay claims.

A simple programme is a bar chart showing activities and their durations. 'Dependencies' or logic links are normally added to these programmes to show the interrelationship between activities by making connections between the starts and finishes of activities. Further sophistication is the use of **critical path analysis**. This calculates **float** of various kinds so that the amount that activities can be delayed or prolonged without adverse effect on other activities can be assessed. Importantly, it produces a **critical path** of activities. It is an understanding of this critical path that is necessary in making and evaluating claims for delay and extra time. Float is quite a complicated concept and is explained further in the section below.

Looking at Figure 16.1, we see a programme for the construction of a garden wall, one brick thick, 12 metres long, 2 metres high, with a gate, and a coping (a capping stone on top of the wall). The terms used in this programme are explained in the next section.

Some programming terms defined

The terms used below are written in 'normal English' and refer to Figure 16.1. Most forms of contract define these terms and use a slightly different terminology that is particular to that form of contract. Hence 'access' may be called 'possession' in some forms of contract such as JCT and ICE7. Similarly there are a number of different terms for 'completion date'.

- The *activities* are to construct the foundations, brickwork, copings and so on.
- The *durations* are shown in days, thus – 2d, 3d, 5d.

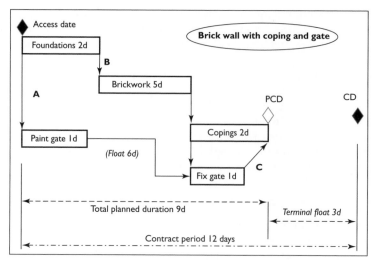

Figure 16.1   Programming terms.

- The total time available is the *contract period*, here it is 12 days.
- The *access date* is when the contractor can start work on-site.
- The *completion date* (CD) is the date when the contractor must finish under the contract terms.
- The contractor can finish earlier and here plans to finish 3 days early on the PCD (planned completion date). These 3 days are called *terminal float* – more on this below. The 3 days represent the contractor's 'risk allowance' in case things go wrong, or activities take longer than the planner thought.
- Connections A, B and C are called *logic links* or *dependencies* and show the logical sequence of activities.
  - A is a start-to-start connection (both activities can start at the same time).
  - B is a finish-to-start connection (one activity needs to finish before the next can start). This is the most used connection.
  - C is a finish-to-finish connection (both activities must finish before the next one can begin, or as is shown here, before a key date is achieved).
  - The start-to-finish connection is not shown. They are little used in practice.
- Looking at the 'fix gate' activity, the *earliest start* is day 7 (2d + 5d) and the *latest start* is day 8, to finish by the PCD.
- The **critical path** is the path through the activities giving the earliest completion date.

The critical path

The critical path is of prime importance in determining contractual claims, and the associated costs and extensions of time that may result. The critical path is the sequence of linked activities through a programme which determines the overall project duration. Many programmes have more than one critical path, each showing a different essential sequence of activities leading to the same planned completion date.

If *any* activity on any critical path is delayed, it also delays the planned completion date. In other words, activities on a critical path have *no float*, so if they are delayed, they also delay the planned completion date. Of course, whether the project itself is delayed will depend on whether the total delays to the critical path use up the terminal float. Once all the terminal float is used up the completion date for the project is delayed. In the critical path is foundations–brickwork–copings. The activities paint gate–fix gate are not critical.

However, determining a *true* critical path can be difficult on a project more complex than the simple example above. A true critical path can vary throughout the contract duration depending on whether non-critical activities use up their **total float** and whether the logic links are correct or need to be reconsidered. Again, to evaluate the delay effects on critical activities, an *up-to-date*, correct and detailed programme is required.

Float

Float is basically a time allowance built into activities or programmes. It is an allowance for delay or slippage, or for activities taking longer than planned. There are a number of types of float of which the main ones are:

- *Free float* is the total time *an activity* can be lengthened or delayed without affecting the *earliest start* of the next activity. So using the assumptions made in this programme, the contractor can paint the gate as soon as site work starts, but the gate cannot be fixed until the wall is finished. Completion of the wall is the earliest start for fixing the gate. This is a *planning assumption* that is a simplification of real possibilities. The *real possibility* is to fix the gate earlier, when the appropriate section of wall is complete. Planning assumptions are usually made to reduce the complexity of programmes, but can generate unrealistic logic links. This is often the case and the existence of planning assumptions should always be considered when evaluating contract claims.

  So with this planning assumption in place, the free float on the gate painting activity is 6 days. This means the gate-painting activity can slip back by up to 6 days without delaying the 'fix gate' activity. Alternatively, the contractor could make a really good job of painting the gate, and take a total duration time of 7 days to do so (though the contractor would no doubt lose money in so doing).
- An activity with zero free float is on the *critical path*.
- *Total float* is the total time an activity can be lengthened or delayed without affecting the earliest finish date of the project. Here, the earliest finish date of the project is the end of day 9. This means that the 'paint-gate' activity has a total float of 7 days, because the 'fix-gate' activity can slip back one day if necessary.

  So unlike free float, total float is dependent on what happens in a sequence of activities, which makes it more complicated. Any activity on a sequence can use up all of the total float, or a number of activities in a sequence could each use up some of it, provided the value of total float is not exceeded.
- *Activity float* (called *time-risk allowance* in NEC3 ECC clause 31.2) is 'spare time' *within an activity*. So if the estimated activity duration for brickwork was actually 4.5 days, and it is shown on the programme as 5 days, then it would

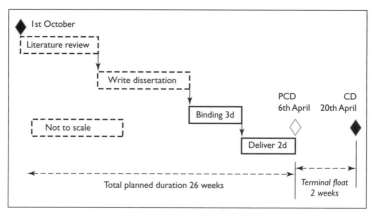

Figure 16.2 The impact of terminal float and time-risk allowance.

have half a day activity float (or 'spare time') within the activity duration shown.

- *Terminal float* is the difference between the contractual completion date and the date when the contractor *intends* to finish (the planned completion date, or PCD in Figure 16.2). So if this garden wall had a contractual completion date of 30 August, where the contractor shows the *planned completion* as the 27 August, then it would have three-days' terminal float.

Terminal float and 'early completion'

Contractors usually produce programmes showing 'early completion'– they are building in terminal float. There is nothing wrong with this, as long as the contract administrator considers the 'shortened' programme to be *achievable*. If the contract administrator has doubts, then the programme must be challenged and the contractor should be asked to provide further justification. The problem with 'over-shortened' programmes is that they can be used as a basis for delay claims. This is because a shortened programme has less 'slack' in it to allow activities to be reprogrammed, or for additional activities (arising from changes) to be added into the programme or for problems to be absorbed within activity durations by using activity float (called time-risk allowance in NEC3 contracts).

If delaying events occur that are the 'contractor's fault', such as late delivery of materials or labour problems, then the contractor has to pay for the time and cost effects of such events. Terminal float is used by the contractor to absorb the time effects without (hopefully) delaying the completion date. However most contracts allow for events that are the employer's risk and compensate the contractor. A typical example is the power of the contract administrator to make changes, typically by issuing revised drawings. Here the employer pays for the time and cost effects of the changes, both for the changes themselves and also their effect on other activities in the programme.

This means that according to the shortened programme, events are more likely to delay the critical path, or produce a new one, thereby increasing the *scope* for contractor claims for financial loss and expense, or for additional time to complete the project. This is known as 'programmanship'. Another technique is to manipulate logic links to show as many critical activities as is possible (or believable). The more critical activities there are, the greater is the possibility of justifying a claim for additional expense for extended overhead costs arising from a delaying event that is the employer's risk. This concept will be explained later.

An example of terminal float and time-risk allowance

Take the example of a university dissertation. Suppose it is allocated to students on 1 October for hand-in on 20 April, and there is a 20 per cent penalty for all work that is submitted up to two-weeks late, and zero marks after that. Only printed and bound coursework is accepted of 10,000 words in total and it has to be handed in, not emailed or posted. In contractual terms the start date is 1 October and the Date for Completion (JCT SBC05) or Completion Date (NEC3 ECC) is 20 April. A programme for this is shown in Figure 16.2. The first two activities would clearly have much more detail in the real programme, but we are focusing on the last two activities.

Suppose the printing and binding centre usually requires about two days; then a wise student will build a small allowance into this activity, making it say three days in case anything goes wrong and delays this activity. Building an allowance *within* this activity is a time-risk allowance but, as we can see, this allowance is not evident on the programme, without further explanation. After binding, there is a further two days allowed in the programme to collect the dissertation and hand-deliver to the university administration office.

To allow for any eventualities during planning and writing the dissertation, a wise student would perhaps build an additional two weeks into the total programme time of 28 weeks. This would be done at the start when the programme is drawn up. If there are no problems and all the activity durations are correct then the dissertation will be completed two weeks early. This is most unlikely however, and the two weeks will provide a 'buffer' for all the delays and other problems that usually occur when writing dissertations.

In other words, instead of planning to use up the whole 'contract period' of 28 weeks, the student would initially plan to use up only 26 weeks, leaving two weeks as *terminal float* such that the planned completion date (PCD) is 6 April. So in this example the student's programme would show delivery to the binding centre 5 days before 6 April, which is 1 April.

Now suppose the university rules are changed such that the penalty for a late submission after 20 April is a mark of zero. A wise student will now build in *more terminal float*, perhaps three or four weeks, to reflect the increased 'risk'. This is exactly what a contractor does when producing a programme. The greater the risk (such as high delay damages), the more terminal float the contractor will wish to allow.

Another example would be deciding on the time to get up in the morning when catching a train or aeroplane. If you were going by train to London for a sightseeing trip, and trains ran every half hour, then missing one is not very important, so you may build in 20-minutes' terminal float when deciding what time to set your alarm to. Of course, you work backwards from the train departure time. However, in deciding what time to set your alarm for your annual holiday from an airport fifty miles away, where being late will cause you a day's delay and extra expense, you will build in a lot more terminal float. It is all about assessing risk, and then making a sensible time allowance for it.

Terms used to describe the impact of events on programmes

There are three main terms used to describe the effect of events on programmes. These terms are delay, prolongation and disruption. These three words are very important, but have different meanings and contractual consequences.

- **Delay** is when an activity cannot start at its intended time, and so is late starting.
- **Prolongation** is when an activity's *duration* is extended. In other words the activity takes longer to complete.
- **Disruption** is where an activity is adversely affected usually by another quite separate activity or some external event.

Delay, prolongation and disruption have very important contractual effects. They describe the result of an event that will cause the completion to be later than planned of

(1) one activity
(2) a sequence of connected activities
(3) the whole project, if the event affects the critical path.

The result of items (1) and (2) may be additional cost directly related to the activities in question. An effect on the critical path, item (3), is much more serious contractually, because it will not only produce the increase in activity cost of items (1) and (2) but also the *cost* of *delay to the planned completion*. This delay to planned completion will mean that the contractor is on-site longer than was intended. The financial effect of this will normally be additional site overhead costs for every day or week of overrun. Additionally, item (3) may entitle the contractor to more time to complete the project. However, whether the completion date is extended or not depends on the form of contract used and whether there is any terminal float available. Delay, prolongation and disruption are shown in Figure 16.3.

Most contracts have clauses which require the contract administrator to grant more time in which the contractor may complete the project. People accustomed to ICE contracts call this extra time 'an extension of time' (EOT) for completion of the project using the words of clause 44. However, in JCT SBC05 the architect/contract administrator 'fixes a later completion date' in clause 2.28, and in NEC3 ECC contracts the project manager 'assesses a delay to the completion date' in clause 63.3. The effect is similar but the phrases used depend on the precise wording of the particular contract.

There is an added subtlety however. In more traditional contracts, it is assumed that 'terminal float' will be used up, but in NEC3 ECC contracts additional time is assessed as the delay to the planned completion date and it is then added to the completion date for the project. The procedure for fixing a revised completion date is an important contractual concept and will be explained further in the chapters on NEC3 ECC and JCT SBC05 contracts (Chapters 21 to 27).

Delay to the start of an activity is fairly easy to demonstrate since it will often be a lack of access or materials supplied by the employer. Prolongation is usually

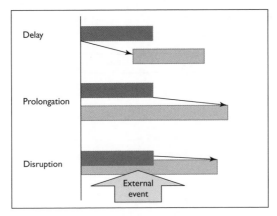

Figure 16.3 Delay, prolongation and disruption.

justified by a comparison between the scope of the original activity, and any revised scope or increase in quantities. Disruption is much harder to prove, and is usually assessed by comparing progress on a 'non-disrupted' activity of similar type; always assuming one exists of course. Unfortunately events on-site often produce delay, prolongation and disruption, and sorting out the consequences and contractual entitlement is never easy.

An example of delay, prolongation and disruption

As an example, imagine two painters painting the walls and ceiling of a lecture theatre in sky blue, using two coats of paint, with activity duration of three days to complete the painting. Just as they have finished laying out the dust sheets, the contract administrator (in this case the architect) arrives and decides to change the colour to light blue. There is a *delay* of one day while they procure the new colour. Three days later they have almost finished when the contract administrator returns, looks at the coverage, and decides that two more coats are necessary. This produces a *prolongation* of three days (the work content has been doubled).

Unfortunately during this three-day period, the electrician (under a different contract) arrives to repair the overhead projector, hung from the centre of the ceiling, and insists that the painters move the dust sheets in 'his area', and change the painting sequence to suit the electrical work. This *disrupts* the painters, and the three-day activity now takes an extra day, *since they cannot work as efficiently*. This reduction in assumed efficiency is the essence of a claim for disruption. These events are shown in Figure 16.4.

So we have
Original activity	3 days
Delay	1 day
Prolongation	3 days
Disruption	1 day
Total	8 days

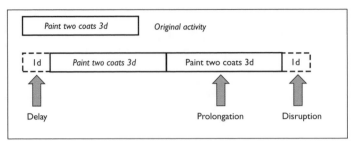

Figure 16.4   Delay, prolongation and disruption.

Under most contracts the above situation will be a 'claim', or a 'compensation event' as it is called in NEC3 ECC, and the contractor will be entitled to extra costs for the painting activity, not only for the extra paint, but also the extra time costs of labour and any plant involved for the additional five days.

If the painting activity was on the *critical path*, then the contractor would be on-site longer, and would be entitled to site overhead costs in addition. Site overheads in this example would be small, but on a normal contract would include such things as offices, welfare facilities, storage, security and fencing, supervision, small tools and equipment. On a normal contract, these overheads often amount to thousands of pounds a week, or even tens of thousands, so the decision about whether a delaying event is on the critical path has a major significance.

It is interesting to note that the obvious cost (the *cost* of extra paint) is a very small proportion of the total extra cost. This is usually true of construction contracts on any scale; the material cost is dwarfed by time and delay costs.

Under most contracts the painting contractor would also get a five-day extension to the completion date which would delay the imposition of any potential liquidated damages (delay damages) by five days as well. Delay damages are written into many construction contracts and represent the loss to the employer of every day or week that the contractor is late. This employer loss might be additional supervision cost, interest on capital borrowed or loss of rent or income. Delay damages are covered in detail in Chapter 18. It can be seen from the example above that even minor changes can often have far-reaching consequences.

What does a good programme show?

Carefully considered, comprehensive programmes that are regularly reviewed and updated are essential for the safe and efficient planning and execution of all construction projects. So the next question is what should a programme show?

> Well-considered, agreed, detailed programmes that are kept up-to-date, are in everyone's interests.

The NEC3 ECC contract is particularly detailed in clause 31 where it sets out its requirements of the programme. Other forms of contract are less specific. A good programme should show the following items:

- A bar chart of activities against time, with the logical sequence clearly showing durations of all activities
- Float
- Suitable logic links between activities
- The critical path, or critical paths if more than one
- Health and safety requirements
- All important dates, such as start, finish and access
- The planned completion date (which has a major contractual significance)
- Key dates or milestones – these are important dates or events between the start and planned completion dates
- The resources attributable to each activity (this is often called a 'resourced programme')
- **Time-risk allowances** (see box below).

A **time-risk allowance** is the term used in NEC3 ECC for the period of time built into the duration of an activity to ensure it is completed in that duration. It simply reflects the usual human characteristic of 'building in a little spare time allowance' such as allowing 5 days for an activity when 4.5 are probably sufficient. It may also reflect a mathematical 'rounding-up'. So if for example an activity is calculated by output records as taking 3.9 days, and it is rounded up to 4 days, then 0.1 day is the time-risk allowance. If a programme is produced without consciously noting the time-risk allowances for activities, then it is possible to carry out a retrospective review of each programme activity. Here each activity is reassessed to calculate the quickest time that it could be completed if everything goes according to plan. The difference between the quickest time and the duration shown on the programme is the time-risk allowance. NEC3 ECC does not state how this time-risk allowance should be shown. It may be easiest to insert a vertical text column on the bar-chart programme with the time-risk allowance period shown against each activity.

So in many ways the time-risk allowance has a similar effect to that of float. The difference is that time-risk allowance is 'hidden' within an activity, whereas float is usually shown openly and clearly by most planning software, since it is the amount by which the whole activity can be delayed (unless it is a critical activity). Most planning software will list and describe the various types of float, but cannot list 'time-risk allowance' because it is hidden within an activity. NEC3 ECC is presumably trying to show time-risk allowance in a similar clear way so that all the various floats can be taken into account in a situation of delay, disruption or prolongation.

Issues in producing programmes

Programmes of various kinds are the best planning and monitoring tools that we have. However, they can never be precise, totally comprehensive or logically perfect. During claims meetings to determine contractor entitlement to extra costs or time, programmes are often treated as if they *were* perfect. A few points to remember about the imprecise nature of programmes are that:

- Durations depend on a best estimate of how long an activity will take. Conditions on-site, labour and plant outputs and a host of other factors mean that planned durations can never be more than assumptions.
- Activities can often start part way through another activity but programmes that split activities into numerous sub-activities become unnecessarily complex and unmanageable. Hence the programme shown is often a coarse estimate of reality. In the wall example above, the coping activity could probably start when half the wall is completed, rather than when the whole wall is completed, as shown, and the gate could be fixed when the relevant section of wall is completed. However showing this more realistic level of detail makes the programme more complicated. The planner always has to strike a balance between reality and complication.
- Activities are often shown as being consecutive where they are actually concurrent. This makes a big difference to the overall project duration. This may be due to an assumption about resources that could be altered if conditions or events on-site change. In planning terminology it is called a 'resource constraint' and should in theory be avoided. For example, an activity duration of 20 days assuming one gang of bricklayers could be reduced to 10 days if two gangs were used. Suppose now an employer-risk event delayed the start of bricklaying by three days. In reality, the project does not have to be delayed by three days, with the consequent increase in overhead costs. The answer is simply to use two gangs of bricklayers. This simple example may not always be replicated in practice because specialist resources or equipment are often not available at a moment's notice. However, having a discussion about 'resource constraints' is always worthwhile in a delay situation.
- Logic links shown on a programme may reflect incorrect logic and impose constraints that do not exist in reality.
- Conversely, constraints on activities may be forgotten.
- Initial programmes are produced before construction work begins and often at a remote location from the site, such as a contractor's head office. This means that they rarely reflect actual conditions on-site as the work proceeds. Unfortunately it is usually this initial programme that establishes the *base line* from which later contract claims are produced.
- Programmes require regular updating. Ideally they should be updated every few days as events on-site unfold. However, updating a large programme can take many days and hence the updating activity alone can cost hundreds or thousands of pounds in 'planner time'. In practice updated programmes are rarely produced more frequently than monthly.
- It can take a number of days to update a complex programme. As a result the 'updated' programme is out-of-date the moment it is issued.
- Modern software allows very sophisticated resource allocation and resource levelling. It can even have holidays and working patterns for individuals built into the assumptions. Once again, of course, this can only be a good reflection of reality. So the time allocation software may allow for a six-day week for the contractor's site agent and bank holiday site closure, but it cannot allow in advance for the fact that the site agent may be off for four days with flu.

'Accepting' construction programmes

The accepted programme is always the major instrument used in assessing the impact of changes or other events on the cost and duration of the contract. This is why it is so important to have an up-to-date, comprehensive programme that has been *agreed*. The contractor submits the programme, and the contract administrator's role is to accept the programme, or to ask for amendments such that it can be agreed. Without such a programme there is no base line for assessing the effects of events as the project proceeds on-site.

Many forms of contract expect the contract administrator to *accept* the contractor's construction programme. ICE7 covers the whole matter of programmes in clause 14 which takes up two pages and requires the contract administrator (called the 'Engineer' in ICE7) to accept or reject the programme in writing. NEC3 ECC has a similarly comprehensive programme requirement in clause 31, this time covering a page. Again the programme has to be accepted by the contract administrator (called the 'project manager' in NEC3 ECC). Although a court would no doubt imply a term which retained the contractor's responsibility, despite any acceptance by the contract administrator, NEC3 ECC says in clause 14.1 that 'the project manager's or supervisor's acceptance of a communication from the contractor or of his work does not change the contractor's responsibility to provide the works or his liability for his design.'

In many ways, building contracts are more complex than civil engineering ones and could be expected to have more detailed programme requirements. However, JCT SBC05 covers the programme in a very cursory way in just a few lines in clause 2.9. Neither does JCT SBC05 require the architect/contract administrator's acceptance or approval of the contractor's programme. JCT Minor Works Contract 2005 has no requirement for the contractor to submit a programme at all, although the architect/contract administrator is obliged to make an extension of time for completion where justified. This approach of JCT is most curious to contract administrators accustomed to civil engineering contracts. Indeed, to emphasise the importance of the programme, NEC3 ECC in clause 50.3 provides for a deduction of one-quarter from payments due to the contractor until the contractor has submitted the first programme for acceptance to the project manager. However, despite this different approach in JCT SBC05, it is suggested that the architect/contract administrator should always check the contractor's programme and raise any queries about its comprehensiveness or achievability.

Fundamentally, the contractor is responsible for the construction of the works, the construction methods and construction sequences, except where any of these are specified in the contract by the employer (or design organisation acting for the employer). So for example the contract may specify different access dates to various parts of the site, different completion dates for sections of the work, and possibly essential working sequences. However it is good practice for the designer to give the contractor as few constraints as possible in programming and sequencing the work in the most efficient way to suit the contractor.

> The construction programme should be logical, comprehensive and *achievable*.

If the contractor administrator raises doubts we are led to the matter of revised programmes. As an example NEC3 ECC requires revised programmes whenever:

- Instructed by the project manager (the NEC3 contract administrator)
- When the contractor chooses to do so
- At intervals that may be stated in the contract.

JCT SBC05 requires a revised programme after acceptance by the architect/contract administrator of a 'relevant event'. These are basically events which cause delay or prolongation or disruption to the contractor. NEC3 ECC is very clear in stating that it is only the *accepted* programme that is used for assessing compensation events (delay and expense claims) notified by the contractor.

Checking construction programmes

Before accepting a programme, the contract administrator should check it very carefully, and discuss any questions or observations with the contractor. When checking a construction programme, the contract administrator should:

- Ensure that there is enough detail shown (in terms of numbers of activities) for the size and complexity of the project.
- Check that activity headings are sufficient and go into enough detail to establish clearly what work they cover.
- Check activities for pessimistic or optimistic production rates.
- Check that overlapping activities really *can* be carried out concurrently.
- Ensure activities relating to temporary works are included where necessary.
- Check that logic links are reasonable – false logic links can produce extra critical paths that may not exist in reality.
- Ensure that the programme looks logical and consistent.
- Ensure that the planned completion date looks reasonable, and there is not too much terminal float.
- Check compliance with any specified requirements for phasing or possession of areas of the site.
- Ensure that all key dates are shown (this is particularly important).
- Check that the programme shows *interface* dependencies and durations for items outside the main contract but which affect it, such as service diversions, power supplies and work by other contractors.
- Ensure that a resource schedule is included. Each task should have resources allocated at a suitable level of detail. For example in the wall example we need to know if two or four bricklayers are envisaged – this makes a 100 per cent difference to productivity rates and hence the activity duration may be five days for two bricklayers and two and a half days for four bricklayers.
- Ensure that lead-in time to delivery of plant or equipment is shown.
- Ensure that design/production/delivery schedules, linked to the on-site construction programme, are provided where the contractor provides plant or equipment. This is so that off-site checks on progress can be made for critical items.

Meetings to review construction progress

Progress against the construction programme is usually reviewed at site progress meetings. On most contracts these meetings are held monthly and provide an essential opportunity to review progress and discuss any issues or problems which have affected progress or which might affect it. They normally have formal minutes taken, and it is good practice for the contractor's progress against every activity to be reviewed, agreed and recorded. This recorded progress will often form the basis of a subsequent calculation of the effects of any delaying events, so it is important that it is accurate and agreed by the contract administrator.

People attending these site progress meetings will always include the contractor, the contract administrator (who may also be the project manager) and any site supervisory staff (such as the supervisor, resident engineer or clerk of works). Other people attending will often include a representative of the designer, subcontractors or suppliers, a planning engineer and the quantity surveyor.

Where the contract administrator and the project manager are different people, then it may be beneficial for the project manager to attend the progress meeting as well, rather than simply receiving *reports* on progress. However in this case the project manager will not have powers under the contract to issue instructions to the contractor: this is the role of the contract administrator. The project manager's role is a *coordination* role; the contract administrator's role is to *administer* the contract according to its terms. Unfortunately, under an NEC3 ECC contract, the *title* of the contract administrator is the 'Project Manager', but the *role* in NEC3 ECC is primarily a contract administrative one.

In traditional contracts such as JCT and ICE7, the monthly site progress meeting is the main forum at which problems and issues are discussed. However, in NEC3 ECC contracts there may be other meetings, called 'risk reduction meetings'. These result from 'early warnings' given under clause 16.1. The idea is that possible delaying events or problems are reviewed as quickly as possible and that sensible actions are agreed between the contractor and the contract administrator.

The agenda of site progress meetings

These meetings are very important and should normally discuss and record:

- Progress (recorded as percentage completion against each activity).
- Any changes in programme logic or assumptions.
- Any further amendments to or reissue of the programme.
- Whether any key dates or the completion date may not be achieved.
- Where delay is predicted, the actions (and costs) that the project participants can take to minimise time and cost overrun.
- Any early warnings given under an NEC3 ECC contract (see later chapters on NEC3 ECC) and any risks in the risk register which are now passed or completed.
- Any new risks which need adding to the risk register.
- Any health and safety issues or incidents which affect the programme.
- Progress against the environmental plan, and any environmental issues or incidents.

- Any interfaces with the public, operational staff or statutory authorities such as the Environment Agency.
- Any visits to site by any regulatory bodies such as the HSE, police and so on.

Who pays for the time and financial impact of events?

This very important issue is complicated and is dealt with in detail in the chapters on NEC3 ECC and JCT SBC05 (Chapters 21 to 27). However at this point it is worth touching on the subject in respect of programmes. We need to remember of course that the parties to the construction contract are normally the employer and the contractor, not the designer. In the painting example above, there will be additional cost and also a delay to the completion date if the delayed work is on the critical path. This is why the critical path is so important. The contract administrator has to decide what the contractor's entitlement is, under the terms of the contract.

We assume here that the contract allows the contract administrator to make changes (called 'variations' in traditional contracts). Most contracts that allow changes also give some guidance about how they should be valued. So in the example above, the contract administrator has made legitimate changes, they have cost the painting contractor more money, and ultimately the *employer* must pay. This is an important concept – in most contracts the contract administrator decides whether the event is reimbursable and if so the *employer pays*. The contract administrator does not pay for anything; the contract administrator's role is to administer the contract fairly between the parties (the employer and the contractor) according to the terms of that contract.

There are a few contracts such as PPC 2000 (the standard form of Project Partnering Contract published by the Association of Consultant Architects) which do provide for all parties to bear some of the cost. However such contracts are not common.

Of course, if the employer has to pay extra or receive a late project, then the employer (or its auditors) may want the contract administrator to *explain every action* and relate it to the terms of the contract. So while the contract administrator does not pay directly, his or her actions should always be legitimate, professional, within the terms of the contract and *justifiable*. The employer or its auditor is likely to require the contract administrator to give a full explanation of the contractual basis for all actions and the way in which any additional time or payment was calculated. This explanation may be required weeks or often months after the construction works are complete. This is because the final account and payment to the contractor are usually due at least a year after construction completion. Contract administrators will often become wrapped up in contract matters relating to the hurly-burly of construction. The contract administrator can easily forget that the employer's auditors may come along after contract completion and want a retrospective justification; this can be Nemesis.

Chapter summary

- The main way of controlling time is the programme.
- 'Being late' usually costs either the employer or contractor a lot of money.
- Hence programmes are of fundamental importance in all construction projects, not only to the contractor but also the contract administrator and the employer.
- Programmes allow all the work required to be planned and monitored for delivery of the project within the timescale prescribed.
- They also allow the employer to arrange access as required and to make provision for paying the contractor at intervals set out in the contract.
- There are three main programmes in construction projects. These are the design programme, the construction programme and the Master Project Programme. The Master Project Programme encompasses the design and construction programmes and adds in all the other essential activities, constraints and requirements to ensure satisfactory and efficient completion of the project.
- There are a number of programming terms such as free float, total float and so on.
- One of the most important concepts is the critical path. The critical path is the sequence of linked activities through a programme which determines the overall project duration. Hence if any activity on the critical path is delayed, then the project is delayed by the same amount, unless some form of corrective action is possible.
- Not only is the critical path of great importance to the contractor for planning the work, but it is also of prime importance in determining contractual claims, and the associated costs and extensions of time that may result.
- The three main impacts of events on a programme are delay, prolongation and disruption. These impacts will almost always mean that the project is completed later than planned and will usually cause additional cost and expense to the contractor. It is the job of the contract administrator to examine the terms of the contract and to decide whether the delaying event is one which gives the contractor more time and also entitlement to costs and expense.
- Another major use of programmes is to allow the effect of any changes to be modelled, to see what impact they might have on the project, and whether there are any alternative courses of action which could be considered.
- Once a change or unexpected event has occurred, the programme can be used to estimate the effect and hence determine any additional time or money that is justified under the contract terms.
- There are a number of drawbacks with programmes and at best they represent something like reality. Programmes have to be *simplifications* of real life otherwise they would be unmanageably complicated. So the planner makes assumptions. It is important to examine these planning assumptions when determining contract claims.

17

Contract law

Aim

This chapter aims to provide a basic understanding of contract law and how it relates to construction contracts.

Learning outcomes

On completion of this chapter you will be able to:

>> Relate the need for a contract.
>> Describe how a contract is formed.
>> Discuss the relevance of contract law to the work of a construction professional.
>> Explain the main features of a valid contract in English law.
>> Describe factors which may make a contract invalid.
>> Differentiate between terms and representations.
>> Differentiate between conditions and warranties.
>> Discuss the meaning of 'performance' and its relationship to construction contracts.
>> Describe the remedies available for breach of contract.

A note of caution

This chapter describes the main principles of contract law, but it is not a legal textbook and the author is a chartered engineer, not a lawyer. Design contracts and the diverse range of contracts associated with construction are all contracts under English contract law. Since construction professionals are all involved in some way with these contracts, a basic understanding of contract law is very important. Significantly, construction contracts such as NEC3 and JCT contracts are usually administered by a construction professional, not a lawyer. It is not unusual for inexperienced construction professionals to make mistakes or errors of judgment that can lead to very expensive claims. This is often through lack of knowledge of the way that contracts and the law actually work. Hence one of the most useful outcomes of this book will be to alert construction professionals to the need to 'take advice'. This advice might be from a colleague or a more senior person and, where warranted, a lawyer or legal professional.

Contracts

Contracts do not have to be formal and they certainly do not have to be written by lawyers. In fact most construction professionals write terms of contracts much of the time they are at work. This does not have to be as formal as modifying a form of contract such as the New Engineering Contract NEC3 ECC or one of the Joint Contract Tribunal's JCT contracts. A term of a contract is a statement of what a party has agreed to do or give under that contract. Hence we may have terms which define payment, the time available, and numerous terms relating to quality in the specification and to dimensions and layout on the drawings.

The people or organisations that make an agreement or contract are called the parties. In construction they will usually be some of the employer, design organisation, contractor, sub-contractors, suppliers, other professionals and so on. It is fairly rare in construction for individuals to make contracts, except in the case of house improvements, extensions and renovations, when one party is likely to be a domestic client and the other party could be a building company or an individual builder or tradesman such as a painter or plasterer.

In a modern economy everything depends on the sale and purchase of goods and services. In such an economy, these activities may be very complex. People and companies make agreements to manufacture, supply and sell goods and services. They need assurance that if they comply with the terms of their agreements, then they will be paid. Contract law is a means of supporting all these agreements or 'bargains' and giving a **remedy** if the agreement is not carried out in accordance with its terms. The usual remedy is the payment of damages (money) as a result of a court judgment. Most of the time well-administered contracts run smoothly and so recourse to the courts is unnecessary. Additionally, many construction contracts have dispute resolution procedures in them. These provide another mechanism to aid the settlement of disputes without court action.

For the construction professional, a basic understanding of contract law is essential to answer the following questions:

- How do we ensure that there is a valid contract in the first place?
- What is the difference between a term and a representation?
- What terms are incorporated in the contract?
- If a term is broken or not fulfilled, how important is it?
- How and when is a contract completed?
- What are the remedies for a **breach of contract** (when a contract is not honoured)?
- Why are previous cases quoted?

These are all sensible and logical questions. The law is sensible and logical, in the same way as are the formulae that a construction professional uses. Unfortunately, because the law covers the multiplicity of possible dealings between people it can become complex and this sometimes obscures the underlying logic. There are many times where slightly different facts differentiate one case from another, and lead to a judgment we may not expect at first. What the judge is trying to do is to decide whether the law or the **precedent** created by previous legal cases applies to the case in question. The judge will be persuaded in this judgment by the facts and the arguments of the legal people on both sides.

Cases and precedent

It is important to have some understanding of how the law works with regard to statute law (made by Parliament) and precedent, which is law made by previous judgments of a court of sufficient status. This is because decided cases (*even very old cases*) are often quoted to amplify or support an argument in law. Decided court cases are referred to as for example *Smith* v. *Jones* (1959). This means that in a court of first instance, Smith was the claimant and Jones the defendant, and the case was published in the 1959 Law Reports (although it may have been tried in a year previous to this).

A longer reference would be *Smith* v. *Jones* (1959) 1 QB 67, which means the first volume of the reports from the Queen's Bench Division (a division of the High Court), the report commencing on page 67.

Much of the law in England is based on previous judgments on cases brought to court. This is called precedent. Law made by the government is called statute law. Case law is built up out of precedents. The judgment in a court is written down by the judge and published. These judgments are often dozens of pages long, sometimes hundreds of pages. Such judgments may be binding or persuasive on another court in a similar case, and lawyers on both sides will usually seek to find cases that they consider to be similar to the case in court, such that the judge can be persuaded to treat the outcome in the same way. The two types of precedent are called:

- binding precedents
- persuasive precedents.

Whether a precedent is binding or persuasive largely depends on the level of court in which it was decided. Thus judgments made in the highest court, the House of Lords, are binding on all lower courts, but judgments in the lowest courts, the county courts and magistrates' courts, are not even reported. The House of Lords was the ultimate court in English Law and many of the most important cases went to the House of Lords on appeal. Cases were usually heard by five 'Law Lords'. However, the Constitutional Reform Act 2005 has established the replacement for the House of Lords, which is now called the Supreme Court of the United Kingdom and it started work on 1 October 2009.

A further complication is that a judgment has two parts and they are:

- The *Ratio decidendi* which is Latin and is taken to mean the principle of law used by the judge in making the decision and the reasons for doing so.
- The *Obiter dicta* are statements of the judge which are made 'in passing', and do not have binding force on other courts, but they may be persuasive.

Ratio decidendi establish binding precedent (depending on the level of the court), but *obiter dicta* do not. Cases can be appealed to a higher court. Where this happens, the decision of the higher court overrides the lower court and becomes the new precedent. The English court system is complicated. A very simplified view of precedent is given below:

- The House of Lords and now the Supreme Court (the highest court) is bound by its own decisions.
- The Court of Appeal is bound by decisions of the House of Lords/Supreme Court and its own decisions.
- The High Court is bound by the decisions of the House of Lords/Supreme Court and the Court of Appeal. A division of the High Court is not bound by but 'persuaded' by High Court judgments made in other divisions. However a division is bound by decisions in the same division of the High Court. These divisions are known as the Queen's Bench Division, the Chancery Division, and the Family Division.
- Magistrates' Courts and County Courts are bound by all higher courts, and their own decisions are not reported.

The use of Latin

The law uses many Latin terms such as *ratio decidendi* and *obiter dicta*. The reason is that certain words and phrases have come to mean very particular things in law over the last few hundred years. English law has its origins in the court system of Henry II where judges started to write down judgments. This system began just after the Norman Conquest of 1066, just under a thousand years ago. Latin is used in law because it was the language of educated people after the Roman invasion of Britain, over two thousand years ago, and continued to be used in scientific writing until the last few centuries. Thus Isaac Newton's famous theories, with which engineers are familiar, were written in Latin, in the *Philosophiæ Naturalis Principia Mathematica*, published in 1687.

As we know, the law seeks as much certainty as possible, so having universally understood words and phrases is clearly a good thing. Hence to a lawyer *ratio decidendi*, **volenti non fit injuria** and so forth have a definite meaning, rather like the formulae of the construction professional. Other English phrases have a similar place in law, such as 'in the course of his employment'. Since much of our English language is based on Latin, the meaning of many of these phrases can be deduced. So for example *volenti non fit injuria* sounds as if it has to do with *volunteers and injury*. It is a phrase used in the tort of negligence and means 'to the willing no harm can be done'.

There are currently moves to reduce the number of Latin phrases used by the courts, but of course this will not affect the massive volume of previous cases that form precedent.

Essentials of a valid contract

A valid contract in English law has a number of essential elements. The elements of a valid contract are:

- Offer and acceptance (this is the agreement)
- **Consideration**
- An intention to create a legal relationship
- Genuine consent (no **vitiating factors, discussed below**)
- The parties must have **capacity**
- Legal formalities (where required).

Offer

When we look at offer and acceptance, we are seeking to ensure that there has been a true agreement. The person or organisation making the offer is called the offeror, and the person or organisation accepting the offer is called the offeree. This may seem simple at first, but in practice there may be statements made before the agreement, which are called **representations**. Many agreements are made after a series of negotiations, probably including representations, so determining the true intent of the parties may not be as easy as it would appear.

An offer is an undertaking by the offeror to be bound in contract by the terms of the offer, if there is a proper acceptance of it.

What may appear to be an offer for acceptance may in fact be an **invitation to treat.** An invitation to treat is not an essential element of a valid contract, but it occurs in many commercial transactions.

Invitation to treat

An offer is where party A offers to do or sell something, and party B accepts. In contrast an invitation to treat is where party A is wishing party B to make an offer which party A is then free to accept or reject. There are many examples of invitations to treat, such as goods on the shelves of a supermarket or in a shop window, advertisements in a newspaper or magazine, lots at an auction and so on. Hence an electric guitar priced at £500 in a shop window is not an offer, it is an invitation to you to go inside the shop and make the offer. You might offer £500, or possibly less, and the shopkeeper is at liberty to reject your offer, or suggest an alternative price. Because the window display is an invitation to treat, you cannot walk into the shop and 'accept'. It is possible that the shopkeeper has made a mistake on the price tag, or that he is unwilling to sell that particular guitar, having similar ones in stock.

An invitation to treat may often be part of the process of agreement, but it does not have to be so. Hence many contracts are made simply with offer and acceptance. An example might be if you said to someone, 'I'll sell you my guitar for £500 if you want it'. If the other person says 'yes, agreed', then the two of you have a contract based on offer and acceptance and consideration (the £500). There was no invitation to treat and there does not need to be. This also indicates the point that most contracts do not have to be written down. Clearly it is a good idea for construction contracts to be written down, because they are usually much more complicated than this example. If contracts are written down, then their terms (what the parties intend to do) become much clearer and less liable to subsequent argument.

The invitation to tender

For the construction professional, the most important aspect of the above discussion is that an employer's invitation to tender (an invitation to tenderers to return a price) is normally considered to be an invitation to treat. The returned priced tender made by the tenderers (usually contracting or building firms) is the offer. This means that the employer does not have to accept any offer at all. This does happen sometimes, where all returned prices are above the employer's budget, or perhaps where the project has had to be cancelled or re-scheduled for some reason. Clearly the tenderers will not be very happy, because of the time and expense of tendering. Certainly if employers behave like this on frequent occasions, they may find few contractors willing to tender.

Whilst the employer does not have to accept any tender, the employer does have to *consider* all properly submitted tenders. The legal case of relevance here is *Blackpool and Fylde Aero Club* v. *Blackpool BC* (1990). The Aero Club submitted a valid tender for the concession to operate pleasure flights, but due to an administrative error it was never opened. The Court of Appeal found in favour of the Aero Club and said that the Borough Council should have *considered* all tenders.

It is possible however for an employer to state in an invitation to tender that the lowest offer will be accepted (most employers would not consider doing this), in which unusual event, the invitation to tender is actually an offer. This situation arose in the case of *Harvela Investments Ltd* v. *The Royal Trust Co of Canada Ltd* (1985), where the defendant invited offers for the purchase of shares and undertook to accept the highest offer. The case was complicated by a bid which offered $101,000 over any other bid, which is why the case went to court. However the principle that the contract was made by the highest valid bid in this unusual 'tendering' process was affirmed.

In order to avoid any doubt on this subject, many invitations to tender contain the words 'the employer does not bind himself to accept the lowest or any offer'; or a similar arrangement of words.

Certainty

When we start to think about the validity of offers we can see a number of possibilities. How clear must an offer be? Who is it made to? Does it last forever? Can it be withdrawn? An offer has to be clear, so that the offeree knows what is intended. In other words, its words or terms have to give certainty. Where the words of an offer are too vague then the resulting 'contract' cannot usually be enforced.

In *Scamell* v. *Ouston* (1941) the defendant refused to supply a van, and the statement 'on hire purchase terms' was deemed to be too vague to create certainty and the court decided that there was no contract to enforce. However courts will go to lengths to try and find a valid contract and may use devices such as 'custom' or 'previous dealings' to find suitable assumptions that can lead to a valid and enforceable agreement.

An offer must be communicated

An offer also has to be communicated. Clearly this is not usually a problem in construction where contractors, sub-contractors and suppliers will all ensure that their offers are communicated properly. They are also likely to make their offers in writing so that there can be no doubt as to their interest and there will also be a record of the date of the offer. Frequently they may attach further terms (such as their standard terms of business), which can lead to the 'battle of the forms', discussed below.

How does an offer end?

It would be unreasonable to expect offers to go on forever. Offers can end in a number of ways:

- An offer ends with a proper acceptance, which together with a number of essential features makes the agreement (the contract).
- An offer can lapse with time.
- An offer can be revoked (withdrawn).
- An offer can be rejected.
- A counter-offer 'kills' the original offer.

An offer will usually lapse within a fixed period if one is stated, or if not after a reasonable time. A court would have to determine what was reasonable in any given case. So where prices fluctuate regularly, as they do with some commodities, a court would probably decide that a short time lapse was reasonable. In the case of *Ramsgate Victoria Hotel* v. *Montefiore* (1866) it was held that Montefiore's offer to buy shares (making a part payment of a shilling) could not be enforced when the shares were supplied five months later.

An offer can be revoked by the offeror but this withdrawal has to be before the offer is accepted and it has to be communicated to the offeree.

If the offeree rejects the offer, or makes a counter-offer, then either of these actions 'kills' the original offer. This principle does sometimes occur during the tendering process, where a contractor may tender a price and then in subsequent discussion reduce or even increase it. A case illustrating this is *Hyde* v. *Wrench* (1840), where the defendant offered to sell his farm for £1,000. The claimant offered £950 (this counter-offer 'killed' the original offer). When the claimant subsequently attempted to 'accept' the original £1,000, the farmer refused to sell. It was found that there was no contract, as the £950 counter-offer had 'killed' the original offer of £1,000 and replaced it with a new offer by the claimant at £950 which the farmer was entitled to refuse to accept. However the law draws a distinction between new terms in a counter-offer, and simple enquiries for information. It all depends on how they are phrased. Another point to make here is that this case was over 150 years ago, but has established a precedent and is still 'good law'.

Acceptance

Once we have an offer, we need a proper acceptance of that offer, plus **consideration** to make a valid contract. Acceptance need not be in writing, it can also be verbal or 'by conduct', which means the way we act. An example of 'by conduct' might be where you offer to buy some bricks, and the supplier never actually 'accepts' your offer but simply delivers the bricks with an invoice for payment. Delivery of the bricks would be acceptance by conduct. However if the offeror stipulates a certain method of acceptance such as 'in writing' then only this method (or an equivalent one) will produce a valid contract.

> Acceptance is an unconditional agreement to all the terms of the offer, communicated to the offeror and made with the intention of accepting.

Just as we saw with offers, this definition produces a number of possibilities, which are:

- Acceptance must be unconditional.
- Acceptance must be communicated.
- The method of acceptance.
- Silence is not acceptance.

Acceptance must be unconditional

For the acceptance to be unconditional it must correspond with the offer in all respects. This is sometimes called the 'mirror-image' rule. Hence if you were buying a car and said 'I'll take it if my partner likes the colour', then you would not have accepted because there is no certainty in your statement. However, this rule can be modified in the so-called 'battle of the forms' discussed below.

In normal consumer dealings, terms of offers are usually fairly simple. In construction however, the employer will normally use a form of contract (with many detailed terms) and the contractor or supplier may respond on their own terms of business. This is particularly true of suppliers. So both parties are trying to impose their own terms, which can often lead to difficulty.

Acceptance must be communicated

Acceptance must be communicated to the offeror by the offeree (or someone with the offeree's authority – this will often be the case in large employer organisations). We need to remember that when responding to a tender invitation, the contractor is the offeror and the employer the offeree.

This raises the question again of: who is 'the employer' in a large organisation? Many large employers will have a scheme of delegation for contracts, which has been approved and signed off by their directors. Such a scheme may require the managing director to sign contracts over twenty million pounds, directors to sign between five and twenty million, and the relevant senior manager to sign contracts below five million pounds.

The method of acceptance

Acceptance is subject to a number of further rules, because there are many ways of accepting in a valid manner. Acceptance of simple contracts will often be verbal. 'Yes', or 'I agree', or 'We have a contract' or even 'Done' would all be valid means of acceptance.

However the legal rule for postal acceptance can create problems. Acceptance by post is when the acceptance is *posted*, not when it is received. This rule applies where acceptance by post has been specified, or where it is a reasonable and appropriate means of acceptance. So the rule applies even if the acceptance is delayed in the post, but it must be properly addressed. However, if the acceptance is posted on the eve of a well-publicised postal strike then it would not be a reasonable means, and the postal rule would not apply. This can create difficulties, since the post may take one or two days. In that time, the employer may have accepted another offer, conveyed by a quicker means, such as delivered by hand. A properly posted acceptance is valid even if it is lost in the post.

Where the method of acceptance is specified, then that method should be used. There have been cases relating to a requirement to complete a standard form of acceptance for example. To be absolutely certain of the outcome it is also wise to specify that no other method will suffice. This avoids a later argument that an equally appropriate method of acceptance was used.

Silence is not acceptance

Silence is not acceptance. Hence it is not possible to make a contract by using a phrase like 'if I have not heard from you by Monday 5 December, I shall assume that you have accepted'. The famous case on this subject was *Felthouse* v. *Bindley* (1862), another very old case. This case concerned the sale of a horse, and a similar phrase was used to attempt to force an acceptance. The case failed – there was deemed to be no acceptance, since it cannot be forced on one party by deeming that silence represents acceptance.

'Subject to contract'

In construction contracts, an 'acceptance' will normally be made 'subject to contract'. As we have seen above this in itself is not an acceptance, but it is an intention to create a formal acceptance. The reason that this is done, is to avoid many of the uncertainties and pitfalls that we see above. It is one thing buying goods or materials for a few hundred pounds, but when an employer wishes to award a contract for a few million pounds, the last thing that the employer wants is a disagreement about acceptance by post, or whatever.

Acceptance 'subject to contract' is not an acceptance in itself but requires a contract (usually a signed one) to be completed by the parties. A construction case illustrating this is *Regalian Properties Plc* v. *London Docklands Development Corporation* (1995). Here Regalian offered London Docklands Development Corporation (LDDC) £18.5 million for a licence to build on their land. LDDC accepted 'subject to contract'. The property market then collapsed and it was not financially worthwhile for LDDC to proceed with the contract. Regalian tried to recover from LDDC the £3 million which they had spent in preparation for the proposed contract. Regalian failed because there was no proper acceptance of its offer.

Deeds

Originally agreements were only recognised if they were contained in a **deed**, and some still require a deed, such as the sale of land. However, as society developed and the pace of commercial affairs increased it became necessary to have a faster mechanism for forming commercial agreements. This led to the contract as we know it, which also requires **consideration** (this usually means the payment of money in return for goods or services).

A deed does not require consideration. Originally a deed had to be 'signed, sealed and delivered'. The seal was in wax, whereupon a ring or similar object would be placed to make an impression. Sealing as such was abolished some time ago, but a deed has to clearly state that it is a deed, and it must be validly executed. This execution normally requires a signature in front of a witness. The main advantage of a deed is that it increases the **limitation period**. This is a complicated subject, but in simple terms claims for breach of contract can be made within a limitation period of six years from when the breach occurs in a normal contract, but twelve years for a deed. Thus having a contract signed as a deed is of considerable advantage to the employer.

Standing offers

A different type of offer and acceptance is often used and is called a standing offer. In construction these standing offers are generally called 'framework agreements'. These are discussed in detail in Chapter 12 but the general principle is set out below.

In a standing offer, an overall contract framework is set up and each individual order is made by giving specific works details, and referring to the terms of the original framework agreement. It is wise to have an agreed layout and format for the order, and a clear name such as 'package order' or 'works order', so that there can be no doubt as to its meaning and relevance. An example of a standing offer is the case of *Great Northern Railway* v. *Witham* (1873). GNR advertised for tenders for the supply of railway stores. Witham submitted a tender offering to provide the required articles for 12 months. GNR placed orders, all of which were carried out by Witham *except for the last one*.

Witham argued that he was not legally obliged to supply the goods, and so was not in breach of contract. The court decided that the agreement to supply for a year was a standing offer which was *accepted each time an order was placed*. Witham lost the case.

A standing offer arises where one party to the contract agrees to supply goods or services as and when required (ordered) over a specific period.

A standing offer can be revoked at any time before acceptance of an order.

Placing a particular order generates a valid contract for those goods/materials.

The 'battle of the forms'

This is a phrase used to describe the common circumstances where there is a series of counter-offers. The situation is further confused by the fact that some correspondence may not represent a counter-offer, but is merely a request for further information. What can make matters worse is that a letter from either party might contain a mixture of the two.

The 'battle of the forms' can occur when contractors return tenders followed by a period of correspondence before signature of the contract, but it usually happens with suppliers. Most suppliers have their own terms and conditions printed on their quotations. Hence the employer (or contractor) sends out a written contract with perhaps payment terms of 60 days, and the supplier returns an offer with 30-days' payment terms as their standard terms. There is therefore a mismatch. This is just one example of what can be many mismatches. Strictly this change to 30-day payment terms would be a counter-offer which 'kills' the original offer. Where contractors send in amended tender documents a simple approach used by many employers is to ask the contractor to rescind the modifications or withdraw from the tender process. This is less easy with suppliers, who are more accustomed to doing business on their own terms and conditions.

The courts may recognise that many commercial contracts are concluded after a series of discussions, letters and negotiations. The courts will also try and find a valid contract whenever they reasonably can, rather than the reverse, but they will look for certainty. Their usual approach will be to take all the terms of the latest accepted counter-offer as the terms to be incorporated in the final agreement. This is based on the 'mirror-image' view discussed above. However this can result in the so-called 'last shot', where both parties continue to send back their standard terms in the hope of being the 'last shot' before acceptance. Additionally the 'all or nothing' result of the mirror-image view can produce a result that was not intended by the parties in the course of their negotiations, where they thought that they had finally agreed on a compromise set of terms that they both find acceptable. The way this situation could be approached is to look at each term and take the latest version of it once the final agreement is made, provided the latest term is properly incorporated in the final contract. Let us take a simple example:

- The employer invites tenders from contractors to build a small wall with red rustic bricks.
- The lowest tenderer returns an offer to build the wall for £3,000 with blue smooth bricks.
- The employer writes back asking the tenderer to comply with red bricks.
- The tenderer agrees, but increases its price to £3,100 in a subsequent letter.
- This tender is still lower than the other tenders.
- The employer accepts the contract referring to the related correspondence in the acceptance letter.
- What does the employer get?

We see that the above correspondence has been properly incorporated (referred to) in the contract acceptance letter. Firstly the employer pays £3,100 (the last statement on price), and receives red bricks (the last statement on colour), but the employer also gets smooth bricks which may not really be wanted, because the word 'smooth' was never modified back to 'rustic'. Hence the employer receives red, smooth bricks for £3,100. Let us hope that rustic bricks were not a condition of the planning consent!

A much sounder solution to the problem of the 'battle of the forms' is for the parties to rewrite parts of the final contract with the agreed set of terms, before final signature. Unfortunately this takes time, at the outset at least. However, the result is a clear final contract which is agreed and that can be administered in an efficient way.

A less satisfactory but more common solution is to incorporate in the contract all the letters and correspondence which show the agreement of amended terms to the final version (red, smooth bricks for £3,100 in the above example).

The first solution takes longer at the outset but produces a contract which is easy to follow during implementation. The latter can produce such a mass of paper where practitioners have little hope of knowing exactly what the contract includes. The author had experience of a major construction contract, signed after weeks of discussion and correspondence, mainly relating to specification and process issues. The original contract and drawings filled five bankers' boxes, and the 'incorporated correspondence' filled three large lever arch files. This was not an easy contract to administer.

Consideration

The third element of the agreement is consideration. Consideration is not required for deeds. Consideration represents the 'bargain' notion of contracts.

Each of the parties to a legal agreement must give and receive something of value; this is consideration. Hence an exchange of consideration whether it be financial or of another sort between the parties to a contractual arrangement is essential for the agreement to be legally enforceable. Paying (or promising to pay) money in return for goods or services is the most common form of consideration for the construction professional. Goods will normally be materials or manufactured items, and services will usually be design or advisory services and of course construction services (actually building the asset). However, consideration has had various definitions over the years and can include 'some right, interest, profit or benefit accruing to one party or some forbearance, detriment, loss or responsibility given, suffered or undertaken by the other.'

In construction contracts the payment of money for construction materials, manufactured items or services is the most common form of consideration.

There are a number of rules concerning consideration and there are exceptions to many of them. However most have rare application to construction contracts. The rules with general application to construction are:

- Consideration must be sufficient but it need not be adequate. So a seller has to receive a certain sum, but it might be much less than the goods are actually worth.
- Past consideration is not good consideration. This could be a service previously performed for example.
- Performing an obligation or a duty that is already owed is not good consideration.
- Consideration must move from the promisee. This is another way of saying that a person to whom a promise is made can only enforce the promise if he or she has provided consideration in return.

The law tries to create certainty as far as possible, so it will not overturn a 'bad bargain'. If I sold my cherry sunburst Gibson Les Paul guitar for £100 when it was worth £2,000, I could not subsequently go to court and overturn the contract. In other words, consideration is sufficient (for legal purposes) but it is not really adequate, because I should have charged a lot more money. If I could go to court and cancel the contract then the courts would be full of people who thought they had made a bad deal, and nobody could be certain of the outcome of a contract.

An example of past consideration would be where a service or activity is carried out with no promise of payment, such as a gardener clipping a hedge in a back garden. If a subsequent contract is made for that activity it would not be enforceable because the consideration was in the past. However a separate agreement to clip a front hedge for £50 would be enforceable, since it now has good consideration.

Where a duty is already owed, then performing that duty is not good consideration. There are a number of old nautical cases on this topic, such as *Stilk* v. *Myrick* (1809) where a sea captain offered to share the wages of two deserters among the remaining crew. After the ship returned to port the captain refused to pay and the court (perhaps a little harshly) decided that the crew were already bound to sail the ship and hence provided no extra consideration for the deserters' wages. A typical legal subtlety is evident in the later case of *Hartley* v. *Ponsonby* (1857) where there were so many deserters that the remaining crew had to sail an unseaworthy ship. This extra duty was deemed to be good consideration since they were clearly giving something well above their original contract.

In a construction contract the issue of consideration is generally fairly simple. Typically the employer undertakes (promises) to make regular payments to the contractor in return for the value (or cost) of work actually done by the contractor in that period. In a design contract, the employer undertakes to make payments for design work and drawings carried out and produced by the design organisation. In a supply contract, the employer, contractor or sub-contractor undertakes to pay the supplier for goods and materials delivered to the site.

Intention to create a legal relationship

The intention of the parties to be legally bound is an essential prerequisite of a contract. Again for the construction professional this is not usually an issue because of the law's presumptions which are:

- In commercial agreements there is a strong presumption that they are intended to be legally binding.
- Domestic and social agreements are presumed not to be legally binding.

So for example, if I told my partner that I will paint the garden fence today if she cooks an Indian meal for me, tomorrow at 7 pm, this would not normally constitute a contract. However, if I went into the local *Indian Restaurant* and asked them to deliver a particular meal to me for 7 pm the following night, at a price of £30, then this would be a binding contract because it is a commercial arrangement.

Hence contracts between employers, contractors, sub-contractors and suppliers will be assumed to be legally binding unless they say clearly and specifically that they are not. An example in the construction industry of an agreement that is not usually legally binding would be one of the many forms of partnering agreements available. As explained in Chapter 11, although partnering can have a contractual basis, it is *usually a non-legally binding agreement* to cooperate, share information and work together on the basis of mutual objectives and trust. The binding agreement is the contract itself; this would probably be a construction contract chosen from the JCT or NEC suite of contracts. The construction contract itself provides the legally binding terms and conditions. In fact the ICE Partnering agreement states in clause P8 that it is not legally binding, whilst the title of the JCT version is the Partnering charter (non-binding) 2005.

Finally, an *intention to create a legal relationship* is not usually assumed in the case of advertisements. They are usually seen by the courts as a 'mere puff', to use an old-fashioned term.

Consequently, with some of the exceptions quoted above, it is usual to assume that parties to construction contracts do intend them to be legally binding.

Genuine consent (no vitiating factors)

We must remember that the law looks for true agreements between the contracting parties. Hence we may have offer, acceptance and consideration, but for a valid reason the contract should not be enforced, or one party ought to have a right to compensation. These barriers to a proper contract are often called **vitiating factors**. Where one or more of these are present the aggrieved party may rescind (cancel) the contract or obtain damages depending on the nature and severity of the vitiating factor and possibly the effect of later circumstances that may have occurred.

A contract that *may* be cancelled is called 'voidable', whereas one that *must* be cancelled is called 'void'. A void contract never legally existed in the first place.

As an initial example let us assume you go into a shop to buy a one-year-old second-hand computer. In discussion with the shopkeeper, he tells you that it has a certain size of hard drive, video card, processor speed and memory. You are persuaded that this computer will be much faster than your existing one and will play the latest games and so you buy it for £500. However, suppose the shopkeeper knowingly told you lies, and the computer is much slower than you had been led to believe. We have offer, acceptance and consideration, but is this a fair contract? Of course not, and you have a right to compensation or even cancellation of the contract. This is an example of fraudulent misrepresentation, which is a vitiating factor that would allow you to choose between cancelling the contract and claiming damages as compensation. There are four principal vitiating factors and they spell DIMM. They are:

D	Duress and undue influence
I	Illegality
M	Mistake
M	Misrepresentation

Duress and undue influence

Here a contracting party subject to duress or undue influence can take the matter to court and have the contract declared void. Duress could be the threat of violence or intimidation and undue influence is concerned with the exercise of unfair power by two parties in a particular type of relationship, such as parent and child, or a doctor and patient. This vitiating factor is unlikely to be of relevance to construction contracts.

Illegality

This could include contracts to commit a crime, or a fraud, or prejudicial to public safety or to promote corruption in public life for example. All would be void. Again, these areas are unlikely to concern us.

Mistake

This is quite a difficult area, because the courts do not encourage people or organisations to avoid properly made contracts because they have made a 'mistake' (such as the price charged) as we commonly mean the word in everyday speech. The courts apply a more restricted meaning.

However there are a number of areas of mistake which are recognised by the courts. 'Operative mistakes' are those that are fundamental to the making of the contract such that it was only made as a result of the mistake. There are three possibilities:

1. A common mistake is where both parties make exactly the same mistake.
2. A mutual mistake is where both parties are mistaken but are at cross-purposes about the substance of the contract.
3. A unilateral mistake is where only one party makes a mistake and, by implication, the other party knows and seeks to take advantage of it.

An example of a common mistake would be where the subject matter (here specific goods) does not exist at the time of contracting. A well-known example is *Couturier* v. *Hastie* (1852) where the contract was for the sale of a cargo of grain in transit by ship, which both parties believed to exist. In fact the grain had fermented during the voyage and had been sold by the captain in a port, before the voyage was completed.

An example of a mutual mistake is the case of *Raffles* v. *Wichelhaus* (1864) which concerned a contract to buy cotton being transported on a ship named 'Peerless'. In fact there were two such ships sailing from Bombay Harbour, and the seller was selling the cargo on one ship whilst the buyer thought he was buying the cargo on the other. The contract was void for mistake.

An example of unilateral mistake is *Hartog* v. *Colin and Shields* (1939). The contract involved the purchase of 30,000 hare skins from Argentina. The price was stated as being per pound (in weight) whereas the usual practice was to sell skins by piece (an individual skin). Since there were about three skins to every pound in weight the difference was very marked and should have been very evident to the buyer. The buyers tried to enforce the contract but the sellers counter-claimed that the offer was wrongly stated as would be common knowledge in the trade. The court declared the contract void for mistake.

In a similar way, a unilateral mistake could occur in construction contracts during tendering. The mistake could be operative (making the contract void) if party A has made a significant and material mistake, party B ought reasonably to have known of the mistake and party A is not at fault in any other way. Returned, priced tenders are usually checked by the design organisation, or occasionally, the employer. Design organisations should be experienced and competent in this area, and one would hope any employer carrying out the task themselves would be equally competent. It is usual to bring any apparent mistakes to the attention of tenderers.

If a contract involved piling for example, and a proper price would be about £50,000, it is apparent that a price of £5,000 is almost certainly a mistake. It is good practice to notify the tenderer of such errors. The usual method of dealing with such errors is to ask the tenderer to stand by the price or withdraw its tender. A less usual method is to allow the tenderer to correct the error. Unfortunately the latter procedure is likely to upset other tenderers, who see one tenderer as having 'two bites at the cherry'.

Another area of unilateral mistake is that of identity, where the identity of the party must be of crucial importance. There have been cases of people misspelling their names to resemble a famous one. This is unlikely to be of particular relevance to construction contracts.

Documents mistakenly signed

This is an area of law which could certainly be relevant to construction professionals. The usual rule is that contracting parties are presumed to have read and understood the documents even if they have not.

> As a general rule someone signing a document is deemed to have read and understood the document.

Hence the moral for contractors, design organisations, sub-contractors and suppliers is to ensure that they have read and understood all contract documents, before signing a contract based on them.

However where someone has been induced to sign a document by fraud or **misrepresentation** (see below) the transaction will be voidable. An example is *Foster* v. *Mackinnon* (1869) where a senile man with poor eyesight was induced to sign a document which he thought was a guarantee, whereas it was a bill of exchange. However, again to avoid valid contracts being set aside, the courts insist that such a mistake should be fundamental and the signatory must not have been careless.

Misrepresentation

Misrepresentation, or 'misrep' as it may be called, is an important area of law to the construction professional. Representations are statements made before the contract is formed, often to provide more information or to induce the other party to enter the contract. In comparison, terms are contained within the contract itself. It is possible for a representation to be false or incorrect and for someone to enter into a contract as a result.

Misrepresentation is a very common vitiating factor, because it is concerned with the behaviour and statements made by the parties, before the contract is concluded. Many contracts involve a period of discussion and clarification, before the actual agreement or 'bargain' is made. During this period statements are made and documents or other material may be produced, which induce a party to enter into the contract. The law provides protection against untrue or misleading statements which produce a contract based on false promises. This is misrepresentation. An example might be a holiday let in Spain. The agent shows you a number of photographs of the area, the beach, swimming pool and holiday apartment. You book a holiday for two weeks, only to arrive in a 'building site' that bears little resemblance to the photographs that induced you to book the holiday. You should have a valid case for misrepresentation.

It is worth noting that representations can be incorporated into the contract, or written into the contract terms before it is finalised. They then become normal terms of the contract and the usual contractual remedies apply. This would be a sensible course of action where it becomes apparent that a representation is *fundamental* to the contract and ought to be incorporated therein as a term, thereby giving it full contractual status.

However there remain all the other representations which may induce someone to enter into the contract. If any of these other representations are falsely made, then the 'injured party' should have some protection in law. This is the concept of 'misrepresentation'.

The consequence of misrepresentation, if it is proved in court, is to allow the 'injured party' to cancel the contract (if they wish) or to receive damages. Misrepresentation thus produces 'voidable' contracts. There are three categories of misrepresentation:

- Fraudulent misrepresentation refers to statements that are made knowing they are untrue.
- Negligent misrepresentation is where the person making the statement has reasonable grounds for believing it to be true.
- Innocent misrepresentation is where the person making the statement has no reason to think it is untrue.

In construction contracts misrepresentation can easily occur during the tender period, when tenderers may make enquiries which are answered by the employer's staff, or members of the design organisation, acting for the employer. Suppose a construction professional (CP), acting for the employer, was showing a prospective tenderer around a low-lying sewage works, where a new tank was to be constructed. Unusually, a ground investigation had not been carried out. The tenderer enquires whether there had been previous groundwater problems. Suppose the CP had constructed a new inlet works there previously and knows the conditions are in fact artesian, and ground water could be a severe problem. If the CP replies that there are no groundwater problems, this would be clearly untrue. If this statement induced the tenderer to provide a price that did not account for severe groundwater conditions, then the statement by the CP would be fraudulent misrepresentation, and the employer could later be liable.

Another construction example might relate to the location of the site. The employer's staff may already operate assets or sites in the vicinity and will know of any special risks of theft or vandalism. Again, such issues would affect the tender price, or even willingness to tender, because of increased security costs and possibly insurance considerations. It would be wise for the employer's contract documents to make a statement as to these risks. Again, false statements about such risks upon which tenderers rely could be actionable later for misrepresentation.

In conclusion, it is worth remembering that the employer or the design organisation working for the employer will supply tender documents based on engineering issues such as the drawings and specification. As professionals, they will no doubt take great care over the preparation of these documents. Additionally the employer is obliged to provide 'pre-construction information' under regulation 10 of CDM 2007. This pre-construction information is likely to contain data and information from the site investigation, including the ground investigation. It is particularly important that this information is clear, comprehensive and factual. If the information is not factual or is misleading, then the successful tenderer may have a later claim for misrepresentation.

This can be a particular problem with ground information. There are frequently two reports arising from the ground investigation, which is usually commissioned by the employer. These reports will be the factual report containing soil parameters, structural properties, grading and identification and so on, and the interpretive report, which will be a commentary upon the facts, deductions and opinions of the geotechnical engineer. The factual report should not pose a future problem, but the interpretive report may, since it could be construed as 'opinion'. Where information containing opinion is given to tenderers, such as the interpretive ground investigation report, it would be wise for the employer to make this clear and possibly insert a disclaimer.

Capacity

It may be better to see this section as 'lack of capacity' since the normal assumption in construction contracts is that the parties have capacity. So lack of capacity may be found in contracts made by minors, the mentally deranged or those so drunk that they do not understand the quality of their actions at the time.

Legal formalities (where required)

As discussed above, some contracts have to be made by deed. However for most contracts there are few if any legal formalities such as the requirement for witnesses (though the parties may insist on some of their own, such as a place of signature). Most contracts can be:

• Formal or informal
• Written or verbal.

However, since large sums of money are usually at stake, construction contracts are almost always of a written, formal nature, based on a form of contract, and many drawings, specifications, schedules and other documents. They are also normally signed using a standard form of agreement as discussed above.

Terms and representations

Representations

Before a contract is made it is necessary to look at the discussions between the parties, and any written material to ensure that the contract has been entered into on proper grounds. These discussions and materials will usually be seen as **representations**.

> Representations may be made before a contract is formed. The contract itself contains the terms.

Representations may be discussions as to the employer's general expectations, the level of site supervision to be provided, general information about the site and locality, how the contractor might be expected to behave and so on. Some very simple contracts are made without representations; they are not an essential prerequisite of a valid contract. Representations were also discussed above in the section on misrepresentation.

Terms

In a typical construction contract the **terms** will be everything contained in the specification, any schedules, the information shown on the drawings, any other information provided to the contractor and incorporated in the contract and of course the conditions of contract such as NEC3 ECC or JCT SBC05. It is very important to remember that the terms of a contract are not just the conditions of contract but all the many other items of information contained in the documents listed above.

Hence contract terms are not just obviously important items like the completion date and the method of payment, but also statements in the specification and lines and dimensions shown on drawings. So if a brick wall is shown as two metres high on a drawing, and the specification gives a maximum construction tolerance of 10 mm, then a wall constructed to a height of 2,050 mm is not in accordance with the contract, and a term has been broken. Clearly in this case the contract administrator will need to decide whether this small difference matters and whether the wall can be accepted as satisfactory. However, strictly speaking, the contractor has broken a term of the contract. There have been legal cases over the constructed depth of swimming pools. We can imagine that a pool 1,950 mm deep instead of 2,000 mm deep may not matter, but if the depth under a diving board is insufficient or the length of a pool designed for Olympic competition is incorrect then we have a rather different issue on our hands and extensive reconstruction may be necessary. This is because an important term of the contract or 'condition' has been broken, and the employer is not getting a benefit from the contract that he or she had a right to expect.

Occasionally the main terms are so uncertain that the contract may be void. However, in commercial contracts, the court may be prepared to enforce an ostensibly vague agreement by reference to trade custom, or previous dealing between the parties. We would not expect this to happen in a typical construction contract, since many months or even years will be spent developing a project until it reaches the stage of going out to tender and becoming a formal contract.

The difference between terms and representations

Figure 17.1 illustrates the difference between terms and representations, and will be explained further in this chapter.

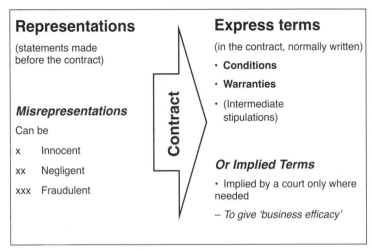

Figure 17.1 Terms and representations.

Differentiating between terms and representations can sometimes be difficult especially where there is no written contract. The courts attempt to deduce the intentions of the parties and there are a number of rules which the courts will apply. Where a party making a statement asks the other party to check it, then the statement will usually be a representation. However, where one party has superior experience and skill relative to the other party, the court will usually construe the statement as a term. This might happen in a discussion between a car dealer and a potential buyer for example. Another concern is the time lapse, thus where there is a distinct interval of time between the statement and the conclusion of the contract then the statement is more likely to be a representation. Finally, the importance of the statement will be taken into account. If the party is unlikely to have entered the contract had it not been made, then the statement will be construed as a term.

Another difference between terms and representations is the remedy available. Where a representation proves to be false then the remedy is a claim of misrepresentation, which was discussed above as a vitiating factor. Where a term of the contract is broken, the remedy is an action for breach of contract, which can lead to damages, or to the contract being rescinded (cancelled).

Terms of a contract can be express terms or implied terms

Express terms

The general principle of contract is that the parties to the contract agree on mutual rights and duties which they would not otherwise have in law. The parties may agree any terms they wish provided that they are lawful. These are express terms – the terms agreed in the contract. They are usually, but not always, written down. It is good practice to ensure that express terms covering all important aspects of the construction contract are comprehensive, clear and written down. This leads to much greater certainty. Where important express terms are missing, and a dispute ends up in court, then the court may **imply** terms. The outcome of such a case is obviously less certain than having clear express terms in the first place.

Types of express terms

Express terms are so fundamental to contracts that they are further divided into three categories depending on their importance. Important terms are called **conditions**; less important terms are called **warranties**. The latter is a legal word, and is not the same as a 'twelve-month warranty' on goods which we purchase such as a washing machine or a kettle. The difference between conditions and warranties becomes important if a term is broken or not fulfilled. The three types of term are listed below, together with the consequence of breach:

- Conditions are important terms that go to the heart of the contract. Hence breach of a condition entitles injured parties to treat themselves as discharged from the contract, and gives them the right to sue for damages.
- Warranties are less important terms. Hence breach of a warranty only gives the injured party the right to sue for damages.
- There is a third category, **innominate** terms, which can be either of the above depending on the effect of the breach. This gives the courts some flexibility when looking for the intentions of the parties.

The court will decide which category a statement falls under by considering the intentions of the parties. Use of the word 'condition' in a contract will often result in the statement being treated as a condition, but not always.

The parol evidence rule

This old rule meant that where one party to a written agreement tried to show that it did not represent the whole agreement, then this would fail. The logic presumably was that if parties made a written contract, they would write down everything that they wished to agree. However, over the years there have been many cases where the court has accepted that a part-written and part-verbal agreement was really intended at the outset. Items that caused problems would usually be verbal statements but could also be other pre-contract written matter such as draft contracts or correspondence. Parol may have a similar derivation to the Italian word 'parola' which means word. In order to avoid any potential problems it is best to properly incorporate by reference all correspondence and other matters that the parties wish to include in the contract. This is done simply by saying, 'we accept your tender of the 5 September 2011 together with your letter of the 10 September and our reply of the 14 September' and so on.

Implied terms

These may be implied by the court in the absence of a suitable express term, to give the contract 'business efficacy'. There are three categories of implied terms:

1. Terms implied by the courts to give business efficacy in the absence of a suitable express term.
2. Terms implied by custom or regular usage in a particular trade or profession in respect of which the contract is silent.
3. Terms implied by statute (law made by the government).

The classic modern statement on the doctrine of implied terms is that of Lord Simon in *BP Refinery Ltd* v. *Shire of Hastings* (1978), where he stated that for a term to be implied, the following conditions (which may overlap) must be satisfied:

- It must be reasonable and equitable.
- It must be necessary to give business efficacy to the contract, so no term will be implied if the contract is effective without it.
- It must be so obvious that 'it goes without saying'.
- It must be capable of close expression.
- It must not contradict any express term of the contract.

Examples of terms in a construction contract could be:

- *Express terms*: bricks shall be laid in stretcher bond, concrete cover shall be 40 mm, the contractor will not be given site possession until 25 March 2012, and the completion date shall be 25 August 2013.
- *Implied terms*: to give the contract business efficacy, where the parties missed something out; so if not actually stated, the courts would imply that the contractor will use reasonable workmanship, complete in a reasonable time and so on.

How does a contract end?

We now need to understand how a contract ends. Normally it ends or is 'discharged' as lawyers say, *by performance*. Performance simply means carrying out a contract in accordance with the drawings, specification and any other terms. However performance is only one of four ways in which a contract can end. These four ways are:

1. Performance
2. Breach of contract
3. Discharge by agreement
4. Frustration.

1 Performance

If both parties perform their obligations under the contract, then it is discharged (completed). This of course is what the parties want. However this raises yet another question – what is performance? The legal starting point for the concept of performance is that it must be truly complete performance of all obligations under the contract. An old case representing this rather harsh aspect of the law is *Cutter* v. *Powell* (1795), where a seaman agreed to serve on a ship from Jamaica to Liverpool for 30 guineas, payable on completion of the voyage. He died half way. His widow went to court for some payment and failed. The court construed the contract as 'entire', probably because there was a single payment payable on complete performance. There had been no complete performance so no payment was due.

'Entire' contracts such as *Cutter* v. *Powell* are rare. Entire contracts are ones where all the obligations are seen as a single transaction that cannot be broken down into constituent parts. Another simple example would be the purchase of a bar of chocolate for 50 pence – you 'contract' to get the whole bar, and it cannot be delivered in parts with payment on delivery of each part.

Most construction contracts are 'divisible' which means that they can be seen as being made up of a number of parts. If each part can be performed separately then it can be enforced separately, and the strict and narrow old rule of *Cutter* v. *Powell* need not apply. A typical example of a divisible contract would be payment in instalments, as is the case with the monthly payment terms in most construction contracts. Hence most contracts for work or materials (such as construction contracts) are divisible.

Another mechanism used in construction contracts to ensure reasonable payment for work done is the concept of **substantial performance**. Because most construction projects comprise many activities carried out over a long period of time, it would be unrealistic to expect them all to be absolutely complete before the asset could be taken over and used by the employer. It would clearly be unreasonable to say that a multi-million pound construction contract was not complete because the hinges on the gate were not painted.

The other point is that the employer will frequently be wishing to take over the new asset, and will not want that take-over to be delayed by very minor defects. Construction contracts usually have a period of a year after completion for defects to be corrected at the contractor's expense, so the employer is not exposed to any particular risks in taking over an asset that is 'almost complete'. However before a completion certificate is given, the contract administrator must ensure that the asset is operable, safe and any necessary training has been completed.

The idea of substantial performance is the norm in construction contracts. **Substantial completion** is the term used in ICE7 contracts, and **practical completion** is the term used in JCT contracts. NEC3 defines completion in a way which facilitates completion being granted, with minor work still outstanding, thereby having a similar effect.

2 By breach of contract

As we have seen above, the contract can be ended for breach of a condition. The party who is the victim of the breach can choose to continue with the contract and sue for damages, or consider the contract as 'repudiated' and effectively have it cancelled by a court.

3 Discharge by agreement

A contract is formed by agreement, thus it is logical for it to be possible to end it by agreement as well. If a contract is discharged by agreement then there should be consideration for the discharge, in order to make the discharge a valid contract in itself. Consideration for discharge would usually be a payment of some kind.

4 Frustration

Under the doctrine of frustration a contract may be ended if some event makes further performance impossible or illegal or if because of changed circumstances the contract becomes essentially different from that originally envisaged. An example would be *Taylor* v. *Caldwell* (1863), where Caldwell was to rent a concert hall to Taylor. After the contract was signed, the concert hall burnt down. Under these circumstances, the court released both parties from their obligations under the contract.

Remedies for breach of contract

'Breach of contract' means that one party has broken a term of the contract. There are four main remedies that the court can apply to cases of breach of contract, or in the event of an actionable vitiating factor such as mistake or misrepresentation. However the most usual remedy is **damages**.

1 Damages

Damages are the usual remedy for a breach of contract or actionable vitiating factor. There are two types of damages. Those granted by a court are **unliquidated damages**, although they are usually simply called damages. However, it is possible to pre-estimate damages for a specified breach, and write these in the contract; these are called **liquidated damages**. The specified breach in construction contracts is usually 'being late', in other words completing the contract after the completion date, or any extensions to it. Liquidated damages are very important in construction contracts and are covered in detail in Chapter 18.

Unliquidated damages (now referred to simply as damages) are awarded by a court with the intention of placing the claimant in the same financial position as it would have been if the contract had been properly performed. They are not intended to punish the defendant.

An important rule used by the court for determining a reasonable level of damages is that of 'remoteness'. The idea is that it would not be reasonable for a party in breach of contract to be exposed to all possible losses arising from that breach. When parties enter into contracts they are likely to do so after considerable thought and **contemplation**. They will no doubt weigh up the benefits, and also the difficulties and risks. One such risk is that they may not perform a term of the contract, and thus be in breach. It is reasonable to expect the parties to think about these issues which the law then says are 'in the contemplation of the parties' when the contract is made.

A famous case illustrating the principle of remoteness is *Hadley* v. *Baxendale* (1854) which involved late delivery of a mill wheel and the losses arising for the mill owner. The two principles of the case are often called the 'two limbs of Hadley and Baxendale'. They are that damages to be paid for breach of contract should only (1) be in respect of losses that arise naturally from the breach of contract or (2) be in the contemplation of the parties as being the probable result of a breach at the time that the contract was made.

Another well-known case is *Victoria Laundry (Windsor) Ltd* v. *Newman Industries Ltd* (1949) which is easier to understand. Here Newman Industries had a contract to deliver a large boiler to Victoria Laundry. Newman knew that Victoria Laundry wanted the boiler for immediate use in its business. The boiler was delivered five months late. Victoria Laundry sued Newman for the loss of profits during that period, both on normal business and on a very profitable special contract. Victoria Laundry won the case and the judge had then to decide on an appropriate level of damages. The judge decided that Victoria Laundry could recover for loss of profits on the ordinary laundry business. However, they could not recover damages for the loss of profits that they would have made on especially profitable dyeing contracts, because Newman did not know of these contracts at the time it entered into the contract. The potential losses on these lucrative contracts were therefore 'not in the contemplation of the parties' when the contract was made.

> The loss must arise naturally from the breach, or be in the contemplation of the parties at the time they made the contract.

If additional consequences of breach are made known to the other party when the contract is made, then it is quite likely that that party will increase the price for the increased risk. Hence if Newman were aware of the potential losses on the lucrative dyeing contracts, then it would have been liable for those losses if they were in breach. The commercial consequence of this is that Newman are now facing a much riskier proposition and will probably take additional steps to ensure delivery or factor the risk into their price.

Finally, when looking at what could be expected to be reasonably in the contemplation of the parties, the courts will consider what a reasonable person may have contemplated, not the particular person in question. In so doing the law is attempting to be as objective as possible.

The case of *Balfour Beatty Construction (Scotland) Ltd* v. *Scottish Power Plc* (1994) illustrates the application of the principle of remoteness in a construction setting. Balfour Beatty were employed to construct the roadway and associated structures of a section of the Edinburgh city bypass. They built a concrete batching plant near the site, and an agreement was made with Scottish Power for a temporary supply of electricity. Part of the construction work was a concrete aqueduct to carry the Union Canal over the road. During construction the batching plant broke down, because of the failure of the fuses provided by Scottish Power in their supply system.

Watertight construction of the aqueduct required a continuous concrete pour. Once power was restored, attempts were made to continue the construction by cutting out the old concrete and adding fresh concrete. Despite these efforts the contractor was unable to meet the specification for a watertight aqueduct. They were instructed by the Engineer to Contract to demolish the structure and start again. In an action against Scottish Power, Balfour Beatty claimed damages for breach of contract. The case went on appeal to the House of Lords against damages of £229,102.53 plus interest and the Lords confirmed that:

- Damages should either arise naturally from the breach or have been in the contemplation of the parties at the time they made the contract.
- What one party knows about the other's business is a question of fact.
- The demolition and reconstruction of the aqueduct was not in the contemplation of Scottish Power (since they did not know, and were unlikely to know, that a continuous pour was required to make the structure watertight).

2 Specific performance

Specific performance is an order of the court, compelling the defendant to perform its part of the contract. It will only be awarded if damages are an inadequate remedy, and is subject to other limitations (for example in contracts of employment).

3 Injunction

An injunction may be granted to enforce a negative stipulation (a promise not to do something) in a contract, where damages would be an inadequate remedy. This might be used for actors or singers who wish to work for someone else whilst under a contract for sole services, to ensure that they do not work for that someone else.

4 *Quantum meruit*

This Latin phrase simply means 'as much as is deserved'. This means a reasonable payment for work performed, where the other party is in breach of contract. Such a breach may well be where one party has prevented the other party from performing contractual obligations. This remedy is an alternative to claiming damages.

Limitation periods

It is a fundamental principle that someone should not be exposed to the risk of court action for an unlimited period of time, because of not only a consideration of fairness but also the difficulty of acquiring evidence. In construction the person or organisation exposed to such action will normally be the contractor but it may also be the design organisation, architect, quantity surveyor, building surveyor or other professional. This leads to the concept of limitation periods. A limitation period is a period in which a legal action can be brought. After this period, a claim can no longer be referred to a court. The periods are different in the various areas of the law. Limitation periods are almost all governed by the Limitation Act 1980 as amended by the Latent Damage Act 1986.

In contract

In contract, the limitation period is:

- Six years from when the **cause of action** accrued for a **simple contract**
- Twelve years from when the cause of action accrued for a **deed**, or a contract under seal.

Many construction contracts are simple contracts, that is, they have offer, acceptance and consideration. Deeds have to be signed 'as a deed' before witnesses. This very formal procedure is used by many employers for larger contracts. One reason is the extended limitation period for actions in law.

A cause of action is the right of a party to bring legal proceedings. A cause of action accrues from the moment the breach of contract occurs, not from when the contract was made or from when it was discovered. It is to be hoped that most construction problems are resolved during the construction period or if not during the period for the correction of defects, which is usually one year after completion of the construction work. This period is called the **maintenance period** or **rectification period** in contracts other than NEC3 ECC which sets a defects date by which defects should be corrected.

However the limitation periods really come into their own for **latent defects**, which are not evident during the construction period or the period for the correction of defects. Latent defects come to light after everyone thinks that the contract is finally completed and that all defects have been corrected. However the latent defect will usually be the result of some serious problem (a breach of contract) during construction that was not evident at the time. Hence in practical terms the six or twelve years will run from the point in time during the construction period when the breach of contract occurred.

Let us take an example. Suppose the columns below the lowest two floors of a tower block were specified to be of a higher strength concrete than columns for the upper floors, because of the higher loading. By an oversight, the contractor's site engineer negligently ordered normal strength concrete for the four columns at the north of the building and they were poured on 6 May 2011. The building was completed in July 2012. The employer took over the building. However it was not until August 2014 that the employer installed a mainframe computer on the first floor, heavy but within the design loading of the structure. Four columns failed and severe damage was caused to the north of the building. Subsequent tests on the columns revealed their inadequate strength. The cause of action accrued from 6 May 2011 (the date of the negligently poured concrete) and so even though the problem was not discovered until over two years from take-over, the employer has a right to take legal action against the contractor because the limitation period on this contract (under seal) is twelve years and the defect is a serious one.

Chapter summary

- All construction projects are subject to contract law, and have 'parties' to those contracts.
- The law sees agreements between the parties as 'bargains' which have offer, acceptance and consideration (usually the payment of money).
- There are many legal rules around the definitions of offer, acceptance and consideration.
- We must differentiate between offers and invitations to treat.
- The parties must also have an intention to create a legal relationship and capacity to make a contract. Formalities may also be required.
- Words and phrases in the contract itself are called terms of the contract.
- More important terms are called conditions, and less important terms are called warranties.
- Where the parties forget to include an essential term, then a court will insert one, called an implied term, to give the contract 'business efficacy'.
- Terms must be distinguished from representations, which are documents and words used to persuade someone to enter into a contract.
- If such representations are false or untrue they can lead to damages or the contract being cancelled, for 'misrepresentation'.

- Misrepresentation is one of a number of 'vitiating' factors which usually lead to cancellation of a contract or damages. The others are mistake, duress and undue influence and illegality.
- A contract can end in a number of ways, but the usual one is by performance.
- In construction contracts performance is rarely absolute, and is often defined as substantial completion or partial completion.
- If a contract is broken there are a number of legal remedies available to a court. The most common one is damages.
- Damages are not intended to punish the party breaking the contract. They are intended to restore the 'innocent party' to the position they would have been in if the contract had been performed.
- Damages imposed by a court can be called unliquidated damages. Damages written into a construction contract as a pre-estimate of loss are called liquidated damages and are discussed in Chapter 18.

18

Liquidated damages

Aim

This chapter aims to explain the meaning and operation of liquidated damages, which are used in many construction contracts.

Learning outcomes

On completion of this chapter you will be able to:

>> Describe breach of contract.
>> Outline the remedies for breach of contract.
>> Define the term liquidated damages.
>> Differentiate between liquidated damages and damages awarded by a court.
>> Explain the advantages of liquidated damages.
>> Calculate liquidated damages.
>> Explain the link between liquidated damages and extensions of time.

Breach of contract

As we have seen, contracts are made between the parties (such as the employer and the contractor) and they contain terms which give the parties both rights and obligations. If a term is broken, then this is likely to be a breach of contract. In a construction contract these terms will normally be written down and can be found in the conditions of contract, the specification, the contract particulars, the works information, details and dimensions shown on the drawings and so on. Hence there will usually be many thousands of terms.

In a simply drafted contract (with fewer terms than our contracts) where a term is broken, the parties may decide to use a **dispute procedure** or go to court to have the matter determined. Dispute procedures are described in Chapter 28 but usually include one or more of **arbitration, adjudication, mediation and conciliation**. However, since the advent of the Housing Grants, Construction and Regeneration Act 1996 (HGCRA), statutory adjudication has been available as a right. HGCRA gives a party to a construction contract the right to refer a dispute arising under the contract to adjudication as defined in the Act, even if this right is not set down in the contract. Since HGCRA applies to *all* parties to a construction contract its procedures can be used by the employer, the contractor or a sub-contractor.

In a simply drafted contract there may be no procedures *within the contract itself* for resolving difficulties. However because construction is a complex affair, with thousands of terms in the contract (including dimensions shown on the drawings and so on), it is very easy to break a term. There would be no sense in construction professionals having to go to court over such breaches. As a result construction contracts usually contain procedures for resolving the day-to-day difficulties and problems that arise on most projects. These procedures will be covered in detail in the chapters on NEC3 ECC and JCT SBC05 contracts (Chapters 21 to 27). However examples are:

- If the employer is late giving access to the site.
- If the contract administrator issues instructions changing the drawings or other contract information.
- If the contractor does not comply with the requirements of the specification or drawings.
- If the contractor is late in completing the work.

If the contractor has been delayed or suffered a financial loss, then the contractor normally makes a 'claim' (called a 'compensation event' in NEC3 ECC) under the relevant terms of the contract. The contract administrator (called the project manager in NEC3 ECC) decides the matter, usually after correspondence, meetings and the provision of supporting information by the contractor. Resolution by the contract administrator and acceptance of the decision by the contractor is almost always cheaper and quicker than going to court. This is why larger construction contracts tend to have more comprehensive contract terms to try and provide for all likely eventualities. The contract administrator uses the terms of the contract to make a decision on items such as those above. Therefore the parties do not have to use a dispute procedure or go to court unless they disagree with the contract administrator's decision or unless an event arises that is not covered by the terms in the contract.

Remedies for breach of contract

Dispute procedures such as mediation and conciliation are intended to help the parties in dispute to agree an acceptable outcome between themselves with the help of a third party called a mediator or conciliator. They are relatively quick and informal methods. The outcome and decision may then be formalised in a specific contractual agreement between the parties.

Arbitration and adjudication are more formal procedures with some of the features of court actions such as clear statements of each party's case and the disclosure of documents. In particular in arbitration the specified timescales stretch over many weeks. In adjudication and arbitration, an 'award' is made by the adjudicator or arbitrator. This award is similar in principle to damages ordered by a court. Because it is more formal, there are very few grounds upon which a decision in arbitration can be challenged in court. A decision in adjudication is valid and can be enforced, but it can be challenged in a court action. In practice, this rarely happens.

A court has a number of remedies, but the usual one is called damages, which is the award of money paid to the innocent party by the party who did not fulfil its obligations. The court award is intended to put the innocent party in the same position that it would have been in if the contract had been carried out properly. Damages are not intended to *punish* the party who is in breach of contract.

Difficulties in taking an issue to court

Taking an issue to court is not a cheap or easy option, and that is why contracts such as ours have provisions for resolving issues and problems within the contract itself. Where these provisions fail, then a dispute procedure is the normal next step before court action is contemplated. Some of the difficulties of court action are listed below:

- It can be very expensive in terms of the fees of solicitors and barristers.
- Additional costs arise from the time taken by the staff involved on the part of the contractor and the employer. These staff costs can amount to more than the legal fees.
- It can take a long time, often months or even years. This is not simply the time taken to have the case booked into the court diary, but the time taken in meetings, document retrieval and preparation and general administration.
- Court decisions are published. So the judgment is public. This may not be acceptable for commercial or marketing reasons to either party.
- Court is risky. Legal people will *predict* an outcome, but it is always an opinion as to the likely result based on the legal professional's judgment.
- A case can be decided in court, and then appealed to a higher court, with all the associated problems of the original case.
- Legal costs of the successful party are not always paid in full.
- The unsuccessful party must pay their own costs usually together with all (or most) of the costs of the successful party.
- Losses have to be proved to a court. In many cases this can be difficult through lack of records or other evidence.
- Losses which are too 'remote' may not be recovered (see Chapter 17).
- A court case may have a *long-term damaging effect on future work* with the employer in question as described below.

The employer's main concern

For most employers the real concern is that the contractor is late completing the contract. If contractors are running late, they will incur the ongoing site costs such as safety, accommodation and welfare. However, these costs may not be high and the contractor may be content to incur them for a few weeks, possibly because their attention is focused on new and larger projects. There is thus a potential imbalance of motivating forces between the employer and the contractor. If there were no alternative, employers would have to take the case to court with all the above problems and it can take many months or even over a year to get an outcome. The answer to this problem is to use **liquidated damages** which the employer writes into the contract. These recompense the employer for the costs of a late completion, and since the contractor pays them, they provide a real incentive for contractors to complete their contracts on time.

Additional problems of court for a contractor

Court action has some additional problems for contractors. A real dilemma for *them* is that they actually involve the *employer* in any court action. The contractor will often see the employer as a client or customer, and will almost certainly be hoping for further work in the future. Let us restate the logic, and extend it to a court action. The construction contract is between the employer and the contractor. It is usually administered by the contract administrator (who is often a named member of the design organisation). All day-to-day relationships are between the contractor and the contract administrator. The employer is rarely concerned and, after signing the contract, may have no involvement until completion and handover. This is particularly true of large employers such as utilities who may have thousands of construction contracts a year but are really in the business of running a successful water, gas or electricity company.

Let us suppose a problem arises on-site and the contract administrator makes a decision that the contractor does not accept. After further discussion, and a final decision by the contract administrator, the problem may go to a dispute procedure if there is one named in the contract. Statutory adjudication under HGCRA is also available whether named in the contract or not.

If the matter is not resolved the contractor must take the *employer* to court, not the contract administrator. It is the contractor and employer who have the contract. So imagine the reaction of directors of a large utility, who are happily running operational functions and suddenly find that they are 'in court' over a contract they may know little or nothing about. This is not good for business relations and future work for the contractor. Most contractors are extremely wary of going to court for this very reason.

Liquidated damages (delay damages)

For the *employer* one very good way around the problem of going to court for damages is to use **'liquidated damages'**. Liquidated damages are sometimes known as 'liquidated and ascertained damages'. Their abbreviated name in the industry is LDs or LADs. The NEC3 ECC contracts have changed the expression to 'delay damages'. At first sight, the word 'liquidated' may seem inappropriate or confusing. It is thought to come from the Latin word 'liquido' meaning clearly 'like a liquid'. In other words, liquidated damages are clear and apparent to both parties when the contract is made.

Liquidated damages are the estimated cost to the employer of a specified breach by the contractor. The contractor pays liquidated damages to the employer, or the employer deducts them from payments due to the contractor.

> The contractor pays liquidated damages to the employer, never the other way round.

Liquidated damages are calculated (not guessed) and written into the contract. So, at tender stage, contractors know exactly what their liability will be for that specified breach. The normal breach specified is 'being late'. This means delivery of the completed contract after the specified completion date, or any extension to it.

- Damages are those fixed by a court after a breach of contract.
- Liquidated damages are *pre-determined*, and written into the contract.

Liquidated damages are usually specified as being at a rate per day or per week. It is also possible to specify a limit to liquidated damages so that the contractor's risk is effectively capped at that limit. The limit is an amount in pounds that marks the contractor's maximum liability. Once liquidated damages reach this level, no more are paid. This can be a good idea, because the contractor will probably reflect its slightly lower risk in a less increased tender price.

Liquidated damages are usually deducted by the employer from monthly payments otherwise due to the contractor (but deducted by the project manager in NEC3 ECC contracts). However liquidated damages can be the subject of a court action, to have them enforced. Clearly, going to court to have liquidated damages enforced rather defeats the object, so deductions from payments are the usual mechanism. There are usually contractual provisions such as written notice before proper deductions can be made, which are described below.

Most forms of construction contract contain carefully drafted clauses for the imposition of liquidated damages, and associated procedures to ensure that a court will uphold them, if they are challenged. Where there are liquidated damages in a contract, they remove the right of the employer to go to court *for the specified breach of contract* but not for other breaches of contract. If the liquidated damages are about 'being late', then the employer can deduct LDs for the contractor's lateness, but the employer would have to go to court for different breaches of contract by the contractor. The other important point is that, for the specified breach, *the liquidated damages are paid regardless of the actual costs.*

So let us suppose we have a contract for the refurbishment of an office building, and the employer's staff have to move into alternative accommodation at say £5,000 per week, then this would be the assessed liquidated damages (or at least part of them). As the contract proceeds, the employer finds alternative accommodation at £3,000 per week. The contract overruns due to the contractor's default. The employer is entitled to deduct £5,000 per week as the original LD rate, even though it is now only costing the employer £3,000 per week. This may seem unfair, but the deducted rate of LDs would still be £5,000 per week if the employer ended up having to pay £7,000 per week for accommodation. The whole point of LDs is that they are a genuine estimate made at the time of tender, of the likely cost of a delay to the contract. Both parties enter into a contract in full knowledge and acceptance of this rate of LDs.

'Pricing the risk'

Another excellent feature of liquidated damages is that the contractor knows the rate to be paid and can thus 'price the risk' at tender. The author had personal experience of a pipe jack crossing (a form of small-diameter tunnel) below a railway, with a contract period of four weeks and a tender price of about £400,000. During this time the railway line would have to be closed because of the fear of settlement and possible derailment. If the contract overran, the railway company would charge £50,000 per day as their costs due to rescheduling trains and potential loss of business. The rate of £50,000 per day was accordingly inserted as liquidated damages in the construction contract. So tenderers faced a very high risk and were obliged to put an allowance in their rates to cover some of it. If they allowed too much they might not win the tender. The contract was completed on time, and in a later conversation with the contractor's managing director, the author enquired how the contractor had priced the risk. The reply was that they had 'put in three days' or in other words an additional £150,000 as a best guess as to what might happen; and in the event had made a healthy profit. This example also illustrates the effect of very high (and genuine) liquidated damages. Tender prices are likely to be increased to reflect the high risk.

Challenging liquidated damages

Once liquidated damages are applied by the employer, the contractor can challenge them in court. This is unusual because it is expensive and it will be bad for future business relations. The courts are likely to uphold a *challenge* to liquidated damages if they can be shown to be a *penalty* rather than a genuine pre-estimate. The court will have to decide whether the calculation of LDs was a genuine pre-estimate, or whether it was designed to punish a party in breach. If the latter is the case then LDs will be overturned and the court will automatically relegate the employer to a claim for general damages in the normal way.

The advantages of liquidated damages

In summary, the advantages of liquidated damages over going to court are:

- There are no legal costs and associated staff costs.
- There is no long delay in getting a court hearing.
- There is a private outcome, unlike a court decision which is public with the associated publicity that brings.
- The outcome of a court hearing can never be predicted with accuracy.
- An appeal to a higher court would result in further delay and cost.
- Calculation of LDs is quick and simple, unlike trying to prove actual losses in court. Courts usually require extensive and clear evidence.
- All proper costs can be included in LDs. If a case came to court, the court might find some heads of cost to be too **remote** and disallow them. The term remote is legal language for outside the contemplation of the parties to the contract when the contract is signed.
- The LD amount is clear to both parties when the contract is signed, which brings certainty.
- This certainty means the contractor can 'price the risk' if necessary, although this will increase the tender price of course.
- Deduction of LDs from payments due to the contractor is very simple.

Calculating and applying liquidated damages

Some simple rules for the calculation and application of liquidated damages can be deduced, based on good practice following court cases decided over the years:

- Liquidated damages must *not be a penalty*.
- Liquidated damages must be a *genuine pre-estimate* of the loss likely to be suffered, calculated at the time when the contract is made.
- It is advisable to keep a copy of the liquidated damages' calculation.
- Liquidated damages are frequently calculated by the design organisation as part of the preparation of tender documents. However, the calculation is contractually made by the employer. Consequently it is good practice for the employer to instruct how LDs should be calculated. A thorough employer will insist that the calculation is also signed off at a senior level in the employer's organisation.
- Details of the liquidated damages must have been properly entered in the contract itself.
- For liquidated damages relating to time (which is the normal case), the contract must have a clear completion date.
- In this case the contract is normally expected to have extensions of time (EOT) provisions for breach of contract by the employer.
- All appropriate contract certificates must have been given properly, especially those relating to time.
- Any special notifications required by the contract must have been given.
- Where *concurrent delay* exists (that is a combination of employer and contractor delay), the courts may view the deduction of liquidated damages with suspicion, unless the contractor is given an extension of time sufficient to cover the delay for which the employer is responsible.
- Where a contract provides for **sectional completion** (see below), it is advisable for liquidated damages to be calculated separately per section.
- Additionally the court will apply the ***contra proferentem*** rule against the employer. So if any contract provisions are ambiguous and have two different interpretations, then the interpretation against the employer's interests is taken. Roughly translated, *contra proferentem* means *against the person proffering the contract clause*.

A **sectional completion** is written into the contract to define a particular part of the contract works. Sections must be clearly identified and described in the contract documents and should have their own completion date. Sectional completions are often used on larger contracts, where a completion of a section of the works in a shorter time is reasonable and the employer wants to take over that section earlier. All contract periods, including those for sectional completions, must be set with sufficient time to complete the works with safety; this is a requirement of the CDM regulations. So for example in a two-year contract for the construction of a production facility comprising a factory, two warehouses and an administration building, the employer may want the administration building and associated access road completing in ten months to enable early transition of staff to the facility. This could be achieved by means of a sectional completion, inserted in the contract and clearly described. Most forms of contract designed for more complex projects contain provisions for sectional completion and the associated wording of the liquidated damages clause to include sectional completions.

So, in summary, liquidated damages will be upheld by the court if they have been estimated properly, a clear end date has been fixed, extensions of time provisions exist for employer default and the relevant terms of the contract have been administered properly.

Determining an appropriate rate for liquidated damages

There are a number of ways of calculating a rate for liquidated damages. One of the main areas of loss to an employer is that the money for the project will have been raised, or borrowed, but the project is late, thereby extending the period for paying interest on a loan, with no associated benefit from a completed project. A formula used by a major water company for the 'loan element' was:

LDs per week = 90 per cent of contract estimate × loan interest rate/52

The 90 per cent assumed that, at the time of liquidated-damage deduction for late completion, 90 per cent of the project would be complete. So assuming a loan rate to the employer of 5 per cent, and a £10,000,000 project, this calculation would give £8,563 per week. This is a very substantial sum.

Another usual area of loss is the cost of extended contract supervision by staff based on-site. Site supervision would traditionally comprise a resident engineer, and possibly clerk of works and inspectors. They would usually be employed or provided by the design organisation, and paid for by the employer as part of the overall payments to the design organisation. So here, let us assume two supervisors on £25,000 per annum and £50,000 per annum with employment on-costs (national insurance, holiday, sickness and so on) of 40 per cent. Their cost would therefore be £2,019 per week.

For these two items alone, of 'loan element' and extended supervision, the liquidated damages would amount to £10,582 per week. However, there are a number of other justifiable costs which an employer might add to the liquidated-damages calculation. These could include:

- Rent or rates on present premises
- Rent or rates for alternative premises
- Other site charges (that are the employer's responsibility)
- Extra payments to directly employed staff
- Insurance costs
- Additional administrative costs
- Other professional fees.

Once we get into the more commercial area of an employer's business, care must be taken to be able to justify liquidated-damages calculations if they are challenged. Such items might include:

- Costs of a delayed start to further contracts
- Loss of rental income

- Loss of profit on manufactured goods
- High anticipated profits for seasonal goods or sales.

An example of high anticipated profits might be a toy retailer expecting to be open before Christmas.

Because of the operation of the legal principle of remoteness, it would be prudent to bring the inclusion of any of these more commercial items to the attention of tenderers. There would then be no doubt that they were 'in the contemplation of the parties' when the contract was made.

A domestic example of liquidated damages

Suppose a family are having a major house extension done by a builder. It is so potentially disruptive that the family have to move out for six months whilst the extension is constructed. They have used an architect and a JCT contract. The extension is costing £80,000, and the time for completion of the contract is six months.

Suppose the builder is 8 weeks late in completing the extension, and it is entirely the builder's fault. The builder is in breach of contract for failing to meet the completion date. Remembering that the family have taken out a large bank loan and are paying interest but have had no benefit so far:

The family's costs per week are

• Loss of interest on £80,000 capital at 7 per cent bank loan	£100
• Hotel bill for the family of four	£1000
• Extra cost of site visits by the architect at £300 per visit	£900
• Extra transport cost of taking their children to school by car	£50
• *Lost rent for hire of two rooms to local tennis club*	£250

So the total cost to the family is £2,300 per week for 8 weeks. This is **£18,400**, which is a surprisingly large amount of money.

Without liquidated damages the family would have to go to court and sue for their money. This could take a year or so and could cost them more in legal fees and other expenses than their cost of £18,400. If they win they should recover most of their legal costs from the contractor, as part of the court settlement. However, first they have to win their case. The other problem is that their loss must not be too 'remote' (in a legal sense). So they would be unlikely to be awarded their *lost rent for hire of two rooms to the local tennis club* for administration of the summer competition unless the builder knew about it when the contract was signed. It is something out of the ordinary, and 'not in the contemplation of the parties' when the contract is signed.

A much better alternative would be

- They get their architect to check their estimate for costs of delay (we've just done it at £2,300 per week).

- The architect inserts this £2,300 per week into the contract as liquidated damages.
- When tendering, the builder now knows the risk for being late (this brings 'certainty' which is a good thing).
- If the builder is late, the family get their losses without going to court.
- The architect has to certify that the builder is late, and the family deduct the LDs from their monthly payments to the builder.

The liquidated damages clause in construction contracts

Most forms of contract in the construction industry have clauses providing for the inclusion and application of liquidated damages. Even some of the forms for small contracts have a liquidated-damages provision, for example clause 2.8 of the JCT Minor Works building contract. Some clauses such as ICE7 clause 47 are very procedural and this covers almost two pages of text, whereas the use of LDs in NEC3 ECC requires the inclusion of option X7 (delay damages) which is a fairly short clause. This is how these liquidated-damages clauses *usually* operate:

1. Where the contractor is delayed (late access, late drawings, instructions, changes and so on) the contractor makes a claim for more time to complete under the relevant clause of the contract.
2. The contract administrator considers the claim for more time and will want to see evidence and justification. Discussions and correspondence usually follow.
3. The contract administrator certifies all justifiable extensions of time (EOTs) or in other words revisions to the contract completion date for 'employer risks'.
4. This results in a new contract completion date, often called the revised completion date. If further circumstances warrant it, the contract completion date can be revised any number of times.
5. If the contractor is still late in completing the contract, the contract administrator informs the employer.
6. There may be other procedural requirements such as in JCT SBC05, where under clause 2.32, the contract administrator must also issue a 'non-completion certificate' to the contractor and the employer must also notify the contractor of whether deduction of LDs or payment of LDs is required.
7. The employer deducts LDs from any sums due to the contractor as a separate item on the interim payments certificate, so the contractor can see exactly how much has been deducted. In NEC3 ECC contracts, the project manager deducts them, rather than the employer.
8. If at a later date, after some LDs have been deducted, the contract administrator makes a further revision to the completion date, then the employer repays any applicable LDs (with interest in ICE7 and NEC3 ECC). So if for example the contractor was ten weeks late, and had been granted a four-week revision to the completion date, the employer could start deducting LDs from monies due to the contractor on contract certificates. If later the contract administrator grants a further two-weeks' revision to the completion date (six weeks in all) then the employer repays two-weeks' LDs (with interest in ICE7 and NEC3 ECC).

Provided all the above conditions are met, the employer may deduct liquidated damages regardless of the actual loss, or even if no loss has been incurred. However, a contractor may challenge the legality of the LD deduction and go to court, if the contractor considers that the contract has not been administered properly particularly in respect of any extensions of time provisions (which produce a revised contract completion date).

Liquidated damages and a revised completion date

Liquidated-damages clauses should always be related to clauses giving additional time to complete the contract, for various stated risk events. This link is very important because, without it, the contractor could be penalised for 'being late' when in fact it was the 'employer's fault'; in other words the result of employer prevention or default, or a risk that the employer had accepted when the contract was made. A court would view liquidated damages with great suspicion if they were imposed on a contractor without a proper extra time allowance for these items of 'employer fault'. It is often said that extensions of time provisions are for the contractor's benefit where in fact they are also for the employer's benefit. Without extension of time provisions the deduction of liquidated damages would almost certainly be viewed as a penalty by the courts, and would be overturned.

In ICE7 these stated risk events are listed in clause 44 and are called 'extensions of time' (or EOTs as they are known). An EOT effectively extends the date for completion of the contract. Contracts such as JCT SBC05 and NEC3 ECC speak of 'fixing a new completion date'. The effect is the same.

In JCT SBC05 the events giving the contractor more time are listed in a section entitled 'relevant events' in clause 2.29; whereas events entitling the contractor to payment for the financial loss and expense are called relevant matters and are listed in clause 4.24. In ICE7 events giving more time are summarised in clause 44, whereas additional payment is covered in a number of clauses. Hence JCT SBC05 clearly separates time from money and ICE7 generally separates the two in different clauses. Normally a revised completion date (extension of time in ICE7 terminology) will generate a valid claim for expense from the contractor, since the site overhead costs will be extended. At a basic level there would always be the additional costs of security, welfare, office facilities and supervision for the extended period on-site. So when the contract administrator certifies additional time, this usually produces a valid reason for an additional payment claim by the contractor. It is not unknown in these contracts for something of an unfortunate game to be played, where the contractor 'only wants more time' and once granted then makes a subsequent claim for the resulting loss and expense.

NEC3 ECC however does not separate time and money, but calls all such events 'compensation events'. Most of these are listed in clause 60.1, but some additional compensation events form part of the main option clauses A to F. The main reason that traditional contracts like ICE7 and JCT separate time and money is to give more flexibility around risk events. Hence in these contracts 'exceptionally adverse weather conditions' (the phrase used in JCT SBC05 in clause 2.29) give additional time, but are not listed in the clauses which give the contractor payment of loss and expense (called a 'relevant matter' in JCT SBC05 clause 4.24). The phrase that is often used to describe this state of affairs is 'there's no money for weather'. The advantage of this separation of time and money is that both the contractor and the employer suffer for exceptionally adverse weather conditions. The contractor suffers because there is no payment for such conditions, and the employer suffers because the project will be delivered late, and liquidated damages cannot be imposed. So this becomes a 'shared risk', as it is sometimes known, since both parties accept some of the consequences. This affects the risk balance and hence the prices inserted in the tender by the contractor.

NEC3 ECC attempts to be more scientific about weather and substitutes a 'probability assessment' based on weather measurements at a specified 'place where weather is to be recorded'. This would usually be a Met Office weather station. Because NEC3 ECC does not share this risk in the same way as many traditional contracts, many employers strike the clause out and simply put the whole of the weather risk on the contractor. Of course, this action is likely to increase the tender price.

Most contracts also cover the situation where the contractor has overrun the extended completion date and is paying liquidated damages, when the completion date is extended again. This frequently occurs because of further evidence or information being produced, or possibly a decision of an adjudicator or often simply by the contractor convincing the contract administrator that more time is due for a particular event. The normal contract provision is for the contract administrator to extend the completion date further, and for the employer to repay any affected liquidated damages (with interest in ICE7 and NEC3 ECC contracts).

Liquidated damages and early take-over of parts or sections

Another possibility covered by most contracts is take-over of **part** of the works by the employer before the whole of the works are completed. This is not the same as completion of a **'section'** of the works, which is pre-determined and described in the contract when it is signed. Sometimes for operational reasons, or possibly for rent or lease, the employer wishes to take over a part of the works which has been completed. There are a number of contract formalities that need not concern us here, but clearly there would be an impact on the rate of liquidated damages due, which were originally calculated for the whole of the works. Most contracts reduce the rate of liquidated damages due by the proportion that the completed part bears to the whole of the works. ICE7 does this in clause 47(1), JCT SBC05 in clause 2.37 and NEC3 ECC in clause X7.3.

Where *sections* of the works are used, they must be clearly identified and described in the contract. ICE7, JCT SBC05 and NEC3 ECC all make provision for stating the rate of liquidated damages that apply to each section. ICE7 and JCT SBC05 automatically include this provision for sections, but in NEC3 ECC sectional completion must be added as option X5.

Poor completion of the *rate* of liquidated damages in the contract

The rate of liquidated damages, in pounds per day or pounds per week, is inserted into the contract, usually by the design organisation after consultation with the employer. There have been problems in traditional contracts that are less likely to occur in NEC3 ECC, since inclusion of delay damages is a specific act by inserting option X7. The question is: what happens if a rate of 'nil' is inserted, or no rate at all is inserted or a dash is inserted? The lack of insertion of a rate of LDs can happen quite easily since the design organisation probably completes the contract data over a period of time as information becomes available and decisions are taken. The design organisation may be waiting for an LD rate from the employer, and simply neglect to complete it when the tender documents are sent out. In the event of a dispute, a court would have to decide what the intentions of the parties appeared to be. There have been a number of cases on this issue over the years and the courts seem likely to decide that in this event no damages at all are available for the specified breach. In other words neither liquidated damages nor general damages would be due against a contractor who completed late. This should be a warning to all construction professionals who write contracts. Great care is needed at all times.

Some commercial issues with liquidated damages

Clearly the deduction of liquidated damages costs the contractor money. It may also affect reputation and tendering opportunities, which is why many contractors will fight very hard to have them repaid, often as part of a final settlement with the employer. Most employers carry out financial checks on tendering contractors. They may also have a number of competence or 'quality' criteria as part of the selection process. Such quality criteria around performance could be questions about the number of arbitrations per annum or other dispute processes, or there could be a question about the number of contracts over say the last three years where liquidated damages were deducted. This latter question would indicate contractor default against time for other clients. Hence contractors do not like the deduction of LDs 'on their record'.

Quite frequently certain events or costs remain 'in dispute' after the contract administrator has certified all time and money that the contract administrator considers due. If the contractor is dissatisfied, there is always recourse to a formal disputes procedure such as arbitration or adjudication. Many contractors will start by seeking a meeting with the employer. It is often possible for the contractor and employer to reach a **commercial settlement** over outstanding issues thereby avoiding a formal disputes procedure which can be time-consuming and expensive for both parties. Liquidated damages that have been properly deducted can be a useful negotiating item for the employer who will often agree to repay them as part of an overall commercial settlement.

A commercial settlement is the phrase applied to a final agreement that avoids the need for further action, be it a dispute procedure or court action. Here, once the contract administrator has made all final decisions under the contract, the contractor may still be dissatisfied and consider that there may be justifiable grounds for taking the matter to dispute or court. The employer may well choose to take further advice (both contractual and legal) and make an 'offer' to the contractor of further payment. This does make a lot of sense and avoids the costly business of dispute or court action, with their less certain outcome for both parties.

Chapter summary

- A party to a construction contract can refer a dispute to statutory adjudication under the Housing Grants, Construction and Regeneration Act 1966.
- A party to a contract can also take the other party to court for breaking a term of the contract. However going to court is very time-consuming and expensive and it can have an uncertain outcome.
- Damages awarded by a court are called damages, general damages or unliquidated damages.
- A very good alternative for the employer is the use of liquidated damages.
- Liquidated damages (or LDs) are called delay damages in NEC3 ECC contracts.
- Liquidated damages are a pre-estimate of the employer's losses (usually for late completion by the contractor) and are inserted in the contract as a rate per week (or per day). This brings much more certainty to both the employer and the contractor.
- If the contractor is late completing, the employer simply deducts the rate of liquidated damages per week (or per day) from payments otherwise due to the contractor. No court action is necessary.
- There are a number of safeguards for the contractor, particularly that the contract must have provisions for employer default and that the contract administrator must have abided by all these provisions and given all extensions to the contract completion date that are properly due.
- There must also be properly applied provisions for staged completion such as sectional completion and part completion.

19

Negligence

Aim

This chapter aims to outline the tort of negligence and to explain its significance to construction professionals.

Learning outcomes

On completion of this chapter you will be able to:

>> Describe the basic principles of the tort of negligence.
>> Differentiate between tort and contract law.
>> Identify how negligence could apply in your own professional practice.
>> Relate the concept of employer's liability to potential work situations.

Tort

The law of tort is separate and distinct from the law of contract:

- Contract law is concerned with parties who enter into a legal agreement with rights and obligations. They do this deliberately in order to create a legal relationship.
- By contrast, people will often incur liability in tort, without intending to.

Hence tort is concerned with the interactions between individuals who do not *usually* have a contract relationship, although they may have. Where a contract relationship exists as well as a potential action in tort, the lawyers will often advise their clients to seek an action in both. If they are successful however, the claimant can only be compensated once.

Torts generally are concerned with maintaining the status quo, and compensating those who have been disadvantaged. The law tries to strike a difficult balance between one person's rights and freedom to act, and the effect it might have on the rights and freedoms of someone else.

Since tort is so wide ranging, particularly the tort of negligence, it is important for the construction professional to have some knowledge of the subject. Indeed, construction professionals may commit negligent acts in a work context. Their employer will usually be liable but of course will not be very pleased with the situation. In particular, design organisations pay a great deal of money every year for professional indemnity insurance which is designed to protect them in the event of a claim for negligent design. The premiums for this insurance depend largely on the turnover of the design organisation, but for the larger organisations will be millions of pounds per year. Hence negligence and its consequences on the construction professional are no minor matter.

> It is important for all construction professionals to have some understanding of the law so they 'know when to take advice'.

There are frequent occasions when a problem or an issue may develop into a legal situation, possibly with far-reaching consequences. It is advisable for all professionals to recognise such situations and either consult their manager, an expert within their company, or if necessary an external expert. This expert may be a legal professional such as a solicitor or barrister, but most consultations begin with one of the many organisations who specialise in construction issues and claims. These firms often comprise experienced engineers and quantity surveyors who have extensive experience of the industry. They often have legal qualifications in addition, but are primarily construction professionals rather than legal ones.

The main torts

The word tort is thought to be a surviving relic of old Norman French and probably simply meant 'wrong'. There are a number of torts of which the main ones are:

- Negligence
- Trespass to the person (assault, battery and false imprisonment)
- Trespass to land
- Interference with goods
- Economic torts such as deceit
- Breach of statutory duty such as those under health and safety legislation
- Nuisance
- *Rylands* v. *Fletcher* (this typically concerns the escape of water brought onto land, such as from a reservoir)
- Defamation.

We will only consider negligence.

Negligence

Negligence is one of the most significant torts and usually compensates people for physical injury, or damage to their property arising from another person's negligence. Mental injury can be compensated as well as physical injury. It should be noted that the tort of negligence imposes stricter rules on recovering damages where the loss is only **pure economic loss**. Pure economic loss is financial loss where there has been no harm done to the person's body or property. This is discussed later in this chapter.

Many of the cases quoted in this chapter such as *Donoghue* v. *Stevenson*, discussed below, were tried in the House of Lords. This was the ultimate court in English Law and many of the most important cases went to the House of Lords on appeal. Decisions of the House of Lords are binding on all lower courts, which is why they are so important. Cases were usually heard by five 'Law Lords'. The Constitutional Reform Act 2005 established a replacement for the House of Lords which is now called the Supreme Court of the United Kingdom; it started work on 1 October 2009.

Donoghue v. *Stevenson*

Today, the tort of negligence is of prime importance and forms many of the cases going through the courts. However this was not always the situation. It is said that the case of *Donoghue* v. *Stevenson* (1932) really established the modern concept of negligence. The case is sometimes referred to as the snail in the bottle of ginger beer. Donoghue's friend bought her a bottle of ginger beer. The bottle was opaque. After drinking some ginger beer, her glass was topped up from the bottle and allegedly the remains of a decomposed snail slid into the glass from this bottle. Donoghue became ill. Because the drink had been bought for her by a friend, there was no contract between Donoghue and the seller of the ginger beer. As a result Donoghue sued the manufacturer (Stevenson).

It was quite evident that this case could have far-reaching consequences and as a result it was ultimately referred to the House of Lords on appeal, as many important cases used to be (they would now go to the Supreme Court). There was no contract between Donoghue and Stevenson because Donoghue's friend had bought the drink. However, the House of Lords found the manufacturers (Stevenson) liable in negligence. They decided that even in the absence of a contract a manufacturer can be liable to the ultimate consumer where a product has caused physical damage. The case established four key principles:

1. Negligence was accepted as a separate tort in its own right.
2. Negligence would be proved if a three-part test is satisfied.
   - The existence of a duty of care owed to the claimant by the defendant.
   - A breach of that duty by falling below the appropriate standard of care.
   - Damage caused by the defendant's breach of duty that was not too remote a consequence of that breach.
3. A manufacturer would owe a duty of care to consumers or users of that manufacturer's products not to cause them harm.
4. The method of determining the existence of a duty of care is the so-called **'neighbour principle'**. Lord Atkin, one of the judges involved, said 'You must take reasonable care to avoid acts or omissions which you can reasonably foresee would be likely to injure your neighbour. Who then in law is my neighbour? Persons who are so closely and directly affected by my act that I ought reasonably to have them in contemplation as being so affected when I am directing my mind to the acts or omissions in question.'

Damages

If the claimant wins the case, the court will usually award a sum of money in compensation, called damages, payable by the defendant to the claimant. Damages are generally awarded to compensate the claimant; in an attempt to put them in the position they would have been in, had they not been 'wronged'. This is not directly possible in many cases of personal injury, but the court will attempt to compensate for pain and suffering, loss of future earnings, loss of enjoyment of life and so on.

Damages are usually awarded for the *injury or loss* (and this may include mental distress where it is directly linked) and the *resulting physical damage*. So in a motor accident resulting from negligent driving we might see damages awarded for physical injury and the cost of repairing the damaged vehicle owned by the claimant. However, the law of negligence is more restrictive in relation to pure economic loss. This is financial loss alone, which is not accompanied by other damage. An example of pure economic loss might be a financial loss resulting from negligent advice on pensions. The main area where pure economic loss can be recovered is in relation to 'negligent statement' typically in relation to the giving of professional advice. This is clearly of interest to construction professionals and is described in the next section.

The steps required to prove a case in negligence

There are a number of steps required to prove a case in negligence and they are summarised in Figure 19.1 and explained further in the sections below.

Duty of care

As the law on negligence developed it became recognised that certain relationships would give rise to a duty of care. Generally, a duty of care arises where one individual or group undertakes an activity which could reasonably be seen to harm another individual or group, either physically, mentally or economically. This includes for example such common activities as driving (where physical injury may occur). The duty of care has two important elements:

- Firstly the question of whether a duty exists in the first place.
- Secondly the *standard* that should be applied to the performance of that duty.

Does a duty of care exist in the first place?

This is a difficult question. In answering it the court will use previous judgments and it will also have an eye for the likely consequences of its own judgment in either direction. In other words it will have some regard for the 'policy' that it is creating. Society and its values change with time, and this can have an effect on what a reasonable policy might be. There is no doubt that in the years since *Donoghue* v. *Stevenson* there have been a number of cases which have overturned previous judgments or modified them considerably. As a result, statute law (that made by the government) has sometimes been enacted to clarify the law which has changed as a result of precedents set by cases themselves.

In principle the claimant has either to prove that his or her case is within one of these existing 'duty situations' or to persuade the court to recognise a new

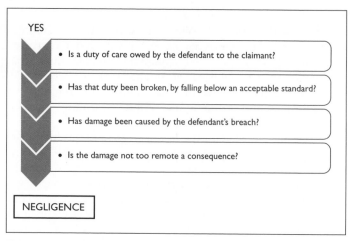

Figure 19.1 Negligence – the steps to proof.

duty situation. The standard of duty is as far as possible an objective one, that of a 'reasonable man' (the 'man on the Clapham Omnibus' as a judge once remarked), not the actual defendant.

However this standard of duty, defined by the courts, can be changed in later cases. An example is the famous case of *Anns* v. *Merton LBC* (1978). Here the Merton LBC, a local authority, had failed to ensure that building work complied with the plans (it is a local council's responsibility to check that all buildings comply with the Building Regulations). A subsequent owner, Anns, succeeded in a case claiming that damage caused by inadequate foundations was the responsibility of the council. Clearly this case had wide-reaching implications for councils and their building inspectors.

However, a later similar case, that of *Murphy* v. *Brentwood DC* (1990) was decided against the claimant and the council were not found liable. These cases came under some scrutiny because of their apparently different result, and today a three-fold test is preferred. This was set out in *Caparo* v. *Dickman* (1990) where shareholders in a company bought more shares followed by a successful take-over bid on the basis of an auditor's accounts. They subsequently found that the company was worth less than they thought and sued the auditors. The House of Lords decided that the auditors owed no duty of care to people who were potentially to take over the company, and company accounts should not be relied upon for this purpose. In *Caparo* the House of Lords established what is known as the 'three-fold test' (a series of three factors), which is that for one party to owe a duty of care to another, the following must be established:

- The harm which occurred must be a reasonable foreseeable result of the defendant's conduct.
- A sufficient relationship of proximity or neighbourhood exists between the alleged wrongdoer and the person who has suffered damage.
- It is fair, just and reasonable to impose liability.

The real difficulty that the courts have is keeping the range of persons to whom a duty is owed within sensible limits, whilst still trying to compensate individuals who have suffered and are within the range of reasonable foreseeability of the negligent individual. This is illustrated by Lord Roskill who said in his judgment on *Caparo* v. *Dickman* 'the submission that there is a virtually unlimited and unrestricted duty of care in relation to the performance of an auditor's statutory duty to certify a company's accounts, a duty extending to anyone who may use those accounts for any purpose such as investing in the company or lending the company money, seems to me untenable.'

Establishing the standard to be applied to the duty of care

Once a duty of care is established in any given situation, it then has to be decided whether the individual concerned has fallen below an appropriate standard. The standard applied is that of a 'reasonable man'. This has been defined in many cases, but a well-known definition is found in *Blyth* v. *Birmingham Waterworks Co* (1856) thus:

> Negligence is the omission to do something which a reasonable man, guided upon those considerations which ordinarily regulate the conduct of human affairs, would do, or doing something which a prudent and reasonable man would not do.

A court will also expect that the standard of care should reflect the potential risk. Hence in the case of *Bolton* v. *Stone* (1951) the claimant was standing in the road outside a cricket ground when she was struck by a cricket ball hit over a five-metre high fence. There was evidence that balls had been hit out of the ground only a few times in the last thirty years. The court decided that the risk was so low that the cricket ground was justified in taking no further measures to prevent a recurrence. In a similar way, the practicability of precautions will be taken into account.

However, people who hold themselves as having a particular profession or skill must attain the level of a reasonably competent person exercising that profession or skill. Thus one would expect from builders the degree of skill appropriate to a reasonably competent member of their trade and from accountants or solicitors also an objective standard of competence. Such people may, therefore, be negligent even though they do their best. The standard of duty applied to professionals such as doctors, engineers and architects is judged by the standard of a reasonably competent member of that profession, and inexperience is no excuse. However the level of competence to be expected will also vary according to the level of the risk. Hence a surgeon performing a heart bypass operation will be expected to reach a higher standard of skill than one performing a simple and minor operation.

One difficulty is deciding who should set that standard. The rule normally applied is to look for the standard generally accepted as proper by members of that profession and the body of knowledge of that profession. The 'Bolam test', as it is known, is often used. The case was *Bolam* v. *Friern Hospital Management*

Committee (1957).The claimant agreed to undergo electro-convulsive therapy (ECT) during which he suffered a fracture to the pelvis. The issue was whether the doctor was negligent in failing to give a relaxant drug before the treatment, or in failing to provide restraint during the procedure. Evidence was given of various medical practices. One body of medical opinion favoured the use of relaxant drugs, but another body of opinion said that they should not be used because of the risk of fractures. The action failed. It was held that a defendant is not negligent if he or she acts in accordance with a practice accepted at the time as proper by a responsible body of professional opinion skilled in the particular form of treatment.

Today the established practice is to have expert evidence from a member of the profession in question who will give evidence on what is reasonable and responsible conduct for the profession concerned. The judge must also be satisfied that the expert's opinion is reasonable and responsible.

Causation

As we have seen, the claimant first has to show that a duty of care existed and then prove that it has been broken by falling below an appropriate standard. However, once this is achieved, the claimant still has to prove that the defendant's negligence actually *caused* the damage. In determining the issue it is usual to apply the 'but for' test. In other words the claimant would not have suffered damage 'but for' the defendant's negligence. The test can also be used to eliminate those issues that have not caused the harm or damage.

An example of the 'but for' test is *Barnett* v. *Chelsea and Kensington Hospital Management Committee* (1969) where three night-watchmen went to a hospital casualty ward at 5.00am on New Year's Day complaining of vomiting and stomach pains. The duty doctor by telephone (in clear breach of a duty of care) told the duty nurse to tell them to consult their own doctors in the morning. A few hours later one of the men died; it was established through arsenic poisoning. The court found the hospital not liable for failure to treat, because it was shown that the man would not have recovered anyway. Even though the duty doctor was negligent, the failure to treat was not the *cause* of death.

Res ipsa loquitur

The usual principle in law is that claimants must prove their cases. This can be a great burden on someone who has clearly suffered harm, but cannot easily track its origin. Hence we have the principle of *res ipsa loquitur*, which simply means 'the thing speaks for itself'. It is applied in limited circumstances by the courts because of its effect on the defendant's position. Generally it has to be proved that the thing causing the harm was in the control of the defendant; that the incident was of a type that could only have been caused by negligence; and finally that there is no other direct evidence of what caused the incident.

Thus in *Scott* v. *London and Saint Katherine Docks Co* (1865), the claimant was standing near the doorway of the defendant's warehouse when he was hit by several bags of sugar falling from a hoist operated by the defendant's employees. There were many ways in which the incident could have happened from negligence such as incorrect stacking, or poor loading and lifting. The claimant was successful in his case because, although the exact cause of the accident wasn't clear, any of the causes were in the control of the defendant, the accident could only have happened through negligence, and there was no direct evidence of any unrelated cause. *Res ipsa loquitur* is also used frequently in medical negligence cases.

Remoteness

Even if all the other tests are met, the claimant may still fail if the damage is too 'remote' a consequence of the original breach of a duty of care. The concept of 'remoteness' limits the extent of compensation for the consequences of a defendant's wrongful act or omission. The consequences must not be of a kind that are unlikely or unforeseeable. If it were not for this legal restriction then compensation for the consequences of our wrongdoings could be endless. This is why the concept of remoteness is necessary, to limit the extent of liability to something that is reasonable. This was not always so, but has become the norm since the case of *Overseas Tankship (U.K.)* v. *Morts Dock and Engineering Co (the Wagon Mound)* (1961) or simply 'the Wagon Mound' as it is generally known.

The Wagon Mound

The Wagon Mound was a tanker which negligently discharged oil into Sydney Harbour whilst moored at a wharf for the purpose of discharging gasoline products and taking in bunkering oil. During the early hours of 30 October 1951, a large quantity of bunkering oil was, through carelessness, allowed to spill into the harbour. The oil was carried by wind and tide to the Morts timber wharf which was about 200 metres away where welding on another ship, the Corrimal, was being carried out. When they made enquiries Morts Dock were advised that it was safe to continue with the welding operations on its wharf. This was justified because sparks do not usually set fire to oil on water. However, two days later some molten metal fell onto a piece of cotton waste, which was floating in the water. The cotton waste caught fire and ignited the oil. As a result, the wharf and the ships being repaired were damaged in the blaze.

Morts Dock failed in its action against Overseas Tankship for the damage caused by the fire because it was too 'remote' a consequence. The oil also congealed on the slipways and interfered with Morts' use of the slips. The court noted that a claim for fouling of the slipways would have succeeded, since this was foreseeable. However the fact that some of the damage suffered (the damage to the slipways) was foreseeable did not make the defendants liable for the fire damage which was unforeseeable. The test for remoteness of damage was whether the kind of damage sustained was reasonably foreseeable.

Hughes v. Lord Advocate

Another key principle is that the defendant is not liable for all the direct losses. The *kind* of harm sustained by the plaintiff must also be reasonably foreseeable as a consequence. However, it appears from more recent decisions that the precise nature of the harm suffered need not be foreseeable. Liability arises if the injury was of a kind that was foreseeable even if it takes an unusual form. An example of this principle is the case of *Hughes v. Lord Advocate* (1963) which began in the Scottish courts but ended up being finally decided in the House of Lords. Here employees of the Post Office negligently left an open manhole unattended in the street. It was covered by a canvas tent and surrounded by paraffin warning lamps. Out of curiosity two young boys entered the tent and the claimant, a boy aged eight, took one of the lamps in with him. The lamp was knocked into the manhole and caused an explosion in which the boy suffered severe burns. The defendants were held liable. Whilst the explosion was unforeseeable the kind of damage which occurred, burns, was of a type which was foreseeable.

The 'eggshell skull rule'

However, provided the kind of harm is foreseeable, then the defendant will be liable for its extent, even if this is far-reaching. This is often called the 'eggshell skull rule'. It has been used in cases involving claimants with unusually thin skulls, or, for example, someone who suffers from haemophilia, notwithstanding that the defendant could not possibly have known of these medical conditions. Haemophilia is a disease which prevents clotting of the blood, and can hence lead to excessive blood loss and even death as the result of an accident. In the case of *Smith* v. *Leech Brain and Co* (1962) a workman, Mr Smith, received a burn on the lip from molten metal whilst at work because of the defendant's negligence. Mr Smith's lip was in a pre-malignant cancer condition, and the burn activated a cancer which led to his death. The employer was found liable. The question was whether the burn was foreseeable (the *kind* of harm). Once this was proved, the defendant was responsible for the full *extent* of the harm (the effect on the pre-cancerous condition), even though the cancer was not foreseeable. Another legal maxim describing this approach is 'you take the victim as you find him'.

Defences to a claim for negligence

There are a number of defences to a claim of negligence. Some of these are absolute in that they rebut the claim entirely, and some simply reduce the damages that would have been paid. So it is possible to 'consent' to a risk, which means there is no claim. Alternatively a person may 'contribute' to the harm he or she suffers, in which case the damages may be reduced.

Volenti non fit injuria

This means 'to one who is willing no harm is done'. It may also be called the 'doctrine of the assumption of risk'. *Volenti* is a complete defence so the action fails and the defendant pays no damages. In other words the claimant's acceptance of the risk may prevent them winning the case in court for some conduct of the defendant which would normally be actionable.

Hence boxers in a boxing match consent to the high risks of their sport. Similarly, if someone takes part in a game of rugby, he is assumed to accept the scrums and tackles which are a normal part of the game, and any harm would not normally present an opportunity for legal action. However, players can expect the referee to control the game. If the referee does not control it properly, and someone is hurt, then the referee can be sued in negligence. Similarly, sticking a knife into a person would normally be actionable, but if a surgeon does it with the consent of the patient it is acceptable. This is why the risks of an operation are now always explained to the patient, and the patient signs a 'consent' form.

People attending ice hockey games (hit by a puck), and motor racing (hit by a car running off the race track) are notable legal examples of *volenti*. However, the ice stadium or race track owners would need to take reasonable precautions against injuring spectators.

For a defence of *volenti* to succeed, the defendant must show three things:
* Knowledge of the precise risk involved
* Exercise of free choice by the claimant
* A voluntary acceptance of risk.

Knowledge of the precise risk involved

This test is a subjective one, involving the actual claimant. The defendant must show that the claimant had a precise understanding of the risk and was prepared to take that risk. It does no good for the defence to show that the claimant 'should have known'. An example of this is *Stermer* v. *Lawson* (1977), where the young and unlicensed claimant borrowed his friend's motorcycle but was not shown how to use it properly, and so could not appreciate the risks involved. The claimant was injured and sued his friend.

The defendant's claim of *volenti* failed because the claimant did not fully understand the precise nature of the risks involved.

Exercise of free choice by the claimant

In accepting a risk, a claimant must have a genuine free choice. Thus acceptance of risk cannot be induced by threat or fraud which would override a person's free consent. An example is *Smith* v. *Baker* (1891), where the employers of a quarry worker claimed that the injured worker accepted the risk of rocks falling from a crane. The court decided that although the worker knew of the possibility of falling rocks, and continued working, he had not accepted the risk because there was no proper warning of when the crane was in use and so he was thus unaware of the danger.

The other point here is that where employers try to demonstrate *volenti* on the part of an employee, they will rarely succeed. This is partly a matter of policy, but also because the unequal relationship means that the employee does not usually have a truly free choice. However, if the employee breaks the employer's safety rules, then *volenti* may be accepted by the court. This was the position in the case of *Imperial Chemical Industries Ltd* v. *Shatwell* (1965) where an employee was injured in an explosion after acting in breach of the employer's safety regulations. The House of Lords accepted that the claimant was responsible for his own injuries and the defendant was not liable.

Voluntary acceptance of the risk

Volenti will usually apply in particular jobs where there is an inherent danger, such as a test pilot. Where an employee expressly assumes a risk, and is even paid extra for doing so, any harm resulting will rarely be the fault of the employer, unless there is evidence that the employer was negligent and created a risk which was not normally present even in an activity that is inherently dangerous. Contrast this with the case of *Smith* v. *Baker* (1891) above, where a quarry worker does not consent to the risk of being hit by rocks falling from a crane.

Volenti cannot be pleaded by an employer in an action for damages based on his breach of a statutory duty, such as a failure to fence machinery under safety legislation. The reason is that the object of the statute, to protect workmen, cannot be avoided even if there is a private agreement between employer and employee.

The rescue cases

There have been a number of cases relating to rescue, where different rules generally apply. These rescue cases arise when the claimant is injured while intervening to save life or property caused by the defendant's negligence. If the intervention is a reasonable thing to do for the saving of life or property, this does not constitute the assumption of risk (*volenti*), nor does the defence of contributory negligence apply. If the intervention is not a reasonable thing to do, then the defences of *volenti* and contributory negligence could apply.

Novus actus interveniens

This means 'a new act intervenes'. We have seen above the need to prove a breach of a duty of care and a chain of causation leading to the loss or damage. If this chain is broken by an intervening act then damages may not be recoverable from the initially negligent party. This doctrine can be expanded to say that when the act of a third person intervenes between the original act or omission and the damage, the original act or omission will only be held to be the direct cause of the damage:

- if the act of the third person might have been expected in the circumstances or,
- if the act of the third person did not materially cause or contribute to the injury.

There are many difficulties with the concept that are beyond the scope of this book. In *Lamb* v. *Camden London Borough Council* (1981), the council negligently broke a water main. As a result, Lamb's house suffered water damage and the family had to move out. While the Lamb family were out of the house squatters moved in and caused much more damage. Camden BC was held liable in **nuisance** for the water damage but not liable for the further damage caused by squatters. The actions of the squatters were a *novus actus interveniens*. It was not foreseeable. Nuisance is another tort relating to a person's rights to 'enjoyment of land' and also acts of general annoyance to the public.

It is also possible for the defendants *and* the third party to be liable for the damage caused. In *Rouse* v. *Squires* (1973) a negligently driven lorry jack-knifed on the motorway, causing an accident involving many vehicles in which the claimant was injured. In a second accident another driver, negligently driving too fast in the circumstances, then collided with the stationary vehicles. The second accident was deemed to be a reasonably foreseeable result of the first and hence the court decided that the chain of causation was not broken.

Contributory negligence

Sometimes when an accident occurs, both the claimant and the defendant have been negligent and the defendant may raise a defence of contributory negligence against the claimant. At one time a claimant guilty of contributory negligence could not usually recover damages. However, under the Law Reform (Contributory Negligence) Act 1945, liability can be apportioned between claimant and defendant.

Contributory negligence must be differentiated from the other defences discussed above. These other defences are complete defences because, if proved, there is no case to answer. However, if contributory negligence is proved then the claim is not defeated but the damages awarded may be reduced according to the degree of fault of the claimant. In other words, claimants have made their own harm or damage worse by their own actions and hence even if they are successful in their court action, their damages are reduced.

It is important to understand that contributory negligence is not concerned with someone helping to cause an accident, but with the increased severity of the resulting harm or injury to the claimant themselves.

Thus failure by a claimant to wear a crash helmet on a motor cycle may reduce the damages he or she obtains on the ground of contributory negligence. This occurred in *O'Connell* v. *Jackson* (1971) where the claimant was injured on a moped and had damages reduced for not wearing a crash helmet. This decision was in the era before wearing a helmet became compulsory. The Court of Appeal's view was that motor cyclists who fail to wear crash helmets in circumstances where a prudent road-user would do so and who are injured in an accident, may be held in part responsible for the injuries which they would not have received if they had been wearing a helmet. This was held even though the motor cyclist was in no way to blame for the occurrence of the accident.

Similarly, failure by a claimant to wear a seat belt in a motor car will also reduce damages on the grounds of contributory negligence (*Froom* v. *Butcher*, 1975). Since 1983 it has been a criminal offence for the driver and front-seat passenger not to wear seat belts. The law has now been extended to include rear-seat passengers. The courts may therefore reduce still further the damages where a seat belt is not worn. It is thought unlikely that they will refuse to give damages altogether even though the claimant is breaking the law.

There have been many occasions of warnings being given to the victim, either written or verbal. The fact that the victim of negligence received a warning of possible harm will not mean that the damages will be totally eliminated. In *Brannan* v. *Airtours plc* (1999), Airtours put on a party with unlimited free wine, as part of a package holiday. Guests were told not to stand on tables because of danger from overhead fans. Brannan climbed on to a table and was injured by a fan. The trial judge reduced his damages by 75 per cent, but the Court of Appeal said the defendants could have avoided the damage by ensuring that tables were not under fans, and so amended the reduction for contributory negligence from 75 per cent to 50 per cent.

Vicarious liability of employers

Employers have many duties and obligations under employment law and safety law. Another area of law which affects employers is concerned with vicarious liability. Vicarious simply means 'through someone else'. Whilst the person who is actually responsible for the commission of a tort is always liable, sometimes other persons or organisations may be liable although they have not actually committed the tort themselves (such as an employer). In such a case both are liable as joint **tortfeasors**. A tortfeasor is a person who commits a tort. We see this particularly in an employer's potential liability to other people, for any negligent act of an employee. In this situation both the employee *and* the employer are liable because the employer is deemed to be legally responsible for the employee's conduct.

> Employers are usually liable for the negligent acts of their employees.

In some ways this concept protects employees from being taken to court, since most claimants will simply sue the employer (though they have the right to sue both the employer and the employee as joint tortfeasors). Fairly obviously, the reason for this is that most employers have much more capital than employees, and should also have insurance to cover most risks. Whilst negligent employees may thus be protected from court action, they will still have to deal with their employer who is likely to be very unhappy. It is quite likely that in the negligence, the employee has broken one of more of the employer's rules of employment, and could therefore face disciplinary proceedings at work and possible dismissal.

The law used to see vicarious liability in terms of 'master and servant' relationships and these are the words used in many earlier judgments. Masters (employers) are liable for the torts of their servants (employees) committed in the course of their employment. The risk is very wide and so is commonly insured against by employers. Indeed certain insurance such as public liability is a legal requirement for most employers. Additionally employers must insure themselves in respect of vicarious liability for injuries caused by their employees to their colleagues.

Who is an employee?

Employers are usually liable for the torts of their employees but an employer is not usually liable for the torts of an independent contractor. This distinction is important in the construction industry as we shall see. Employees have a contract 'of service'. Most employers in construction contracts use independent contractors who have a contract 'for services'. Since construction is a complex and fairly hazardous activity, the distinction between employees and independent contractors is clearly of great significance.

In simple terms an employee carries out work or services for the employer, and receives payment for that work usually on a weekly or monthly basis. Most employees are paid an annual salary. Many people go to work each day to factories, offices or hospitals and to teach in schools and universities. Today, a contract of employment is a legal requirement for all employees. However, even without such a contract, the law may decide that someone is under a contract of service, rather than one for services. The courts have used a variety of 'tests' to determine the employment status of individuals who are at first sight not employees and have been assumed to be self-employed people or independent contractors. The result of such tests may be to decide that the persons in question are in fact 'employees' under a contract of service. If this is the case, the employer is usually liable for their torts, provided they are acting 'in the course of their employment'.

The question is to determine whether there is:
- A contract for services (an independent contractor)
- A contract of service (an employed person).

The 'control test'

If the employer controls not only the work but also the way in which it is carried out, then this is likely to indicate a contract of service, and the employer becomes liable on the basis that the person is an employee.

As Mr Justice McCardle said in *Performing Right Society Ltd* v. *Mitchel and Booker (Palais de Danse) Ltd* (1924): '... the test to be generally applied, lies in the nature and degree of detailed control over the person alleged to be the servant.'

Today, many employers (such as utility companies or the Highways Agency) seek to employ self-employed people or independent contractors but wish them to appear to be part of their business. Thus the distinction between 'for services' and 'of service' may become blurred where for example an employer wants a nominally self-employed person to wear company clothing, and drive a van with company livery, possibly even follow some of the employer's rules of conduct or safety provisions. The difference becomes very important because of the vicarious liability of employers for acts of their employees, but not generally for those of an independent contractor.

Determining employment status

A good illustration of some of the difficulties is *Ready Mixed Concrete (South East) Ltd* v. *Minister of Pensions and National Insurance* (1968) where the question to be answered was whether an owner-driver of a vehicle used exclusively for the delivery of a company's ready mixed concrete was engaged under a contract of service or a contract for services. Ready Mixed Concrete (South East) Ltd (RMC) decided to introduce a scheme whereby concrete was delivered by owner-drivers working under written contracts. The owner-drivers entered into a hire-purchase agreement with Ready Mix Finance Ltd to purchase a lorry but the mixing equipment on the lorry was the company's property. In 1965 the company asked the Minister of Social Security for a determination of the employment status of one of the owner-drivers, Mr Latimer, whose written contractual terms included the following:

- He was entitled with the consent of the company to appoint a competent and suitably qualified driver to operate the truck in his place but this was subject to the company's entitlement to require him to drive the truck himself unless he had a valid reason for not doing so.

- He was responsible for paying any substitute.
- He had to wear a company uniform.
- He had to carry out all reasonable orders from any competent servant of the company.
- He had to maintain the lorry at his own expense and pay its running costs.
- There was a mutual intention that Mr Latimer was an independent contractor.
- He did not work set hours and had no fixed meal break.
- The company did not tell him how to drive the truck or what routes to take.

The Minister decided that Mr Latimer was employed under a contract of service but, on appeal to the High Court, Mr Justice MacKenna held that he was running a business of his own. He concluded that the contract was not one of service but was one for services, specifically carriage services.

Hence we see a number of themes used to determine the question of employment status. These include the power to appoint, the power to dismiss, the method of payment, the power to delegate, place of work, provision of plant and equipment by the employer and working hours and similar arrangements.

The integration test

Another test is to look at the person's integration into the business. This test was proposed by Lord Denning in *Stevenson, Jordan and Harrison Ltd* v. *McDonald and Evans* (1952) where he said 'It is often easy to recognise a contract of service when you see it, but difficult to say wherein the difference lies. A ship's master, a chauffeur, and a reporter on the staff of a newspaper are all employed under a contract of service; but a ship's pilot, a taxi-man, and a newspaper contributor are employed under a contract for services.' In construction a sub-contractor would not usually be seen as a contractor's employee, but a temporary hourly paid labourer, employed for a few days, probably would be.

'In the course of their employment'

If we turn now to the meaning of the phrase 'in the course of their employment', we see that where the employer authorises the acts of the employee then the employer is clearly liable. Frequently however employees carry out work that they are employed to do but in an unauthorised way or they may even be negligent. Their employer will generally still be liable. Exclusions to an employer's liability have been seen in cases involving workers going off 'on a frolic of their own', or carrying out activities which are nothing to do with the employer's business.

An example might be two junior engineers, sent out in the company van to do some levelling at a site twenty miles away. Suppose that one of them drove negligently and injured a pedestrian:

- On the way to the site, where their employer is probably liable.
- On a minor detour to collect some sandwiches for their lunch, where their employer is probably liable.
- Leaving the site to go on a country drive to a local pub for lunch, where their employer is probably not liable; they are on a 'frolic of their own'.

Employers and independent contractors

There are two meanings to the word 'employer'. In everyday speech the word means someone who employs another person to carry out tasks and duties for payment, often for a salary. Generally in this book, however, 'employer' is used to denote the organisation which places the contract with the contractor (and the designer in many contract strategies). In this sense the employer might be a local authority for example. Hence employers will usually employ and pay their own staff and quite separately place construction contracts with a contractor or builder.

When we examine the responsibility for negligent acts, the law distinguishes between these two different roles of the employer. Generally, all employers are responsible (vicariously) for the torts of their own direct employees (their staff or workpeople), but the employer as a party to a construction contract is *not* generally responsible for the torts of an independent contractor. This independent contractor would usually be the contractor or builder appointed after a tendering process, under a form of contract such as NEC3 ECC or JCT SBC05. The logic is that employers direct the work of their own employees and hence can be held responsible for their actions but the employer does not direct the work of an independent contractor. As we know, under most forms of construction contract the employer specifies what is required in the specification and on the drawings, but it is generally up to contractors *how* they build the project.

Another way of looking at this is to understand that an employee is a person whose work is integrated into the employer's business organisation, whereas an independent contractor does specific work for the employer's business (builds the construction project) but is not integrated into the employer's business.

This concept of an 'independent contractor' is very important in the construction industry, because it is very common. Thus, firms of builders, civil engineering contractors and suppliers under normal forms of contract with the employer are usually regarded as independent contractors. Of course contractors are also 'employers' themselves in terms of the contractor's own staff such as joiners, steel fixers and other trades-people and so contractors would be responsible for their negligent acts because they are the contractor's employees.

Hence under a normal construction contract such as NEC3 ECC or JCT SBC05 as far as the 'employer' (the person or company awarding the construction contract and paying for it) is concerned, the contractor is independent. However, the contractor's employees are 'servants' of the contractor and the *contractor* will potentially be liable for any negligent acts that they commit. So to repeat an important point, an employee works under a contract of service, whereas independent contractors' contracts are 'for services' under which they are to carry out a particular task or tasks.

Let us take a typical construction project using the traditional strategy. The 'employer', say Western Homes, signs a design contract with architect A and a separate construction contract with builder B under JCT SBC05. Builder B would normally be seen as an independent contractor. If a site engineer working for builder B is negligent in the course of employment, then builder B is liable as the site engineer's employer. However, the 'employer' in terms of the JCT SBC05 contract, which is Western Homes, would not be liable. There are exceptions to this for work of special danger, which is discussed below.

Circumstances where an 'employer' can be liable for the acts of a contractor

Since an independent contractor is by definition a person or company whose methods of work are not controlled by the 'employer' it would be unfair to give the employer general liability for the torts of such a contractor. However, there are circumstances in which employers may be liable for the torts of an independent contractor employed by them. These are set out below though it should be noted that they are not true examples of vicarious liability. Instead they are based on the idea that the employers themselves are in breach of a primary duty which they owe to the claimant. Such circumstances could be:

- Where employers are negligent themselves. An example of this would be selection of an independent contractor without ensuring its competence. This is why good tendering procedures always have checks for contractor competence.
- Where the employer authorises the actions of the contractor. This might happen, for example, where an employer approves the transfer of construction waste to land owned by the employer, which is not a registered waste disposal site.
- Where liability for the tort is strict, so that responsibility cannot be delegated. An example would be the tort of '*Rylands* v. *Fletcher*' which imposes strict liability on the owners of reservoirs, for damage caused by the escape of water, even if this was caused by an independent contractor.
- When the work which the employer has instructed the independent contractor to undertake is extra hazardous.

This last point has some wide-reaching consequences in construction contracts. The employer has been found liable for work being carried out by independent contractors in the highway or adjoining the highway which creates a danger to highway users. Similarly demolition work or work involving explosives has been seen by the courts to join the employer in liability for the negligence of an independent contractor.

Liability of professionals (such as engineers and architects)

The law of negligence also applies to the provision of *services* which will include design services and advisory services. In particular it applies to those whose work involves giving *professional advice*. The law has developed mainly in cases against accountants but the principles apply also to lawyers, building surveyors, architects, engineers and similar professionals.

There is a distinction between negligent statements (which can frequently be giving advice) and negligent acts. Negligent statements occur when someone (usually a professional) makes negligent statements to someone who then relies on those statements. This could be a doctor, for example, who advises a patient to take a certain drug (which should not be prescribed to patients with high blood pressure), but who does not inquire as to the patient's history with regard to blood pressure. A negligent act on the other hand could be, for example, a doctor negligently injecting a patient with ten times the normal dose of a drug.

The distinction between negligent acts and negligent statements is important because liability for negligent statements carries the additional legal risk of having to compensate the claimant for pure economic loss. Pure economic loss is a difficult area of law, but, as we mentioned before, an example might be negligent advice on pensions which causes someone to lose a lot of money. The person is not injured, nor is property damaged; he or she has simply lost money. However for negligent statements as opposed to negligent acts, that person can go to court, and should be able to recover losses because in this case they resulted from negligent advice.

In a construction context negligent acts could be such things as not maintaining traffic signals in a highway excavation or not tying a ladder properly, whereas negligent statements could be statements made to tenderers such as 'there is little vandalism or theft in this area', where the person making the statement should have known otherwise. Hence negligent acts are usually carried out by contractors and builders, whereas negligent statements are usually made by architects, building surveyors and engineers.

Liability for negligent statements is an important area of the law, and the number of claims continues to increase. Negligent statements may now have become a more effective cause of actions at law than have negligent acts. This state of affairs continues to have a major influence on the cost of professional indemnity insurance arrangements (PI). Liability for pure economic loss without injury to a person or property can arise in the making of negligent statements.

The importance of the 'duty of care'

The case of importance in this area of negligent statement is *Hedley Byrne & Co Ltd* v. *Heller & Partners* (1964), which came before the House of Lords. Whilst Heller and Partners were not actually found negligent, because of their 'disclaimer', the House of Lords established an important principle which would become precedent for other, similar cases. Advertising agents, Hedley Byrne, needed a reference from a banker as to the creditworthiness of a potential customer. They approached their bankers who sought the advice of merchant bankers who in turn reported to Hedley Byrne. The report was headed 'for your private use and without responsibility on the part of this bank or its officials.' The report said that the potential customer was 'considered good for its ordinary business engagements.' Hedley Byrne proceeded with their contract and because the customer was actually not good for ordinary business arrangements, lost a considerable sum of money.

So although Hedley Byrne were not awarded damages because the bank's disclaimer stating it would not be held responsible was sufficiently clear, it is worth any construction professional contemplating the words of one of the Law Lords in this case, Lord Reid, who said 'a reasonable man, knowing that he was being trusted or that his skill and judgment were being relied on, would, I think, have three courses open to him. He could keep silent or decline to give the information or advice sought: or he could give an answer with a clear qualification that he accepted no responsibility for it or that it was given without that reflection or inquiry which a careful answer would require: or he could simply answer without any such qualification. If he chooses to adopt the last course he must, I think, be held to have accepted some responsibility for his answer being given carefully, or to have accepted a relationship with the inquirer which requires him to exercise such care as the circumstances require.' It is important to note the word 'rely' because this is an important component of this tort. In summary the House of Lords held that:

> A professional man is liable for statements made negligently in circumstances where he knows that those statements are going to be acted on, and they were acted on and he will accordingly be liable for pure financial loss caused as a result.

This important decision is ignored by architects, building surveyors and designers at their peril. This is because a duty of care arises in relation to statements where there is a 'special relationship' between the giver and the recipient of the advice. It is worth noting that this special relationship does not have to involve payment. It is also the reason that academics may be wary of giving references for students, and should always do so with great care. It is no problem stating exam performance which can be evidenced. However saying 'this is a marvellous student and will do well as a design engineer' is a statement that an academic is probably not qualified to make, and may be relied upon by a future employer.

A later case, which followed *Caparo* (which we described earlier in the chapter and relates to the 'duty of care'), went to the Court of Appeal. It concerned negligently prepared accounts by the defendant firm of accountants, Hicks Anderson, who were aware that the claimants intended a take-over bid of a company for whom they were accountants. The case was *James McNaughton Papers Group Ltd v. Hicks Anderson & Co* (1991) and it was held that the accountants owed no duty to the claimant because the accounts were not prepared for them but for the defendant. In his judgment, Lord Justice Neil set out some important concepts for determining liability for negligent statements. They are:

- The purpose for which the statement was made.
- The purpose for which the statement was communicated.
- The relationship between the advisor, the advisee and any relevant third party.
- The size of any class (see below) to which the advisee belongs.
- The state of knowledge of the advisor.
- Reliance by the advisee.

Reid LJ explained 'size of any class' by saying that where there is a single advisee or he is a member of only a small class it may sometimes be simple to infer that a duty of care was owed to him. Membership of a large class, however, may make such an inference more difficult, particularly where the statement was made in the first instance for someone outside the class.

A thread running through the cases regarding negligent statement is that the duty in regard to a negligent statement would be owed only to those persons whom the maker of the statement *knows* will rely on it, and not beyond that to those whom he might *foresee* relying on it. Thus, the test for negligent statements causing monetary loss (which is knowledge) is narrower than that for negligent acts causing physical injury (which is foresight).

Negligent statement could lead a designer into liability for incorrect statements given to a contractor at a pre-tender stage as to the design, or nature of the subsoil, or the possibility of constructing the works in a certain sequence, and so on. Other possibilities might be statements regarding likely ground conditions, or even the likelihood of theft or vandalism in the area. Likewise, the designer could incur liability to the building owner, sub-contractors and suppliers. Although it is usual for an employer's approval of drawings to be obtained, such approval will not usually absolve the designer of a liability for negligence. The employer is usually relying on the designer's skill and care in a particular specialist field in which the employer may have no detailed knowledge whatsoever.

Where a firm comprises partners, as many architects, consulting engineers and building surveying firms do, then the partners are liable for a tort committed by another partner, in the ordinary course of the firm's business, or with the authority of the other partners.

Limitation periods

It is a fundamental principle that someone should not be exposed to the risk of court action for an unlimited period of time, not only because of consideration of fairness but also the difficulty of acquiring evidence. In construction the person or organisation exposed to such action will normally be the contractor but it may also be the design organisation, architect, quantity surveyor, building surveyor or other professional. This leads to the concept of limitation periods. A limitation period is a period in which a legal action can be brought. After this period, a claim can no longer be referred to a court. The periods are different in the various areas of the law. Limitation periods are almost all governed by the Limitation Act 1980 as amended by the Latent Damage Act 1986. In very simple terms in actions for damages for negligence the limitation periods are:

- Three years for claims involving personal injury (even if this is only part of the claim)
- Six years for claims that do not involve personal injury.

In tort, the limitation period generally begins from the date when the damage *occurs*, not when it is discovered. This is very different from the rule in contract law and can lead to difficulties: sometimes it can be very difficult to decide when damage is suffered. This applies particularly to damage which is discovered some time after a building has been constructed for example. This is a complicated area of law and is beyond the scope of this book.

Negligence – some legal difficulties

We have seen that proof of negligence contains a number of key steps. These are the existence of a duty of care, breach of that duty by falling below an acceptable standard, and damage caused by that breach that is not too remote a consequence. The difficulty that reading previous judgments presents is that judges may use certain words in different contexts.

In *Lamb* v. *Camden LBC* (1981) Lord Denning said: 'the truth is that all these three, duty, remoteness and causation, are all devices by which the courts limit the range of liability for negligence or nuisance … it is not every consequence of a wrongful act which is the subject of compensation. The law has to draw a line somewhere.

- Sometimes it is done by limiting the range of the persons to whom duty is owed.
- Sometimes it is done by saying that there is a break in the chain of causation.
- At other times it is done by saying that the consequence is too remote to be a head of damage.

All these devices are useful in their way. But ultimately it is a question of policy for the judges to decide.'

In *Caparo Industries* v. *Dickman* (1990) Lord Roskill said: 'that it has now to be accepted that there is no simple formula or touchstone to which recourse can be had in order to provide in every case a ready answer to the questions whether, given certain facts, the law will or will not impose liability for negligence or in cases where such liability can be shown to exist, determine the extent of that liability. Phrases such as foreseeability, proximity, neighbourhood, just and reasonable, fairness, voluntary acceptance of risk, or voluntary assumption of responsibility will be found used from time to time in the different cases. But, as your Lordships have said, such phrases are not precise definitions. At best they are but labels or phrases descriptive of the very different factual situations which can exist in particular cases and which must be carefully examined in each case before it can be pragmatically determined whether a duty of care exists and, if so, what is the scope and extent of that duty.'

Contribution

The situation where a number of parties may be liable to one party occurs frequently. The law deals with this by using the concept of **contribution**. This is a completely different concept from 'contributory negligence' which was discussed above, and the two should not be confused despite the similar terminology. An example involving several parties in a building contract will be used as explanation of contribution.

Suppose several parties are all liable to hotel owner H in respect of the same damage. We will use the example of a negligent hotel design by architect R, where structural engineer E negligently fails to spot the problem and builder B negligently also fails to note the problem and builds the hotel anyway. The hotel then falls down owing to a major structural failure. Let us also suppose the loss caused is one million pounds.

The position of the hotel owner H

In the above situation H can sue R, or E, or B, or all of them, or any combination of them. Let us say hotel owner H sues all of them, and the judge decides that they were all negligent and so they are all liable to H. The professionals R, E and B are 'jointly and severally' liable to H. In that situation the court will enter judgment against all of R and E and B in the sum of £1 million. Hotel owner H is then entitled to enforce this judgment against whichever of R, E and B in whatever way he wishes. The only rule is that H cannot actually recover more than £1 million in total – so, he could decide that builder B is the richest and try to get the full £1 million from B.

If H sues only B and wins, he gets a judgment for £1 million, and then enforces that judgment and recovers the full £1 million. H cannot later decide to sue R or E because H now has no loss so there is nothing to claim. H is allowed to go after R and E as well, in the event that H loses his claim against builder B, or if H wins but doesn't manage to get all of the money back. Whoever H sues must prove his or her own cause of action (in negligence or breach of contract).

The position of architect R, structural engineer E and builder B

Let us again suppose that H sues only B. H could have also sued R and/or E, but has decided not to bother because he has been advised that he has a very good claim against builder B and that B has a lot of money. In that situation B can make an application to 'join' R and/or E to the proceedings. Builder B's cause of action is not in contract or in tort, but arises under the Civil Liability (Contribution) Act 1978, which says that:

'(1) Subject to the following provisions of this section, any person liable in respect of any damage suffered by another person may recover contribution from any other person liable in respect of the same damage (whether jointly with him or otherwise).'

This is called a claim in contribution, and it operates as an independent cause of action. In order to win in the claim, builder B must prove to the court that R and E were also liable to the hotel owner H. This may seem strange at first because B does not need to show that R and E did anything wrong in relation to *himself* (that is B). B needs to show that R and E wronged the hotel owner H, and that accordingly H *could* have sued and succeeded against R and E too.

In this situation, builder B's claim against R and E operates like a claim in its own right; it does not affect the hotel owner H's position at all. Hotel owner H (having only sued B) can only get a judgment against B. If H loses against B he has no claim against R and/or E unless they change their own claim form to sue too.

Builder B's claim against R and E is for a 'contribution' to B's liability to the hotel owner H. As to the amount of the contribution, that is a matter for the judge to decide on the basis of what he or she thinks is just and equitable in all the circumstances.

So, in this scenario, if hotel owner H wins, he gets a judgment for one million pounds against builder B. Hotel owner H is entitled to enforce the whole of that one million pounds against B and so B has to pay the one million pounds to H.

Let us say B persuaded the court that architect R and structural engineer E were also liable to H. Builder B will then be awarded a judgment against architect R and structural engineer E. The amount of the judgment will depend on who the judge thinks is most at fault. Let us suppose that the court decides that B is 50 per cent responsible, E is 30 per cent responsible and R is 20 per cent responsible. This means that B will get a judgment for £300,000 against E and £200,000 against R.

The cause of action in contribution lasts for two years from the date that the main claimant (in this case the hotel owner H) has to be paid. So, if hotel owner H sues B, then builder B might just wait to see what happens. If B loses and thus has to pay H, then builder B may later decide to start new proceedings against R and E. Builder B does not have to do it by 'joining' architect R and structural engineer E to the original claim brought by H.

Chapter summary

- Negligence is a tort, or a civil wrong.
- The law of tort and hence negligence is separate and distinct from the law of contract.
- Negligence is generally concerned with acts or omissions which cause foreseeable harm or injury to others.
- To prove negligence one has first to show a duty of care, which has been broken, judged by objective standards.
- This act or omission actually has to cause the harm to another person.
- The harm has to be a reasonably foreseeable consequence of the act or omission, and it must not be too remote a consequence.
- There are absolute defences to negligence such as the claimant's consent to the risk (*volenti*) or an intervening act of another which breaks the chain of causation.
- Where claimants are negligent themselves, and they prove their case of negligence against the defendant, then the damages paid to them by the defendant may be reduced for what is called 'contributory negligence'.
- Employers are responsible for negligent acts of their employees, provided the employees are acting in the course of their employment.
- However employers are not usually responsible for the torts of independent contractors.
- Another important area of law for construction professionals is negligent statement. Here a professional can be sued for negligent advice, under most circumstances, provided the recipient is expected to rely and does rely on the advice. The advisor does not need to be paid for this advice.
- Where more than one person is negligent the principle of 'contribution' will apply. This is a way of sharing the damages to be paid between the parties at fault.
- There are time limits to claims in negligence, called limitation periods.

20

Construction contracts

Aim

This chapter lists the important ranges of construction contracts used in the UK and explains where the more common ones would be used.

Learning outcomes

On completion of this chapter you will be able to:

>> List the main contracts in the NEC, JCT and ICE suites of contracts.

>> Describe the features and application of some of the more common NEC and JCT contracts.

Families of contracts

The major publishers of contracts produce a range of forms of contract for different purposes and the various contract strategies. Sometimes these families are called 'suites' of contracts. Since construction is complex and covers such a diversity of projects, there are many forms of contract.

The NEC3 suite of conditions of contract

The NEC3 publishes a wide-ranging suite of contracts for most applications although they are generally used for work of a civil engineering nature. The most extensively used of this suite is the NEC3 Engineering and Construction Contract (NEC3 ECC) which is covered in detail in Chapters 21 to 26. The standard publication in a black cover includes options A to F as supplementary clauses to the main document. To make its use more straightforward, NEC also publishes individual versions for each main option A to F in green covers. In these individual versions, the supplementary clauses are incorporated into the main contract (using bold font for identification). The NEC3 suite of contracts includes:

- NEC3 Engineering and Construction Contract (ECC)
  - NEC3 Engineering and Construction Contract Option A: Priced contract with activity schedule
  - NEC3 Engineering and Construction Contract Option B: Priced contract with bill of quantities
  - NEC3 Engineering and Construction Contract Option C: Target contract with activity schedule
  - NEC3 Engineering and Construction Contract Option D: Target contract with bill of quantities
  - NEC3 Engineering and Construction Contract Option E: Cost reimbursable contract
  - NEC3 Engineering and Construction Contract Option F: Management contract
- NEC3 Engineering and Construction Subcontract (ECS)
- NEC3 Engineering and Construction Short Contract (ECSC)
- NEC3 Engineering and Construction Short Subcontract (ECSS)
- NEC3 Professional Services Contract (PSC)
- NEC3 Term Service Contract (TSC)
- NEC3 Term Service Short Contract (TSSC)
- NEC3 Supply Contract (SC)
- NEC3 Supply Short Contract (SSC)
- NEC3 Framework Contract (FC)
- NEC3 Adjudicator's Contract (AC).

The NEC suite of contracts is intended to be a fully interrelated set of contracts. As far as possible, NEC contracts are all drafted in a similar way, with consistent terminology. The NEC3 Term Service Contract and Framework Contract reflect much modern procurement practice, where long-term arrangements and frameworks are increasingly common. The NEC also publishes guidance notes and flow charts for these contracts, together with advice on contract selection in its publication *Procurement and Contract Strategies* (see www.neccontract.com).

JCT 2005 family of contracts

The JCT 05 family of contracts is a range of contracts comprising main contracts and sub-contracts, together with other documents that can be used across certain groups of contracts. The families cover a wide variety of scenarios which arise in construction but are generally used by the 'building' side of the industry. The JCT family of contracts is reproduced below:

- Adjudication Agreement
- Collateral Warranties
- Construction Management
- Consultancy Agreement (Public Sector)
- Design and Build Contract
- Framework Agreement
- Generic Contracts
- Home Owner Contracts

- Intermediate Building Contract
- JCT – Constructing Excellence Contract
- Major Project Construction Contract
- Management Building Contract
- Measured Term Contract
- Minor Works Building Contract
- Partnering Charter
- Pre-Construction Services Agreement
- Prime Cost Building Contract
- Project Bank Account Documentation
- Repair and Maintenance Contract
- Standard Building Contract.

As an example of subdivision, the widely used Standard Building Contract (SBC) shown above contains the following specific contracts and related documents:

- Standard Building Contract with Approximate Quantities, Revision 2 (2009)
- Standard Building Contract with Approximate Quantities without contractor's design, Revision 2 (2009)
- Standard Building Contract with Quantities, Revision 2 (2009)
- Standard Building Contract with Quantities without contractor's design, Revision 2 (2009)
- Standard Building Contract without Quantities, Revision 2 (2009)
- Standard Building Contract without Quantities without contractor's design, Revision 2 (2009)
- Standard Building Contract Guide, Revision 2009
- Standard Building Sub-Contract Guide, Revision 2009
- Standard Building Sub-Contract Agreement, Revision 2 (2009)
- Standard Building Sub-Contract Conditions, Revision 2 (2009)
- Standard Building Sub-Contract with sub-contractor's design Agreement, Revision 2 (2009)
- Standard Building Sub-Contract with sub-contractor's design Conditions, Revision 2 (2009).

Hence in reality, JCT publishes dozens of versions of their contracts, each for a particular purpose, whereas NEC publishes far fewer contracts, but these NEC contracts have options within them to provide a similar range of contract coverage as JCT.

As this book was going to press, the Joint Contracts Tribunal announced the launch of the 2011 range of contracts.

The Royal Institution of Chartered Surveyors publishes a regular survey of 'contracts in use' in the building side of the industry. Hence NEC contracts are under-represented in the survey, although their use for building contracts is clearly increasing. The survey is available on the website www.rics.org. The 2007 survey indicated that the design and build contract (JCT DB2005) is the most used, followed by the standard building contract (JCT SBC2005). The minor works contract (JCT MW2005) was used for about a quarter of contracts by number, but the total value of these contracts was, unsurprisingly, small.

The ICE family of contracts

The current edition of the Institution of Civil Engineers contract is the seventh or ICE7 for short. Of the range of ICE contracts available, the one most used for construction is the Measurement Version 7th Edition. It was first published in 2003 and amendments have been published since then. ICE7 contracts have a similar history to JCT contracts and have much in common. ICE7 was once the predominant civil engineering contract in the UK, but its use has declined with increasing use of NEC3 contracts. There is now an ICE7 edition designed specifically for target price contracts, which have become very common; this was published in 2006.

In 2010 the ICE announced its intention to cease support for the ICE7 contract. However, it still has followers in the industry and actually provides the basis of FIDIC contracts, which are used internationally. It is likely that another industry association will take over the publishing and updating of ICE7. The Institution of Civil Engineers publishes the following contracts:

- Measurement Version 7th Edition
- Design & Construct 2nd Edition
- Term Version 1st Edition
- Minor Works 3rd Edition
- Partnering Addendum
- Tendering for Civil Engineering Contracts
- Agreement for Consultancy Work in Respect of Domestic or Small Works
- Archaeological Investigation 1st Edition
- Target Cost 1st Edition
- Ground Investigation 2nd Edition
- Amendments to ICE Conditions of Contract.

Most of these contracts were published between 2001 and 2003. The target cost edition is more recent at 2006.

FIDIC contracts

FIDIC conditions of contract are used for much international work. FIDIC conditions are published by the International Federation of Consulting Engineers. FIDIC is substantially based on the ICE Conditions of Contract Fourth Edition (the current edition is ICE7). The other contract that is still often used for international work is ICE7. NEC3 contracts are little used for international work (they are based on more modern principles of project management and contractual relationships).

In a similar way to other major conditions of contract, FIDIC publish a suite of conditions, including conditions of contract for:

- Works of civil engineering construction (red book)
- Design and build and turnkey (silver book)
- Electrical and mechanical work (yellow book)
- Short form of contract (green book)
- Client/Consultant services (white book)
- Joint venture agreement.

The main FIDIC additions are:

- The language of the contract.
- The law which applies to it.
- Currency and rates of exchange.
- A clause that allows the price to be adjusted if variations exceed plus or minus 15 per cent (applied to the excess only).
- Customs clearance and re-export of contractors' plant and equipment.
- Provisions for action if contractor becomes bankrupt.
- Provisions for war, or attack by bomb or missiles.

Specific details of forms of contract from the NEC3 and JCT family of contracts

NEC3 contracts ECC and ECSC for construction

These contracts are intended for use in the contract between the employer and the contractor. The contract between the contractor and sub-contractors are covered below. For the more usual contract strategies such as traditional, and design and build, the initial choice in NEC3 contracts for the employer/contractor contract is between the NEC3 Engineering and Construction Contract (ECC) and the NEC3 Engineering and Construction Short Contract (ECSC). NEC3 ECC is 85 pages long and contains options for different payment mechanisms such as bill of quantities, cost-reimbursable or target. NEC3 ECC is covered in detail elsewhere in this book (Chapters 21 to 26).

By contrast NEC3 ECSC is just 14 pages long and only provides for payment on the basis of a 'price list'. The standard format provided in NEC3 ECSC is basically a bill of quantities, which is similar to NEC3 ECC option B. Hence if the employer wants anything other than a bill of quantities payment mechanism, the only choice is NEC3 ECC.

NEC3 ECSC can be used for the design and build strategy, by specifying contractor design in section 1 of the works information. Beyond that, however, NEC3 ECSC gives little design guidance except for a brief mention in clause 20.2 which relates to acceptance of the design by the employer.

NEC3 ECSC also contains a list of fourteen compensation events, but some are simplified. For example the compensation event relating to 'weather' in clause 60.1(10) is cut down from being based on weather measurement and probability to reference to the contractor losing more than one-seventh of the available working time, as a result of the weather. NEC3 ECSC also has provision for retention and delay damages if these are inserted in the contract data, but all the other secondary option clauses (X options) from the NEC3 ECC are missing.

NEC3 ECSC is well laid out, with clear and simple sections for the contractor's offer, and works information, with brief advice given in a box at the beginning of each sub-section. There are related guidance notes and sub-contract forms, the NEC3 ECSS. It is possible that on some smaller contracts, the employer and contractor may not be familiar with the NEC3 ECSC. In this case it may be worth organising some joint training. This is not only likely to ensure a better run contract, but can be the start of teambuilding and more cooperative relationships.

NEC3 contracts for management contracting and construction management

The use of management contracting and construction management as contract strategies has declined markedly in recent years. They are both described in earlier chapters. The normal NEC3 ECC contract has option F for use with the management contracting strategy. In many respects, management contracting is similar to the traditional strategy.

For the construction management strategy, NEC suggests that the construction manager can be appointed on the professional services contract (NEC3 PSC). The employer then awards separate contracts using the appropriate NEC3 form such as:

- NEC3 Engineering and Construction Contract (ECC)
- NEC3 Engineering and Construction Short Contract (ECSC)
- NEC3 Supply Contract (SC)
- NEC3 Supply Short Contract (SSC).

NEC3 sub-contracts

These contracts are designed for use in the contracts between the contractor and sub-contractors. On most projects, there will be many such sub-contractors. NEC3 sub-contracts are written in a very similar format to NEC3 ECC except that in the sub-contract versions the contractor takes the place of the employer, and the sub-contractor takes the place of the contractor. The NEC3 ECS has similar provisions for programme, early warnings and compensation events as NEC3 ECC. In the section on compensation events it is the contractor rather than the employer who issues changes to works information, instructions and decisions and so on. Again there are the options A to E, with which we are familiar, from NEC3 ECC and a similar list of secondary option clauses (the 'X options'). The full titles of the two forms of NEC sub-contract are:

- NEC3 Engineering and Construction Subcontract (ECS)
- NEC3 Engineering and Construction Short Subcontract (ECSS).

The facility to use key dates in NEC3 ECC makes it possible for the contractor to mirror these key dates in their sub-contracts. It is therefore important that the sub-contract programmes interface properly with the contractor's main programme provided under NEC3 ECC clause 31.

Options A to E in the NEC3 ECS are similar to those in NEC3 ECC. However it is likely that option A (priced contract with activity schedule) or option B (priced contract with bill of quantities) will be used the most in sub-contracts because of their relative simplicity. It is quite likely that the contractor will have an option C contract with the employer. This is the target contract with activity schedule, which is often used where the employer wishes to incentivise the contractor. Here the contractor is paid for a variety of items, the main one being payments to sub-contractors. Those payments to sub-contractors under NEC3 ECS could well be on the basis of a bill of quantities under option B of the sub-contract. Whilst this has the benefit of simplicity and follows established and understood sub-contract procedures, it does rather defeat the employer's intention to incentivise the contractor. Clearly, on larger and more complex sub-contracts, option C in the sub-contract would become a more feasible proposition.

The NEC3 Engineering and Construction Short Subcontract (ECSS) is most likely to be used where work being done by the contractor under the NEC3 Engineering and Construction Short Contract (ECSC) is sub-contracted. It has a similar layout and format, with simple front pages for the works information. NEC3 ECSS is 17 pages long, as opposed to the 83 pages of the NEC3 ECS. Payment is on the basis of a price list, which has a similar format to a bill of quantities.

However, the NEC3 ECSS could equally be used for straightforward, low-risk sub-contracts by a contractor working under NEC3 ECC.

Preparation of the NEC3 ECSS by the contractor is important. There is little point in copying works information from the main NEC3 ECC contract into the short sub-contract, where it is not relevant to the sub-contractor. Once again, joint training events may be very valuable for sub-contractors who are not familiar with either of these NEC3 forms, the ECS or the ECSS.

NEC3 contracts for long-term arrangements

NEC now publishes a range of contracts for long-term arrangements, both for the supply of goods and materials, and for frameworks. Framework contracts are very common today, and are explained in Chapter 12. The relevant NEC forms are:

- NEC3 Framework Contract (FC)
- NEC3 Term Service Contract (TSC)
- NEC3 Term Service Short Contract (TSSC).

The NEC3 Framework Contract (FC)

The NEC3 Framework Contract is just five pages long. It is intended as an umbrella agreement, with other contracts below it. It cannot stand on its own. The framework contract is designed to allow the employer to invite tenders from suppliers to carry out work on an 'as instructed' basis over a set term. The NEC3 Framework Contract is designed as an overarching contract, with one or more contracts below it.

The NEC3 Framework Contract defines the parties as the employer and supplier, where a supplier could also be a contractor or design organisation. When employers want work to be carried out, they select a supplier using the prescribed selection procedure.

The NEC3 Term Service Contracts

These contracts are the Term Service Contract (NEC3 TSC) and the Term Service Short Contract (NEC3 TSSC). The NEC Guidance Note says 'the NEC3 TSC is also essentially different from other forms of contract in the NEC family of contracts. It has been designed for use in a wide variety of situations – not restricted to construction. It is essentially a contract for a contractor to provide a service (not limited to a professional or construction service) to an employer from a starting date and throughout a service period.' It goes on to say 'the service is usually provided on the employer's premises but may not be. The service may include physical work, such as cleaning, painting or other maintenance, but may not do. In the public sector, the NEC3 TSC is designed to be used for all contracted-out services, whether including a physical content or not.'

The NEC3 TSSC is intended for more straightforward work of lower risk where straightforward management techniques will be adequate.

Hence these contracts are intended to be used for the appointment of a contractor who will provide a service for a period of time. Most large organisations have a need for some type of maintenance work; particularly with the trend to outsource work to suppliers rather than using directly employed personnel. So for example they could be used by a local authority for maintenance work to a number of properties or perhaps by a hospital for regular maintenance work to its premises. It is also possible to extend the idea of 'work on the employer's premises' to highway maintenance, or to street lighting maintenance. Some employers use one of these contracts for ground investigation work comprising boreholes and trial pits.

The NEC3 TSC and TSSC are intended for *physical* work on the employer's premises. Where professional services are needed, the NEC3 PSC would be more appropriate, since these services would normally take place in the offices of the design organisation.

The term services contracts are not intended to operate as an umbrella agreement like the framework contract described above. The NEC3 FW contract 'calls off' packages of work with a package order. By contrast, the term services contracts define the work to be done throughout the service period, and the contractor is paid to do this work, usually on a monthly basis.

The NEC3 TSC is drafted in a similar way to NEC3 ECC except that 'service manager' is the equivalent role to the project manager under NEC3 ECC. The words 'employer' and 'contractor' are retained. The contractor provides services over a period of time, called the 'service period', which is stated in the contract data. The NEC3 TSC also envisages work in or to the employer's property, and this is defined in clause 11.2(2) as the 'affected property.' Again, compensation events are retained, but reduced to fourteen in the core clauses. There are three payment options, named A, C and E, which mirror those in NEC3 ECC. Hence option A is a priced contract with a price list, option C is a target contract with a price list, and option E is a cost-reimbursable contract. Secondary options (X options) are available, and again they are similar to NEC3 ECC, but reduced in number.

Unlike many construction contracts, which are often completed in one or two years, the term service contracts may operate for many years, perhaps five or even ten. Hence provision for inflation is more important. This is done by means of option X1. For similar reasons in option X2, 'changes in the law' are likely to be more important on the NEC3 TSC. This is particularly true of changes in employment law, or laws regulating to health and safety or the environment which are likely to affect the contractor's costs of providing the service.

NEC3 Supply Contracts

The Supply Contract (SC) and Supply Short Contract (SSC) are forms of contract designed for complex (SC) and low-risk/simple (SSC) purchasing of goods. They are written in the same format as other NEC documents. They are intended to be flexible enough to apply to any industry or supply situation:

- NEC3 Supply Contract (SC)
- NEC3 Supply Short Contract (SSC).

The NEC3 supply contracts are differentiated in a similar way to the NEC3 ECC and ECSC. NEC gives advice as follows:

- The NEC3 Supply Contract should be used for local and international procurement of high-value goods and related services including design.
- The NEC3 Supply Short Contract should be used for local and international procurement of goods under a single order or on a batch order basis and is for use with contracts which do not require sophisticated management techniques and impose only low risks on both the purchaser and the supplier.

JCT contracts

For traditional contracts with a separate design organisation, the most usual JCT contract would be the Standard Building Contract, JCT SBC05. This is 118 pages long and assumes that the design organisation carries out the design and the contractor constructs it. However, JCT SBC05 does have provisions for definition and use of a 'contractor's design portion'. This is explained in Chapter 27.

A slightly simplified version of this contract for use on less-complex projects is the Intermediate Building Contract, or JCT IBC05. The JCT practice note *Deciding on the appropriate JCT contract* (2008) suggests that JCT SBC05 is for use 'where the proposed building works are of simple content involving the normal, recognised basic trades and skills of the industry, without building service installations of a complex nature or other complex specialist work.'

JCT also publish the Minor Works Contract (JCT MWC05). This is a much simpler contract, 38 pages long and intended for small and straightforward contracts. It is extensively used in the building side of the industry for this type of work. It is a lump sum contract and is not intended for use with a bill of quantities.

JCT IBC05 and JCT MWC05 are also available in a version 'with contractor's design', which is similar to the automatic inclusion of this aspect in JCT SBC05. The JCT guidance refrains from setting price levels for these contracts, preferring to refer to complexity and risk. However, David Chappell, who has written many excellent books on the JCT contracts, suggests an upper limit of £400,000 or twelve months for JCT IBC05 and £150,000 and six months for JCT MWC05 (*Understanding JCT Standard Building Contracts*, Taylor & Francis, 2007: 1). These seem sensible recommendations, and at least give a sense of scale. Final choice should still be based on risk and complexity of course.

Where the design and build strategy is used, the most appropriate contract is likely to be the JCT Design and Build Contract, JCT DB05 which was written specifically for this purpose. In JCT DB05, the employer (often through an architect or other design organisation) produces a performance specification, which forms the employer's requirements which the contractor should satisfy. The contractor produces the contractor's proposals, which explain how they will do this. These requirements and proposals form part of the contract.

The JCT Major Projects Construction Contract is intended for really large and complex contracts. It assumes contractor design, rather like JCT DB05. The JCT practice note says that it is intended 'for major works where the employer regularly procures large-scale construction work and where the contractor to be appointed is experienced and able to take greater risk than would arise under other JCT contracts and where the contractor is not only to carry out and complete the works, but also to complete the design.'

Contracts for other purposes

Design contracts

There is also a range of contracts written specifically to cover 'professional services'. These forms of contract can be used for the appointment of consulting engineers, architects, building surveyors, quantity surveyors and other professionals. In the traditional contract strategy the employer will have a direct contract with the design organisation, whereas in the design and build strategy, the design contract will be between the contractor and the design organisation. These professional services contracts are appropriate for both strategies.

Forms of contract for design services are usually simpler than are the forms of contract for construction, since design services are generally more straightforward, and risks are easier to define and less extensive. For example, adverse weather and unforeseen ground conditions are major risks in construction, which do not exist in a design contract.

The appropriate professional services contract is usually selected from the same suite of contracts used for construction. Hence it is usual to use a JCT professional services contract with a JCT building contract, such as JCT SBC05. The JCT professional services contract is now called 'Standard Conditions for the Appointment of an Architect' (CA-S-07-A), but many employers still use the earlier version called the 'Standard Form of Agreement for an Architect' (SFA/99 updated in April 2004).

For professional services associated with NEC3 construction contracts (of which NEC3 ECC is the most common), it is normal to use the 'NEC3 Professional services contract' or NEC3 PSC as it is usually known. The NEC3 PSC is written in a similar style to NEC3 ECC contracts. Hence it will seem familiar to people used to the NEC3 construction contracts. However NEC3 PSC is very different from the JCT professional services contracts which are generally used for the design of building work.

There are also standard forms of contract for other professional services such as Planning supervisor (under CDM 2007), Project manager, Quantity surveyor and Building surveyor. They have similarity with the 'Standard Conditions for the Appointment of an Architect' (CA-S-07-A), but reflect the different range of services involved in the appointment of other professionals.

As an example of typical terms in a design contract, the list below is abstracted from the Association of Consulting Engineer's conditions:

- Definitions
- Duration of engagement
- Ownership of documents and copyright
- Settlement of disputes
- Standard of design liability
- Normal services required
- Additional services required

- Supervision on-site where included
- Information to be supplied to the consulting engineer
- Payment for **normal** services
- Payment for **additional** services
- **Disbursements**
- Payment following termination or suspension by the client or consulting engineer
- Insurance requirements
- Form of appointment letter.

Normal services cover design, drawings and specification, whereas additional services might include gaining planning consent, or negotiating the purchase of land for the project. A disbursement is a rather old-fashioned word for additional costs such as postage, printing and telephone calls.

Contracts for dispute resolution

Most construction contracts such as NEC3 ECC and JCT SBC05 contain dispute resolution procedures such as adjudication. The adjudicator is an independent professional, appointed and paid by the parties in dispute (often the employer and contractor). Hence a form of contract is useful to clarify the appointment. NEC3 publish the Adjudicator's Contract, known as NEC3 AC, and JCT publish the Adjudication Agreement.

Partnering agreements

Partnering agreements are becoming increasingly common and are described in Chapter 11. A partnering agreement can be legally binding, but generally it is not. So although partnering can have a contractual basis, it is *usually a non-legally binding agreement* to cooperate, share information, and work together on the basis of mutual objectives and trust. The legally binding agreement is the contract itself. Hence a normal construction contract such as NEC3 ECC or JCT SBC05 is still needed. So for example the JCT Partnering Charter states on the cover that it is non-binding, and begins with the following:

- For use with most standard forms of construction and engineering contracts and sub-contracts
- Where the parties do not wish to enter into a legally binding agreement but wish to create a collaborative working environment.

The ICE7 suite of contracts has a similar non-binding agreement called the Partnering Addendum. NEC3 ECC uses option X12 to create a multi-party partnering agreement.

Typical terms in a form of contract

NEC3 ECC is quite a complicated contract and covers many contract strategies and payment mechanisms together with a wide array of special clauses, called secondary option clauses (or 'X clauses').

As a simpler example, in JCT SBC05, which is one of the most used of the JCT family of contracts, a *selection* of the main terms is concerned with:

- Definitions and interpretation
- Contractor's obligations
- Possession
- Supply of documents, setting out etc.
- Errors discrepancies and divergences
- Adjustment of completion date
- Practical completion
- Defects
- Architect/Contract administrator's instructions
- Payment
- Variations
- Injury, damage and insurance
- Assignment
- Termination
- Settlement of disputes.

The differences between terms in forms of contract can be subtle but significant

Whilst forms of contract have much similarity especially ICE7 and JCT contracts, small differences in concepts or words can have very significant effects on the risk balance between the employer and contractor. A good example is the treatment of adverse weather. Since most construction projects are built outside, adverse weather is a major unknown and therefore risk. Hence excessive rain may make excavation difficult or unsafe; it may also make haul roads and accesses impassable. Snow and weather below freezing point will prevent concreting or bricklaying taking place. Many external painting or finishing works cannot take place in rain. We can all expect some adverse weather in the UK and the contractor can be expected to price in for a reasonable amount of lost time. What is reasonable will of course depend on the time of year and location. We would expect much more severe winter weather in the north of Scotland than in Kent.

ICE7 contracts use the concept of 'exceptional adverse weather conditions' and JCT contracts use the concept of 'exceptionally adverse weather conditions', and these wordings are subtly different. Importantly, these two forms of contract can only entitle the contractor to a longer time to complete the works, *but no money*. In JCT terminology they are a relevant event (time) but not a relevant matter (money). The JCT argument for compensating the contractor hinges on whether the weather is adverse (against the contractor's operations), and whether it is exceptionally so. In contrast, NEC3 ECC contracts substitute weather measurements at a designated weather station which are more unlikely than once in ten years. Hence NEC3 ECC tries to be more scientific. The big difference in NEC3 ECC is that if this compensation event is proven, then the contractor gets time and resulting costs. Many employers see the NEC3 ECC clause as too great a risk, and modify it.

Another example is the valuation of variations (changes). ICE7 and JCT contracts use a hierarchy of calculation based on any relevant item in the bill of quantities. However, NEC3 ECC option B (the equivalent bill of quantities option) relies on a quotation from the contractor which is not usually based on the original bill of quantities, but it can be by specific agreement.

Finally, payment in ICE7 and JCT SBC05 is on the basis of completed work so it is retrospective, whereas in NEC3 ECC options C, D and E it is based on the forecast cost to the next assessment date, so it is prospective.

Which form of contract is most appropriate?

Some reasons behind the choice of a particular contract from a suite or family of contracts are:

- Is the contract for design services only?
- Is the contract for construction only?
- Is the contract for special services such as adjudication?
- Is the contract a mixture of design and construction (design and build for example)?
- The size of contract.
- The complexity of the contract.
- Is it a contract with high uncertainty or risk?
- Does the employer want the power to name (nominate) sub-contractors?
- Will the employer use a contract strategy such as management contracting or construction management?
- Does the contract have a bill of quantities as the payment mechanism?
- How certain does the employer wish to be of likely final cost?
- How definite are the employer's requirements, and are changes likely?
- Is it a simple small contract (often called a 'small works' contract)?
- Is it a 'partnering' contract?
- Is it a contract between a contractor and sub-contractor?
- Is it a long-term or framework contract?

Excellent advice on the choice of a form of contract is given in the NEC publication *Choosing the right NEC contract*, by Bill Weddell. Further good advice is given in the JCT Practice note *Deciding on the appropriate JCT contract* which is available on the JCT website (www.jctcontracts.com).

The importance of the reaction of tenderers

In choosing the right contract it is always worth considering the likely reaction of tenderers. Smaller contractors may be used to very straightforward contract conditions, and for domestic clients they may even use their own. So for a small, straightforward project the NEC3 Engineering and Construction Short Contract (ECSC) or the JCT Minor Works Building Contract would be best. These contain the essential terms of their larger brethren but are less likely to worry a prospective tenderer. If you give a tenderer a contract form of over 100 pages for a £20,000 house extension, two things are likely to happen – the price goes up, or the tenderer suddenly becomes 'busy' and declines to tender.

Chapter summary

- All construction projects are subject to contract law, and have 'parties' to those contracts.
- Most construction projects use a standard form of contract, published by either the Joint Contracts Tribunal (JCT contracts) or one of the New Engineering Contract (NEC3) forms. These are published by NEC, which is a division of Thomas Telford Ltd which is a wholly owned subsidiary of the Institution of Civil Engineers.
- Previously the predominant civil engineering contract was ICE7, published by the Institution of Civil Engineers; but its use has declined in recent years as NEC3 has become the usual contract of choice for the civil engineering industry.
- JCT, NEC3 and ICE contracts contain suites of separate contract forms designed for different purposes.
- JCT tend to publish separate forms of contract for different purposes, whereas NEC3 relies much more heavily on selectable options within a much smaller range of standard forms.
- NEC3 in particular publish all their contracts in a similar format and layout, and they are intended to be mutually compatible.
- Forms of contract drawn up specifically for design are also published by JCT and NEC, as well as a number of professional bodies representing other design and support professions.
- Where they are used, partnering agreements are usually signed in addition to a standard form of contract. Partnering agreements are usually not contractually binding.
- Most forms of construction contract cover similar events and issues, but small differences in wording can have far-reaching effects on the balance of risk.

21

NEC3 ECC fundamentals

Aim

This chapter explains the fundamentals of NEC3 ECC such as the format, the options and the steps required to prepare the contract.

Learning outcomes

On completion of this chapter you will be able to:

>> Explain the format of an NEC3 ECC contract.
>> Describe the clause numbering system.
>> Describe the main options A to F and the X options and understand which you might select and why.
>> Complete the contract data part one.

Background to the NEC

The New Engineering Contract or NEC has its origins in a 1985 initiative from the Institution of Civil Engineers to fundamentally review contract strategies. The NEC state that the main reasons for this were:

- A proliferation of standard forms was available.
- Most projects are multi-disciplinary yet most contract forms were single disciplinary.
- There was a high incidence of disputes, and wastage of resources involved in resolving them.
- Most forms of contract were written before modern principles of project management were established.
- There was a perception that clients wanted greater certainty of achieving project objectives.

A third edition (NEC3) was launched in 2005 taking into account user feedback in the intervening 10 years.

The three key characteristics of the NEC

The NEC website www.neccontract.com has this to say about NEC contracts:

'NEC is a family of standard contracts, each of which has these characteristics:

- Its use stimulates good management of the relationship between the two parties to the contract and, hence, of the work included in the contract.
- It can be used in a wide variety of commercial situations, for a wide variety of types of work and in any location.
- It is a clear and simple document using language and a structure which are straightforward and easily understood.'

A 'cultural' difference?

Engineers have a training that mainly concentrates on mathematical and technical subjects, often following from a similar education at school or college. The design element of projects is indeed mathematical, but most other aspects are more about communication, teamwork and project management. Until recent times, UK construction comprised two divided professions – designers and builders (usually called consulting engineers or architects and contractors and construction managers). Engineers or building professionals spent their careers in one or the other profession with little movement between. There was often an element of distrust, and certainly little attempt at collaboration or teamwork. The contract administrator in traditional contracts, called the Engineer to Contract (ICE7) or Architect/contract administrator (JCT contracts), was seen as the administrator and certifier of the contractor's work. In many ways it was a 'master/servant' relationship. Conditions of contract of the time, such as ICE7 and JCT, reflected this relationship, both in their structure and in their language.

> Successful projects are about *people*; what they do and how they interact.

Since the 1990s, following the Latham and Egan reports, and a genuine desire to improve construction in the UK and make it more efficient, attitudes and processes have slowly changed. We are now at the point where collaboration and teamwork based on mutual trust and respect are possible. NEC3 reflects the changed culture in the UK construction industry, and arguably will not work without it. It is debatable that until the social attitudes and the culture of the construction industry in any given country are ready for NEC contracts, then it is unlikely to be a successful choice of contract. Simply changing forms of contract from ICE or FIDIC to NEC3 is unlikely to work well until a construction industry is organisationally and culturally ready for that change. It is necessary for a construction industry to be able to work successfully with the opening clause 10.1 of NEC3 ECC, which says:

- The employer, the contractor, the project manager and the supervisor shall act as stated in this contract, and in a spirit of mutual trust and cooperation.

Structure of the NEC3 ECC

NEC publishes a family of contracts, which are described in Chapter 20. The contract at the heart of this family and the one most used is the Engineering and Construction Contract (NEC3 ECC), which is written for the contract between the employer and the contractor. It has eight main parts:

- The core clauses set out in nine sections.
- The main options A to F (these are largely choices of pricing/payment mechanism) and one of them *must* be selected.
- Dispute resolution options W1 or W2.
- Secondary option clauses X1 to X20 (these are not obligatory).
- Options Y.
- Any additional conditions of contract called Z clauses.
- The schedule of cost components and the shorter schedule of cost components.
- The contract data part one (provided by the employer) and part two (provided by the contractor).

The three selections which produce the contract

The NEC3 ECC contract is produced from the core clauses plus three selections, which the employer should make for every project. These three selections are:

---

1 Selection of one of the options A to F
2 Dispute resolution W1 or W2
3 Secondary options (X options).

---

1 The employer must select a pricing/payment mechanism from six main options A to F

This option selection has a significant effect on the risks and costs of a project. Selection and recommendation of options A to F is likely to be done at a senior level in the relevant organisation.

2 Dispute resolution option W1 or W2 must be selected

Option W1 or W2 must be selected to go with the chosen main option. The dispute resolution option selected will generally be W2 which is appropriate for contracts where the Housing Grants, Construction and Regeneration Act (1996) applies. This act (HGCRA) applies to most construction work in the UK and gives the right to statutory adjudication. Most of section W2 covers details of the procedure to be followed. Option W1 might be used for international contracts for example.

3 Secondary options

The X options are very important in an NEC3 ECC contract. The base contract comprising the core clauses does not include many of the items such as retention and delay damages which are automatically included in more traditional contracts such as ICE7 and JCT. Thus NEC3 ECC is very flexible. Where a simple contract is required, few if any X options will need to be included. However the range of X options available gives the employer the opportunity to tailor NEC3 ECC to the needs of a particular project. The X options are described in more detail below.

The employer must select a Y option, which will usually be option Y (UK) 2 for contracts where the Housing Grants, Construction and Regeneration Act 1996 (HGCRA) applies. Option Y (UK) 2 will therefore be selected for most UK contracts and covers payment in compliance with HGCRA and supplements the payment clauses found in section 5 of the core clauses, and ensures that the contract complies with the HGCRA.

Option Y (UK) 3 should *not* normally be included because it gives rights under the contract to named third parties to enforce named terms. Details of such third parties and rights must be included in the contract data. Where no details are given in the contract data the Contracts (Rights of Third Parties) Act 1999 *does not apply*. Hence if the contract does not include details under option Y (UK) 3 in the contract data then the contract will exclude third-party rights, which is the situation that most employers and contractors wish to see. In the rare instances where employers wish to include third-party rights, they would be well advised to seek legal advice on the completion of relevant items in the contract data.

The Z clauses are intended to cover any additions or changes to other clauses that the employer wishes to make. New clauses and changes to existing clauses should always be made as Z clauses so they are collected together in one section of the document and are thus very apparent to tenderers and people administrating the contract.

The use of Z clauses can have far-reaching consequences and legal advice may be needed. Z clauses should be very carefully drafted to avoid inconsistencies with other clauses in the document. What employers are usually doing is changing the balance of risk between them and the contractor.

For example, some employers remove the compensation event 60.1(13) relating to weather measurements in order to put all the risk of adverse weather onto the contractor. Such changes will almost certainly oblige tendering contractors to increase their tender prices to reflect their increased risk. Tenderers will have to assess the risk, so deletion of this 'weather clause' for a pipe-laying contract in the winter in the north of Scotland would be high contractor risk for example, whereas deletion of the clause for a road junction improvement in the summer in the south of England would be low contractor risk. This is because the actual weather in Scotland is more likely to deviate from the weather statistics taken at the relevant weather station because of its increased variability.

The difficulty for tenderers is pricing a contract with a deleted 'weather clause' between these two extremes, particularly contracts of twelve months or more duration (encompassing weather conditions throughout the year), and contracts with extensive work in the ground, or requiring access over fields, such as sewers, pipelines and highway contracts. Adverse weather can soon make fields impassable to contractor's plant and machinery.

Some Z clauses are routinely inserted by local authorities and government organisations such as 'the prevention of corruption', or clauses to prevent discrimination. Many employers will wish to add clauses relating to confidentiality.

The core clauses

The nine sections of core clauses

The core clauses are common to all the options whether these options lead to priced, cost-reimbursable or target contracts, and whether the contract contains all, some or no contractor design. The nine sections of core clauses which are described in detail in Chapters 22 to 26 are:

---

1 General
2 The contractor's main responsibilities
3 Time
4 Testing and defects
5 Payment
6 Compensation events
7 Title
8 Risks and insurance
9 Termination.

The system of clause numbering

Each of the nine sections of core clauses are numbered 10, 20, 30 and so on, which aids navigation. Hence the first clause of section 2 is clause 20, and the second clause is 21. Sub-sections are numbered 21.1, 21.2, 21.3 and so on.

The clauses in each option A to F are numbered in order *to be added* to the core clauses as appropriate. Hence for example option B (bill of quantities, or BoQ) adds a new clause 55 to section 5 to explain the contractual significance of information in the BoQ. Similarly it adds clauses 60.4 to 60.7 (the core clauses finish at 60.3) as compensation events so as to include mistakes or where the BoQ does not conform to the standard method of measurement. The additional clauses also cover changes in quantities between the billed quantities and the as-constructed quantities.

Another innovation of NEC3 contracts is that as far as possible the whole suite of NEC contracts uses the same structure and numbering system. Thus compensation events are found in clause 60 in NEC3 ECC. Compensation events are also numbered 60 in the NEC3 Professional Services Contract and the NEC3 Sub-contract for example.

Identified and defined terms

NEC3 also introduces a unique identification system. Most contracts have **defined** terms, which usually have definitions in the first section of the contract. It is easy to skip over this section, but if the definitions of the words used in the contract are not properly understood, then misunderstandings and mistakes are easily made.

NEC3 defines the words with general application to all options in section 11. They are listed as 11.2(1) to 11.2(19). Some options introduce new terms such as the phrase 'bill of quantities' in options B and D. Thus options B and D have a definition of 'bill of quantities' in clause 11.2(21) which can be found at the beginning of these two options. NEC3 gives defined terms a capital first letter, examples being Defect and Accepted Programme.

Many important items *particular to any given contract* are **identified** in the contract data. These include all the important dates, the names of the employer, contractor and project manager, and payment information such as the direct fee percentage. NEC3 calls these 'identified terms', and they are printed in *italics*. There are some small areas of inconsistency in this system, but they need not concern us here. Since this book covers other forms of contract and general contract matters, it *does not* use the NEC3 ECC system of italics and capital letters.

Selection of the main option A to F

The main options A to F

These are shown in the box below:

Option A	Priced contract with activity schedule
Option B	Priced contract with bill of quantities
Option C	Target contract with activity schedule
Option D	Target contract with bill of quantities
Option E	Cost reimbursable contract
Option F	Management contract

Overall considerations

These are 'options' in that there is a choice of six, but their *use* is not optional. The employer must select one of them, or there would be no pricing/payment mechanism in the contract. The options A to E are largely pricing/payment mechanisms and option F (management contract) is also a contract strategy.

The design and build contract strategy can be used by specifying contractor design in the works information but only options A and C would be suitable. This is because options B and D use bills of quantities and option E is cost-reimbursable. None of these three would be appropriate for a design and build strategy because a BoQ is usually produced at the end of the detailed design phase by the design organisation, before tender. If the traditional strategy is used then any of the options A to E could be used.

For the employer, selecting one of the options A to E is primarily the choice of a pricing/payment mechanism. However, when considering the best way to procure and implement a project some other factors that should be considered are:

- Is the employer at a stage where it can clearly define what it requires?
- How much design has been completed already?
- Who will ultimately be responsible for completing the whole design?
- Will the contractor be expected to contribute or comment during the design process?
- Is an early start on-site required?
- Are changes likely?
- How does the employer wish to distribute risk?
- Does the employer want relatively simple contract administration?

- Does the employer want a contract which gives the contractor incentives?
- Is a very flexible contract required?
- Do a number of separate contracts require coordination?

These considerations will not only determine the option A to F. They will also indicate how to divide the project into separate contracts where necessary, how to allocate design responsibility and to whom; and which of the X options to choose.

Option A – priced contract with activity schedule

This option is used when the designer can define accurately what the employer needs. The drawings have to be in sufficient detail for contractors to produce activity schedules with lump sum *prices covering all their obligations.* However it is possible to provide less detail than is required for the preparation of a BoQ for options B and D. The contractor is paid a lump sum for every activity which is completed at each assessment date (these dates are usually monthly). This makes the selection of activities, and number of them, very important for the contractor, since it will affect the cash flow once the project gets underway. However the use of a very large number of activities makes the contract cumbersome and difficult to programme and administrate efficiently.

The use of an activity schedule also makes tendering for option A contracts relatively expensive for contractors, because they have to take off accurate quantities and apply suitable rates to them, to build up the lump sum prices against activities in their activity schedule. It also means that *the tenderer takes the risk* that the activities are not comprehensive, or that some work has been left out and not priced.

For the employer, comparison of tenders is more difficult than on a bill of quantities contract, because each tenderer is likely to produce a different list of activities. Some employers produce an activity schedule for contractors to price. This practice does make tender comparison easier but it rather defeats one objective of activity schedules which is to link the activity schedule to the contractor's programme as required by clause 31.4 (found in the specific additions in options A and C).

Another possibility used by some employers, is to produce a 'skeleton' of major activities which must be priced (and can hence be compared) but contractors may add any detailed activities that they wish below each major activity heading.

Thus the activity schedule, which is used in option A as a pricing mechanism and a payment mechanism, gives the employer a tender price – by summation of the prices of all the activities. In common with the other options, this tender price will potentially change as the construction proceeds, if any changes or compensation events occur.

Option A is very suited to contractor design, but can also be used for employer design, or divided design responsibility. This is because the contractor can be paid from the starting date, which can be many months before the access date (the start on-site).

Another advantage of option A (and option B) is that it is much cheaper to administer on-site than options C and D which usually involve complicated checks on the contractor's claimed costs.

Option B – priced contract with bill of quantities (BoQ)

This is the NEC3 ECC equivalent of a 'traditional' ICE7 or JCT contract 'with quantities'. Bills of quantities (BoQ) are explained in detail in Chapter 15 and are usually produced by the design organisation from completed drawings. Payment is made by multiplying the quantities of completed work by the appropriate rate from the bill of quantities. Unlike most traditional contracts however, in option B the BoQ items are not generally used for valuing changes (called 'variations' in ICE7 and JCT contracts). Instead, the contractor submits a quotation.

Option B is similar to option A, in that it can only be used when the designer can define accurately what the employer needs. The drawings have to be in sufficient detail for the BoQ to be prepared.

An advantage of option B is that it reduces the cost of tendering for contractors, since the work in producing a BoQ has been done by the design organisation or quantity surveyor and it also makes tender comparison easy for the employer.

Option B is a remeasurement contract, and the actual quantities are measured and paid for as work proceeds. Unlike option A, items of work do not have to be *completed* for payment to be made, because payment is based on the quantity of work done multiplied by the applicable rate. Again, unlike option A, in option B the 'risk on the quantities' lies with the employer where an item is missed or incorrectly described. Pricing errors remain the contractor's risk of course.

Option B has all the advantages and disadvantages of a normal BoQ contract. Where there is a lot of contractor design, option A is a better choice than option B, because the contractor can include design activities in the activity schedule, and receive payment once they are complete.

Another advantage of option B (and option A) is that it is much cheaper to administer on-site than options C and D which usually involve complicated checks on the contractor's claimed costs.

Option C – target contract with activity schedule and option D – target contract with bill of quantities

These are target price contracts and are usually used where the extent of the work is not *fully* defined. Target price contracts are a development of cost-reimbursable contracts which are described under option E below. However the extent of work should be sufficiently defined for a reasonably accurate target price to be given by the tenderer. The tenderer also submits a fee, which will be applied to payments made to the contractor as the work proceeds. Options C and D are also used where anticipated risks are greater because many risks are effectively shared by the operation of the 'pain/gain' mechanism. Pain/gain is explained in Chapter 14.

It is important to understand that the contractor *is not paid against the activity schedule or the BoQ*. These are pricing mechanisms in options C and D, but not payment mechanisms. They are used for comparing tenders and to establish the target price. The payment mechanism for the contractor is the 'price for work done to date' which is the defined cost plus the fee. This will be explained in detail in Chapter 24.

- Under option C, the target price is adjusted for 'compensation events'.
- Under option D, the target price is adjusted for 'compensation events' and remeasurement of quantities and any inconsistency with the standard method of measurement used.

Options C and D are particularly appropriate where the employer wants to produce a more collaborative contract in the expectation of efficiencies. By adjustment of the 'share percentage' which applies to 'pain/gain', the employer can attempt to maximise the contractor's incentive to produce efficiency. Another attraction of option C is that it can be used for less defined work, since a BoQ is not required. The contractor's concerns about this lack of definition are reduced by the knowledge of the risk-sharing mechanism created by the pain/gain system.

A major drawback with options C and D is that determining contractor costs for payment purposes can be a difficult process, and involve the employer in extra administration costs. Cost consultants or quantity surveyors are often used by the employer for this purpose. It is for this reason that target contracts are generally only worthwhile for larger projects, where their use is now very common.

Option E – cost reimbursable

In this option contractors are simply paid their costs. This option would normally be used where an early start is required but the definition of the work is not developed enough to allow one of options A to D to be used. This is clearly a high-risk strategy for the employer, since the contractor takes minimum risk.

In option E contractors are paid their cost plus the fee inserted in the tender documents. The fee will usually cover overhead costs plus profit. Thus there is little incentive for the contractor to work efficiently to minimise costs compared with the other NEC options. Option E could be suitable for the following types of work:

- Emergency work
- Ill-defined maintenance work
- Work where an early start on-site is needed
- Where the employer needs to be involved in directing the work to be done
- Work of an unusual or experimental nature where day-to-day changes may be required
- Work of high potential risk or unpredictable risk, which contractors might be unwilling to tender under options A to D
- Where the design needs to be developed as construction work proceeds.

Option F – management contract

Option F is used where the management contracting strategy is required. Here contractors manage the work of sub-contractors and do not usually carry out any significant work themselves. The contractors tender their fee, and a lump sum price for any work they specifically intend to carry out themselves. Payment is made on the basis of the cost of work done plus fee. Advantages and disadvantages of management contracting are discussed in Chapter 9.

Secondary options: the X options

The employer doers not have to use any of the X options at all, but most employers will use a number of them, which are described in more detail below. It is important to remember that these X options are agreed in the *original contract* between the employer and the contractor, because many introduce significant pricing or other risks which the tenderers may wish to reflect in their tender prices. X options cannot be added into a signed contract later, by the project manager. The project manager has no power to do this.

> The project manager has no power to add X options into a contract once it has been signed by the employer and contractor.

The X options most commonly included that produce a contract similar to ICE7 or JCT will be X7, X16 and possibly X5. Where the contractor is given design responsibility, it is usual to include option X15, which again produces a result similar to more traditional contracts. The secondary options in full are as follows:

Option X1: price adjustment for inflation (only with options A, B, C and D)

This option is unlikely to be used in times of relatively low inflation. Without option X1, the contractor carries the risk of increasing prices or labour rates, which could be more of a problem with longer contracts, of over a year or so. The employer chooses the indices used for determining the price increases.

Option X2: changes in the law

This option allows the employer to take the risk for changes in the law after the contract date; changes then become a compensation event. Such changes could be new or amended waste or environmental regulations for example.

Option X3: multiple currencies (only with options A and B)

Used for priced contracts where the contractor is to be paid in more than one currency.

Option X4: parent company guarantee

Many larger contracting companies have subsidiary companies. A parent company guarantee is often used to make the parent company liable for the subsidiary company's default or insolvency. The precise terms of the guarantee must be set out in the works information, by the employer.

Option X5: sectional completion

This option allows employers to define specific *geographical* parts of the works along with their own completion date which is earlier than the completion date for the whole of the works. Sectional completions can be very useful for producing earlier completion dates for areas of the works, possibly for employer occupation, or handing over to other contractors for further work. After the date of sectional completion, the contractor has no further rights of access to that section, except for the correction of defects.

An example of a section might be the requirement for the earlier completion of an administration block on a factory complex, for earlier occupation by the employer's admin staff. Another example might be the completion of the car park for a multi-storey office block (so it can be used earlier), or perhaps one floor which is required for a particularly complex fit-out contract.

Where required, sections must be defined very carefully and clearly in the contract data and their completion date must be stated. Where delay damages are included as option X7, they must be calculated separately for each section. Sections should not be used unnecessarily because they increase the complexity of contract administration, and because any revision to the completion date will need calculating for any existing sections as well, as applicable to them.

If option X7 (delay damages) is used, then delay damages need calculating separately and properly for each section.

Option X6: bonus for early completion

This option allows the employer to reward the contractor for achieving earlier completion, if such an event is of sufficient advantage to the employer. In common with any NEC3 ECC contract, what exactly constitutes completion should be defined. This is particularly important for option X6, where a bonus is payable in addition to the normal issues about take-over.

Option X7: delay damages

Delay damages are the NEC3 ECC phrase for liquidated damages. These are frequently included in contracts and are very important. Chapter 18 covers the matter in detail.

Option X12: partnering

The NEC3 ECC contract is a two-party contract between the employer and the contractor. Option X12 is intended to allow a multi-party partnering contract to be set up with objectives and procedures. Such parties might be the employer, design organisation and contractor for example.

Option X13: performance bond

This option would be used where the employer requires a performance bond. Drafting the terms of such a bond requires great care, and legal advice should be sought. If the contractor fails to perform in the way defined, the bond is paid to the employer. It is usual for the employer to pay for the cost of the bond itself, which will be a few per cent of the tender price. In options A and B, the contractor should include the cost of the bond in the tender price. It is usual for the employer to include an item in a BoQ contract, such as option B for the price of the bond to be inserted.

Option X14: advance payment to contractor

Many employers can obtain funds more cheaply than contractors. This option is a way of using such funds at the beginning of a contract to help the contractor finance the work, particularly if there is significant machinery or process plant to be purchased. In theory this should result in lower tender prices. It clearly increases the employer risk and is often used in association with a performance bond. Repayment terms of the advance payment are set out by the employer as instalments in the contract data. The contractor pays these instalments to the employer, or they are deducted from payments due to the contractor.

Option X15: limitation of liability

Where the contractor designs all or part of the works, this option reduces any potential 'fitness for purpose' liability on the contractor to the normal one of reasonable skill and care. This is the standard normally applied to design organisations and is used in the NEC3 PSC.

Option X16: retention (not used with option F)

This option is often used and allows the employer to retain a percentage of payments to the contractor, until completion (when 50 per cent of the retention is repaid) and the issue of the defects certificate (when the remaining 50 per cent is paid). Details of the retention percentage (three per cent is usual) must be inserted in the contract data. The retention gives the employer some security against the contractor failing to complete the works or rectify defects. However it has an adverse effect on the contractor's total borrowing requirement and so may increase the tender price.

Option X17: low performance damages

This option might be applied to the performance of mechanical plant such as pumps. The employer would have to set performance requirements, tests and the low performance damages to be paid in the contract data. It can be very difficult to be precise about the exact conditions that will apply to a test, and such clauses need drafting with care.

Option X18: limitation of liability

This option can be used to place limits on financial or other liabilities that the contractor has with the employer. This would reduce the contractor's risk, particularly for insurance, and should result in lower tender prices.

Option X20: key performance indicators (kpis)

These can set key areas for contractor performance such as health and safety, environmental matters, quality, achievement of time targets and so on. The incentive schedule, set out in the contract data part one by the employer, defines the various kpis, levels of performance and associated payments to the contractor. This option is not used with option X12, which already includes kpis.

People and roles in NEC3 ECC

All contracts have a person in an administrative role. In ICE7 contracts this person is called the 'Engineer' and in JCT contracts the 'Architect/contract administrator'. In NEC3 ECC the contract administrator is called the 'Project manager', but performs a much wider function including working with the contractor to implement the NEC3 contract's project management procedures such as risk-reduction meetings.

There are a number of other specific roles in NEC3 ECC which are explained below. It is very important that everyone concerned with the contract understands the roles, responsibilities, functions and powers of all these specified people.

Competent people performing clearly defined roles and working together are essential to the success of any contract.

Once the contract is awarded the key players are of course the contractor and the project manager. Except for mention in a few clauses the employer, having signed the contract, is not involved, except of course for making regular payments to the contractor. Those named in an NEC3 ECC contract are the:

Employer
Contractor
Sub-contractor
Project manager
Supervisor
Adjudicator
Others.

The employer, contractor and sub-contractors

These three roles are fairly evident from the titles. The employer signs the construction contract with the contractor, who then 'provides the works in accordance with the works information' (clause 20.1). After signing the contract, the employer normally takes a back seat, leaving the administration of the contract to the contract administrator (the 'project manager' as the person carrying out this role is known in NEC3). The contractor must also design any of the works that are specified in the works information (clause 21.1). Other obligations of the contractor are set out in section 2, 'the contractor's main responsibilities'.

Sub-contractors appear mainly in the payment sections of the options A to F. Most contractors sub-contract much of the construction work but are not obliged to do so. Sub-contractors have a contract with the contractor to construct or install part of the works, provide a service or supply plant and materials. Of course the contractor is responsible for any work carried out by its sub-contractors and NEC3 ECC does make this very clear in clause 26.1. The contractor should gain the project manager's acceptance for all sub-contractors under clause 26.2.

The project manager

Under an NEC3 ECC contract, the employer *must* appoint a project manager. The project manager is *named* in the contract data. The role of the project manager carries clearly specified duties and obligations. Under NEC3 ECC, these are not just concerned with instructions, notifications and so on, but also with proactively managing the contract, in cooperation with the contractor. This is the real difference between NEC3 ECC and traditional contracts.

In NEC3 ECC, the project manager could be seen as an agent, acting for the employer, but recent case law has confirmed that when making contract decisions, the project manager must *act impartially*. Hence in NEC3 ECC, the project manager is appointed to ensure that the employer's business needs and requirements are met, but must act fairly between the employer and the contractor, when certifying work or payments or administering the contract. This is subtly different from ICE7 and JCT contracts, where the Engineer/Architect is seen as having a role that is more independent of the employer. One could argue that in ICE7 and

JCT contracts, there is nobody *directly* looking after the business interests of the employer; and this could be seen as a weakness. In any contract, the other party, the contractor, will obviously seek to maximise its own business interests, as any commercial company will do. So perhaps NEC3 ECC achieves a better balance.

If contractors are dissatisfied with a project manager's decision in NEC3 ECC, they do have direct access to the adjudicator. Adjudication is a quick method of dispute resolution that is dealt with elsewhere in this book.

In an NEC3 ECC contract, the project manager will often obtain the views of the employer, to ensure that the project meets the employer's business needs and requirements. If changes are necessary, they can only be instructed by the project manager. The employer has no similar powers under the contract; the change must be issued by the project manager. These changes will be *compensation events*, for which the contractor will be reimbursed.

It is particularly important in NEC3 ECC that the project manager should be experienced in *project management* of engineering projects as well as having expertise in contracts and contract management. Some background or understanding of both design and construction issues is also invaluable. It is also essential that the project manager is not too distant from the work, is on-site regularly (or constantly on larger projects) and has the time to devote to the role required by the contract. This role can be much more demanding than that of the contract administrator in a traditional contract, because of the NEC3 ECC project management procedures that must be followed, such as early warnings and risk-reduction meetings (more on this later).

Issues can sometimes arise between the employer and the contractor, such as delayed access or delayed delivery of material that the employer is to provide, or complaints from stakeholders about noise, dust or other construction activity. The project manager is required and empowered to resolve these issues, but *only within the terms* of the contract.

Most employers wish to devote themselves to the real needs of their business which may be manufacturing, running hospitals, railways or utilities; and expect the project manager to ensure that their construction projects are constructed to agreed time, cost and quality criteria, as set out in the details of the relevant contract.

Who does the project manager work for?

In traditional contracts, the Engineer (ICE7) or Architect (JCT) was usually a senior member of the organisation that designed the project such as a consulting engineer or architectural practice. However, because of the different and more wide-ranging role of the NEC3 ECC project manager, they may come from different organisations. Three common possibilities are:

1. The project manager is selected from a senior construction professional in the employer's own organisation whether the project is designed in-house or not.
2. The project manager is selected from a senior construction professional in the organisation that designed the project where this is not in-house design. The design organisation would normally be a firm of consulting engineers.
3. The project manager is *selected from a quite separate organisation*. Many firms of quantity surveyors now offer their services in the role of NEC3 ECC project manager.

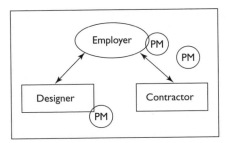

Figure 21.1 The project manager could be chosen from one of three organisations.

Figure 21.1 shows these three possibilities. In each possibility, of course, the project manager administers the contract between the employer and the contractor, regardless of who actually employs the project manager and pays the salary.

All of these options are perfectly legal, provided the NEC3 ECC project manager is named in the contract data. Hence the project manager 'works for the NEC3 ECC contract' and is responsible for its administration. Who pays them is different of course. Ultimately the employer pays the project manager's salary, either as an employee of the employer, or by way of professional fees paid to the design organisation or a firm of quantity surveyors and so forth.

A final complication is the NEC choice of title for this contract administrator. The employer, the designer and the contractor will almost certainly have a person called the 'project manager' as well, and this can lead to confusion.

Project managers and ethical considerations

Where the project manager is also *employed* by the employer (as in many 'client organisations' such as local authorities or utilities), particular care must be taken to ensure impartiality and the separation of 'employer' and 'project manager' roles. This also applies to other named individuals under the contract such as the 'supervisor'.

The project manager can also be in a difficult position where he or she works for the organisation that *designed* the project. This could be as a member of the employer's own staff, or a consulting engineering practice, or an architectural practice. There may be times when the project manager must impartially correct mistakes or ambiguities in designs and documentation which have been produced by *the project manager's own staff*. The project manager may also have to report to the employer on embarrassing cost or time overruns, caused by design errors or late information. In these situations, project managers should act properly and impartially according to the NEC3 ECC contract, and their professional codes of ethics.

Delegation of powers of the project manager

Project managers can (and often do) delegate many of their powers. All delegation must be notified to the contractor under clause 14.2. Clearly, the contractor needs to know the authority levels of any delegated person, and the clauses under which that person can act. This should be clearly set out in the notification. In some cases, the employer may restrict the project manager's powers, possibly with a financial limit on instructions to change the works information. If this is the case, it should be clearly set out in the contract, so the contractor understands the position at tender stage.

The supervisor

Like the project manager and adjudicator, the supervisor is named personally in the contract data. The supervisor's role in an NEC3 ECC contract is primarily the monitoring of quality, and so the supervisor appears regularly in the clauses of section 4, which deals with testing and defects. The supervisor carries out tests and inspections and notifies defects to the contractor, for correction. Ultimately, the supervisor issues the defects certificate on the defects date when all defects have been corrected. This date is usually 52 weeks after the date of completion for the project. The supervisor is not empowered to order changes to the works information, to accept defects or to determine payments. These are the roles of the project manager. Hence in an NEC3 ECC contract the roles of project manager and supervisor are independent of each other and they have separate duties.

The adjudicator

It is assumed that the adjudicator is personally named in the contract data. In many cases the employer will insert the adjudicator's nominating body, such as the Institution of Civil Engineers. Adjudication is the method of dispute resolution used in NEC3 ECC. It is generally a much quicker and cheaper method than arbitration which was often used in traditional contracts. Thus adjudication is used where the contractor and project manager cannot agree on the application of the terms of the contract to a particular event. Such events will usually be compensation events listed in clause 60. Of course, it is hoped that most well-designed, clear, comprehensive and well-administered contracts will never have recourse to adjudication.

The tribunal

This is also identified in the contract data and refers to the next stage after adjudication, if adjudication has not produced a satisfactory result. The tribunal is generally named as arbitration or litigation.

Others

Others are defined in clause 11.2(10) of NEC3 ECC as people or organisations who are *not* the employer, the project manager, the supervisor, the adjudicator, the contractor or any employee sub-contractor or supplier of the contractor. So *others* could be other separate contractors, specialist contractors or professional advisors such as designers, architects, quantity surveyors and so on. They could also be external organisations such as regulators, public authorities, utilities responsible for power, gas or water supplies or diversions, the police and so on.

The importance of others is that the employer generally takes responsibility for them if they do not work within the times or conditions stated in the works information. Hence 'others' should not be seen as a catch-all for people who might affect the contract. When the contract is being prepared, the employer and design organisation should attempt to foresee the activities of others, and ensure that they are defined as far as possible in the works information. Where others do not work within these definitions, their activities may become a compensation event as clause 60.1(5) for which the employer will have to pay.

Designers (the design organisation)

By the word 'designers', again we generally mean the design organisation for whom they work. The design organisation is not part of the NEC3 ECC *contract* between the employer and the contractor. They are of course essential to the investment process. Whilst the designer is not mentioned specifically in the NEC3 ECC contract they have an important role in providing and updating the design information and drawings.

The designer as a person usually works for a separate design organisation such as a firm of architects or consulting engineers. The designer may also be part of the employer's organisation, as is often the case with local authorities. Few contractors employ their own designers directly as employees, and where the design and build strategy is used the contractor will usually engage a consulting engineer to carry out the design role.

The design organisation will normally be separately engaged by the employer (or the contractor in the design and build strategy) using the NEC Professional Services Contract (PSC), which forms part of the NEC3 suite of contracts.

Where the design organisation is separately engaged, as in the traditional strategy, they will often advise the employer on option selection, and the feasibility of options. They will then carry out the detailed design stage. In other words, they will develop the selected option to a point in design where tenders can be invited. The design organisation will frequently advise the employer as to the appropriateness of tenders, and which one to select. The designer's role is to design a project that meets the employer's needs, and communicate this via words and drawings to the contractor.

Consultation with the designer during construction

The designer will have made assumptions as to ground parameters, existing services, water table levels and suchlike, so it is important that the project manager maintains close contact with the designer during the construction period, in case any assumptions need checking or revalidating, or where design changes may become necessary. If the design and build strategy is used, the project manager's first point of contact will be the contractor, who will decide how to involve the design organisation in the discussions.

Preparing the contract for tender

There are *ten main steps* in preparing an NEC3 ECC contract for tender. We have already covered steps one to five above, which are shown below for completeness:

1. Choose the pricing/payment mechanism by selecting one of options A to F.
2. Choose dispute resolution from W1 or W2 (W2 is normal).
3. Choose as many of the secondary options as are required, from X1 to X20.

4. Choose Y (UK) 2 (and possibly Y (UK) 3 where appropriate).
5. Add any additional conditions of contract as Z1, Z2 and so on.
6. Complete the **contract data part one.** This will also include references to the **works information** (drawings, specifications and so on) which should either be included in the hard copy of the tender documents, or be available in electronic form.
7. Refer to relevant **site investigation** information.
8. Include the **contract data part two** for the tendering contractors to complete.
9. Include a bill of quantities for the tendering contractors to complete if option B or option D is used.
10. Include tendering instructions and any articles of agreement as appropriate.

The contract data part one – data provided by the employer

This is information of fundamental importance and *should be completed very carefully.* It is data specific to the particular contract to be awarded. If some items from a previous project are used as a basis (as is often the case), then they must all be checked carefully to ensure that they are applicable to the contract in hand.

Many items will be too bulky to include in the contract data itself, such as the specification, tender drawings or site investigation information. This issue is overcome simply by referring to their whereabouts in the contract data. The employer's standard specification for example may be published on the employer's website. Where reference is made to electronic data then the issue number and applicable date should be included. The contractor cannot be responsible for anticipating possible future changes by the employer! So for example the works information may refer to the 'employer's standard specification S123 issue 5 dated 6th October 2012'. In this way, the parties to the contract are clear as to its basis.

The contract data provided by the employer begins with a very important clause which defines the option A to F which is used, the dispute resolution option, and the secondary options (X options). There follows a brief description of the works and the names and addresses of the employer, project manager, supervisor and adjudicator.

Two very important and often extensive sections at the bottom of the first page of the contract data part one, after the names of the project manager and so on, are the works information and the site information which are defined in clause 11.2 and reproduced below. It is essential not to confuse the two.

Works information

The works information should contain all the details of what is to be constructed, what tests are required and any constraints on the contractor. Works information is defined in clause 11.2(19) as information which either:

- specifies and describes the works or
- states any constraints on how the contractor provides the works and is either
- in the documents which the contract data states it is in or
- in an instruction given in accordance with this contract.

This final provision is to allow the works information to be changed by the project manager, after the contract is awarded, possibly by issuing revised drawings for example.

Whilst NEC3 ECC gives no further advice, the NEC3 Guidance Notes take pains to explain what should be included. The list below is simplified from the Guidance Notes.

Works information

Works information would normally include such matters as:

- A detailed description of the works
- The specification
- The drawings and any reinforcement schedules
- Any constraints such as limited working hours or access requirements
- Any programme requirements that are additional to clause 31
- Any specified sub-contractors or suppliers, such as framework suppliers
- Details of any tests required
- Details of others who may share the working areas
- Details of the contractor's design (if any is required)
- Specific health and safety requirements
- Plant or materials to be supplied by the employer
- Any services required of the contractor such as accommodation for the project manager and staff, access roads, car parking areas and so on
- Details of any operational interfaces on sites which are also occupied by the employer
- A statement as to the title to excavated or demolition materials
- Details of records which the contractor must keep
- Any acceptance or procurement procedure required by the employer
- A definition of what constitutes 'completion' as required by clause 11.2(2).

Site information

Site information is defined in clause 11.2(16) as information which:

- describes the site and its surroundings and
- is in the documents which the contract data states it is in.

Again the NEC3 Guidance Notes provide further detail, which includes:

Site information

Site information would normally include such matters as:

- The ground investigation report
- Any geological data of the area
- Risks of flooding if the site is in the flood plain of a watercourse
- Details of services below the site such as gas, water or electricity services
- Details of buildings or structures adjacent to the site, which could be affected by the construction (condition surveys for example)
- Publicly available information that needs referring to because it may affect the construction of the works.

The final item is very important because it is referred to in clause 60.2. The significance is that it is used in judging a possible compensation event under clause 60.1(12) relating to physical conditions that the contractor could not reasonably be expected to allow for. This is because the contractor is deemed to have taken into account publicly available information, *but only* if it is referred to in the site information (together with information that an experienced contractor could be expected to obtain).

The remaining parts of the contract data provided by the employer

These contain very important data such as:

- The boundaries of the site
- The language of the contract
- The law of the contract
- The period for reply
- The adjudicator's nominating body
- The tribunal
- Any items to be included in the risk register (this is used to alert tenderers to specific risks)
- The important dates such as the starting date and access dates
- Details of when revised programmes are required
- Information on testing and defects
- Information on payment such as the assessment interval
- Weather information, for compensation events
- Limits for insurance
- The completion date
- Key dates and conditions (if required)
- Details of any employer risks (this is referred to in clause 60.1(14) which by definition becomes a compensation event, if it occurs)
- Further insurance requirements
- Share percentages for option C or option D
- Further information relating to any X options required by the contract
- Details needed for option Y (UK) 2
- Additional employer risks
- Any additional conditions of contract required by the employer (the Z clauses).

The contract data part two – data provided by the contractor

Contract data part two: general particulars

The contract data part two begins with the contractor's name and address, and then lists the direct fee percentage and sub-contracted fee percentage, which are applied to the contractor's own work and those of sub-contractors, respectively. The *contractor* defines the working areas, which may add additional areas to the 'boundaries of the site' defined by the employer in the contract data part one.

Key people, their jobs and responsibilities are listed by the contractor, possibly after a specific requirement by the employer in the instructions to tenderers. There is a practical difficulty for the contractor of course. The contractor will have a limited number of key people available, such as site agents/site managers (who run the project on-site). Some of these key people may become available when existing projects actually finish, and some may be needed for other projects that the contractor is tendering for. Predicting exactly who might be available in a few weeks' time (when the tender may be accepted) is very problematic for the contractor. The contractor is therefore likely to keep the list of key people as short as possible.

Where contractors are required to carry out design work they must complete the section referring to the percentage of design overheads and other information in the contract data part two. The contractor must produce any preliminary design details or drawings as required as a response by the contractor to requirements set out in the works information supplied by the employer. Again NEC3 ECC is relatively light in this feature, and relies on the employer making clear provisions in the works information for their design requirements (although NEC3 ECC does not say this). In comparison JCT SBC05 is very good in this regard, and calls this the 'contractor's design portion'. JCT SBC05 sets out a procedure for dealing with the 'employer's requirements' and the 'contractor's proposals'. It also sets out a detailed submission procedure for the contractor's proposals in schedule one, which extends to over a page of text.

Contract data part two: programme and completion date

NEC3 ECC then follows with information of any programme required to be identified by contractors *in their tender*. There have been problems in traditional contracts with tender programmes becoming subsequently incorporated into the contract. They can thus become a term of the contract, which is not the intention because a programme is simply a good approximation that is revised as work proceeds and situations change. It is rarely deliberately intended to be 'cast in stone' as a *term* of the contract. If the programme were to become a term then the contractor would be obliged to follow it exactly and any change to the programme itself would arguably be a change to the works information, which would be a compensation event. This is almost certainly not intended by the contractor or the employer.

However NEC3 ECC may escape such potential legal problems, because the tender programme should become the 'accepted programme' by definition in clause 11.2(1), and the accepted programme can subsequently be changed in accordance with clause 32.

In order to make the situation absolutely clear, the employer would be well-advised to state in the invitation to tenderers that the programme submitted with the tender will become the 'accepted programme' as clause 11.2(1). This statement would clearly bring it within the provisions of the contract as a programme that can be changed and updated, and not one 'set in stone' as a contract term, thereby avoiding any doubt on the matter.

It is also possible for the contractor to insert a completion date if one is not specified in the contract data supplied by the employer, which is the norm. The usual practice is for the employer to insert the completion date in the contract data part one, but there may be occasions when the employer has flexibility over the completion date, and wishes tenderers to select a date to suit themselves and their resources. This could result in tenders with lower prices, but it adds another dimension to tender comparison, where there will not only be the tender prices for the employer to consider, but also potentially different completion dates proposed by tenderers.

Contract data part two: final sections

Items then follow for the contractor to identify the whereabouts of the activity schedule or bill of quantities depending on the option used, and a statement of the tendered total of the prices for options A, B, C and D.

Final sections are information for the schedule of cost components and the shorter schedule of cost components. These schedules can be used to define payment to the contractor or are used for valuing compensation events depending on the option A to F. Their use is described in Chapters 24 and 25.

Commercial and tendering issues

The contract data part two contains essential data that contractors return with their tender. It is mainly pricing information, and so has to be completed carefully. A tendering contractor will also have an eye on the way that other tenderers may complete the information, and the resulting tender prices, if they wish to be successful. There will also be a commercial judgment on the part of tenderers in completing this section because many items such as fee and overhead items will be later applied to work done under the contract, but are not seen in the tender total price itself. Employers too must be wary of this section of the contract because of its potential impact on the selection of a tenderer and the final project cost. The situation is further complicated by the different significance of entries made by the contractor in this section, depending on the option A to F which is chosen.

Let us take an example relating to payment. Payment under NEC3 ECC is discussed in detail in Chapter 24. The point here is the effect on the potential contract outturn price. Payment on an option C contract is made on the basis of the price for work done to date. Essentially this is made up of payments to sub-contractors plus the sub-contracted fee percentage and contractor work plus the direct fee percentage. Both fee percentages are entered by the contractor in the contract data but do not appear in the total of the tender prices which is the total of the activity schedules. The problem in comparing tenders is to estimate the proportion of work which the contractor is likely to sub-contract. Suppose the sub-contracted fee percentage was inserted by a tendering contractor as seven per cent and the direct fee percentage as twelve per cent. If little work is sub-contracted then the overall cost of the project will now be higher, because of the higher direct fee percentage.

Another significant payment item is the percentage addition for 'working area overheads', which is inserted in the contract data by the tendering contractor in options C, D and E. This addition can often be significant. It is applied to payments for the contractor's own work (as opposed to sub-contracted work) in options C, D and E. Additionally it is applied to the cost of the contractor's own work in these options when assessing compensation events.

On a well-defined and well-administered contract there may be few compensation events. However, because of the complexity of the design and construction process, many contracts have dozens of compensation events. If this were the case, it is clear that the working area overhead inserted by each tenderer can be a significant element of the total cost. A similar argument applies to the 'percentage for people overheads' which is inserted by the contractor and applied to 'people costs' when assessing compensation events in options A and B.

These points and others of a similar nature make it quite difficult for the employer to compare tenders, particularly for options C, D and E. As a result many employers have notional figures for the proportion of sub-contracted work, and a nominal estimate of the likely total of compensation events, perhaps five or ten per cent of the total tender price. They then apply the contractor's percentages to these estimates to derive an additional tender comparison factor.

Bill of quantities for options B and D

Where option B or option D is used, the employer should arrange for a bill of quantities to be prepared. This would usually be done by the design organisation or a quantity surveyor. Two copies are usually sent to each tenderer. The tenderers insert their rates and prices in one copy and return it as part of their tender documents, to the employer. The second copy is for the tenderer to keep, as a record. Bills of quantities are explained in Chapter 15.

Tendering instructions and any articles of agreement

It is usual to include tendering instructions which will usually include what must be returned, exactly when and where. There may be other specific requirements for any given employer.

The employer will often want to sign a formal article of agreement with the successful tenderer. This is a short legal document summarising the details of the parties, titles of the contract documents, tender price and a very brief outline of the project to be constructed.

Chapter summary

- NEC3 ECC has eight main parts which are:
  1. The core clauses set out in nine sections
  2. The main options A to F
  3. Dispute resolution options W1 or W2
  4. Secondary option clauses X1 to X20
  5. Options Y
  6. Any additional conditions of contract called Z clauses
  7. The schedule of cost components and the shorter schedule of cost components
  8. The contract data part one (provided by the employer) and part two (provided by the contractor).
- Use of options A to F is not optional, and one must be selected by the employer to determine the pricing and payment mechanisms.
- NEC3 ECC defines a number of key roles such as the employer, the contractor, the project manager, the supervisor and so on.
- The employer completes the contract data part one, which contains fundamental information on dates, payment and so on.
- Two very important sections of the contract data part one are the works information (drawings, specification and so on), and the site information, which describes the site.

- The contractor completes the contract data part two and returns it with the tender.
  - This section contains key pricing data along with other important information such as the working area.
  - The employer would be well-advised also to take this key pricing data into account when comparing tenders.
- The employer should be aware of the commercial issues in the preparation and analysis of the contract data.

22

NEC3 ECC core clauses one and two

Aim

This chapter aims to explain the meaning and operation of the most significant clauses in core clauses one and two of the NEC3 ECC contract.

Learning outcomes

On completion of this chapter you will be able to:

>> Explain in detail the significant clauses of core clauses one and two of the NEC3 ECC form of contract.
>> Apply these clauses to typical contract issues.
>> Manage and administer an NEC3 ECC contract.

Note: compensation events

This chapter will often refer to compensation events, which are covered fully in Chapter 25. Compensation events are those that are the employer's risk. If a compensation event occurs and it is not caused by a fault of the contractor, then the contractor is compensated. The compensation could be more time to complete the contract or more money because of the cost of the compensation event, or both. Compensation events include such items as changes, 'adverse weather', late access being given by the employer and so on.

Core clauses one: general

This section begins with the definitions in clause 11 which are self-explanatory. For people used to more traditional contracts, there are some interesting definitions which have been selected and are listed below.

Definitions, clause 11

Accepted programme

Clause 11.2(1) defines the accepted programme, which is fundamental in this contract to assessing the time implications of compensation events. Programmes are discussed more fully in Chapter 16. The definitions relating to dates are discussed in core clauses, section 3, which deals with time, in Chapter 23.

Completion

Clause 11.2(2) *defines* **completion**. This is a new departure in forms of construction contract, because traditional contracts use the term 'substantial completion' or 'practical completion' without actually defining them. This has presented many practical and legal difficulties. Completion is defined in clause 11.2(2) thus:

> 'Completion is when the contractor has
> - done all the work which the works information states he is to do by the completion date and
> - corrected notified defects which would have prevented the employer from using the works and others from doing their work.'

And clause 11.2(2) goes on to say 'if the work which the contractor is to do by the completion date is not stated in the works information, completion is when the contractor has done all the work necessary for the employer to use the works and for others to do their work.' It is advisable in NEC3 ECC for the employer to state in the works information precisely what is required for completion to be achieved. Where this is not done, the employer must rely on the more general words of clause 11.2(2).

It is hoped that completion is achieved on or before the completion date, or any revision to it arising from compensation events for example. The employer is likely to be unhappy if completion is achieved after the completion date (or revision to it). Of course delay damages will be payable by the contractor, if they were included as option X7.

Equipment and plant

Clauses 11.2(7) and 11.2(12) are significant because they are a departure from the traditional use of the words '**equipment**' and '**plant**', and it is very important to understand the NEC3 ECC use of the words. In traditional contracts 'plant' meant contractor's construction plant such as excavators, cranes and lorries. Materials were used to build the works, and temporary items such as scaffolding, excavation supports and so on were called 'temporary works'.

- Clause 11.2(7) defines equipment as 'items provided by the contractor and used by him to provide the works and which the works information does not require him to include in the works.' So 'equipment' in NEC3 is likely to encompass contractor's plant (cranes, excavators and so on), and also temporary works.
- Clause 11.2(12) defines plant and materials as 'items intended to be included in the works.' So in NEC3 ECC 'plant' does not mean contractor's construction plant (cranes, excavators and so on) but items which will be *included in the works*. Hence plant items might comprise boilers for heating systems, pumps, electrical panels and so on. Materials would include concrete, structural steel, aggregate, reinforcement, bricks and so forth, in the traditional sense of the word. The NEC3 ECC definition of plant and materials is part of what traditional contracts call 'the permanent works'.

Key dates

Key dates are discussed again in Chapter 23 because they are referred to in many later clauses of NEC3 ECC. They may be used to tie the activities of the construction contract contractor to the work of others or the employer for example. If the employer wishes to have geographical areas of the site completed earlier then sectional completions should be used, not key dates. Key dates are defined in clause 11.2(9) as:

> 'A key date is the date by which work is to meet the condition stated. The key date is the key date stated in the contract data and the condition is the condition stated in the contract data unless later changed in accordance with this contract.'

What is important to remember is that setting a key date is relatively simple. The argument that arises is whether a key date has been met, not in terms of time, but in terms of the work that must be complete (the 'condition'). Hence the 'condition' needs defining very carefully in the contract. Failure by the contractor to meet a condition is explained below under clause 25.3.

Risk register

Clause 11.2(14) defines the risk register. The risk register is started by inserting the risks listed in the contract data at the tender stage by the employer or the contractor, or both. Such risks will be listed in the contract data under the headings *'the following matters will be included in the risk register'*. This same heading appears in both the contract data part one (provided by the employer) and part two (provided by the contractor). There may frequently be no such risks identified.

The real purpose of the risk register is for it to be an ongoing project management tool for monitoring, recording and controlling risks as the contract work proceeds on-site. Nonetheless it does make sense for the employer to alert the contractor and the project manager to any significant risks (that are known about at tender stage) that may need monitoring, and also for the contractor to do so. Once the contract work commences the risk register is developed to include other risks which arise and, once determined, the actions taken to avoid or reduce those risks. As a result, most of the content of the risk register will be driven by the outcome of **risk-reduction meetings**, which follow **early warnings**. Thus the risk register is a management tool at the heart of the NEC3 ECC contract. The risk register records risks, management actions taken (and by whom) and resulting costs and time consequences. It is therefore a live document as the contract proceeds. When the contract is complete the risk register will provide a valuable record of important events and actions taken. This could be important if there are any later questions about the administration of the contract, by the employer's auditors for example.

Risks inserted for the purpose of the risk register in the section entitled 'the following matters will be included in the risk register' are *not the same* as those inserted by the employer in the contract data part one, under the heading 'these are additional employer's risks'. This can be confusing because of the similarity of the words used. If any risks are inserted under the heading 'these are additional employer's risks', then they are to be used in connection with clause 80.1 and compensation event 60.1(14). The significance of this issue is discussed in Chapters 24 and 25.

Site information, works information and working areas

Site information and works information, clauses 11.2(16) and 11.2(19), are very important and were discussed in Chapter 21.

Working areas, which are defined in clause 11.2(18), are important with regard to payment and are discussed below under clause 15.

Interpretation and the law, clause 12

This clause begins with the usual legal point that words in the singular also mean the plural and the other way round; and words in the masculine also mean in the feminine and neuter. It also states that the contract is governed by the law of contract.

Clause 12.3 states the important point that the project manager does not have power to change the *contract itself.* Only the employer and contractor can do that by agreement in writing (presumably with consideration to produce a binding contract).

> The project manager can only act within the powers given under the contract.

So for example the project manager can issue instructions and change the works information. However the project manager has no power to instruct an X option for instance, if one is not included in the original contract between the employer and the contractor. Hence the original contract can contain option X5, sectional completion, and define a section in the contract data, so the contractor understands the implication when tendering a price. However, the project manager has no power to put a sectional completion into the agreed contract once it has been signed at the contract date. A similar argument goes for all the X options such as delay damages and retention.

Clause 12.4 states that 'this contract is the entire agreement between the parties'. Eggleston in his book *The NEC3 Engineering and Construction Contract: A Commentary* (Blackwell, 2006) suggests that the words are not written with sufficient force or clarity to exclude common law rights and remedies. He goes on to speculate that the clause may be intended to mean that all listed documents form the written aspects of the contract. Recent cases have confirmed that to exclude the right to an action for misrepresentation, for example, very clear words must be used: the point made by Eggleston. The Guidance Notes to NEC3 are silent on the meaning of clause 12.

Communications, clause 13

Form of communications

Most forms of contract require verbal instructions by the contract administrator to be later confirmed in writing. This is because many instructions have financial consequences, sometimes very expensive ones, and there should be no doubt in the future as to exactly what was required. NEC3 ECC does this in a slightly different way in clause 13.1 which states that each communication which this contract requires is communicated 'in a form which can be read, copied and recorded.' This definition would clearly include telex, fax and of course emails.

Period for reply

The communications clause has a number of features which are unusual in a traditional contract such as the period for reply. This is identified by the employer in the contract data part one, and would typically be two or three weeks. However different periods for reply can be given perhaps as the response period to contractor design proposals, where a longer time would probably be appropriate. The period for reply applies to the contractor, the project manager and the supervisor. The project manager can extend the period for reply, but only with the agreement of the contractor and before the reply is due.

The period for reply is given added weight because it is a compensation event as clause 60.1(6) if the project manager or supervisor does not comply. This is a provision designed to aid prompt and proactive project management. The difficulty of course is to know what constitutes a 'reply'. NEC3 ECC no doubt anticipates a carefully worded reply with reasons and possibly decisions; and for many project manager responses, reasons must be given. At the other extreme is a reply from the project manager which states, 'I acknowledge your communication of the 13 April', which would probably not constitute a valid reply if tested in court.

Notifications

Another important point is made in clause 13.7 which says that a *notification* which this contract requires is communicated separately from other communications. Hence it is very important to look for the word 'notify' in clauses of this contract because of this particular meaning.

Notifications must be communicated separately.

Separate notification is a good provision and ensures that notifications, which are usually important and should be tracked, can always be filed and identified separately. Examples of notifications are early warnings, ambiguities and inconsistencies, illegal and impossible requirements, tests and inspections, compensation events and termination. The operation of clause 13.7 does mean, however, that some communications may have to be given more than once, so that the 'notification' itself can be separated.

The project manager and supervisor, clause 14

Acceptance of work or a communication

Clause 14.1 begins by setting out the normal contractual provision that the 'project manager's or supervisor's acceptance of a communication from the contractor or of his work does not change the contractor's responsibility to provide the works or his liability for his design.' This statement is worth making because this point is often misunderstood by less experienced construction professionals.

The construction contract is between the employer and the contractor, and the contractor is contractually obliged to provide the employer with works in accordance with the works information, unless this works information is changed by the project manager in accordance with the terms of the contract. It would not be right if in some way the project manager or supervisor could diminish or undermine the legal rights of the employer by accepting a communication from the contractor that reduced the contractor's obligations in some way.

Hence, for example, if the supervisor accepts a new bridge deck in terms of quality and subsequently finds it did not comply with the works information, then the contractor must correct the defect at the contractor's expense. A similar provision can be found in most conditions of contract – that the actions of the project manager or supervisor should not reduce the contractor's responsibility to construct the works as originally agreed with the employer in the contract documents. Clearly, this sort of situation can cause friction between the supervisor/project manager and the contractor and should be avoided by proper and thorough checking procedures in the first place.

Power to delegate

Clause 14.2 gives power to the project manager and supervisor to delegate any of their actions to other named people. This is a wider power of delegation than some contracts such as ICE7, which does not allow delegation of some of the more important decisions. The contractor must of course be notified of any such delegation.

Instructions

Clause 14.3 gives a very important power to the project manager, and that is the power to issue an instruction which changes the works information or a key date. This instruction would normally become a compensation event under clause 60.1(1), since instructions often cause the contractor to incur additional expense. The power to change a key date is interesting. Like most contracts, NEC3 ECC does not give the project manager the power to bring the completion date forward (although there are circumstances under which this can be done). However the project manager can change a key date. This is a compensation event under clause 60.1(4) and so the contractor is compensated for the effects.

Unlike ICE7, for example, NEC3 ECC does not restrict such changes to the works information by defining them. Most contracts give contract administrators powers to 'instruct' the contractor, but it must of course be an instruction in accordance with the contract. They cannot instruct anything they wish.

There would also be a legal argument that such instructions should not substantially change the nature of the contract from that originally envisaged by the contractor. Hence an instruction to change dimensions or materials would be acceptable, but an instruction to add a new five-kilometre water supply main to a contract to construct a motorway service station would probably not be acceptable, since the nature of the pipe-laying work is substantially different from the original contract work and would not have been in the contemplation of the contractor when the contract was signed.

It is worth remembering that an instruction must comply with clause 13.1 (communications) in that it must be able to be read, copied and recorded. As with many contracts, the project manager or supervisor may give verbal instructions, but they should as soon as possible comply with clause 13.1. This normally means that they should be confirmed in writing.

Replacing the project manager or supervisor

Finally clause 14.4 gives the employer the usual power to replace the project manager or supervisor after notifying the contractor. This would usually be as a result of one of them leaving to start new employment elsewhere or being transferred to another project.

Working areas, clause 15

Working areas, as opposed to 'the site', are an important NEC3 ECC concept, particularly with regard to payment because the working areas come under the definition for payment in the two schedules of cost components. The *employer identifies the site* in the contract data part one, usually by referring to a drawing, whereas the *contractor identifies the working areas* in the contract data part two, submitted by the contractor at tender. These may include additional temporary land or access which the contractor has negotiated in order to improve access, storage space or whatever.

> The employer identifies the site, but the contractor identifies the working areas.

Clause 15 gives the working area idea flexibility by allowing the contractor to submit a proposal to the project manager for adding an area to the working areas. A reason for not accepting is that the proposed area is not necessary for providing the works or used for work not in this contract. If the project manager withholds acceptance for other reasons not stated in the contract, the contractor's recourse is compensation event 60.1(9). The working area also has significance, as we shall see, in later chapters in terms of payment (in some options) and also legal title to goods and materials in the working area.

Early warning, clause 16

This is a very important feature of NEC3 and is a new departure from traditional contracts which do not have this excellent provision. It has been called the 'jewel in the crown of NEC'. Construction is complicated and difficult, and on many contracts events arise or conditions are met that nobody expected. What tends to happen on traditional contracts is that the contractor notifies the contract administrator of a 'claim event' with suitable explanation and the two parties usually begin to discuss liability and potential cost, rather than focusing on trying to solve the problem. ICE7 calls this process a claim, and JCT SBCC05 refers to such items as relevant events (time) or relevant matters (money) under the overall heading 'claims for loss and expense'.

> The early warning procedure is an important innovation in NEC3 ECC.

The contractor may be claiming (usually with justification) financial loss and expense (a change to the prices) and probably more time to complete the contract. This process pitches the contractor and contract administrator directly into a contractual process of establishing liability for the costs and extra time for completion rather than looking for possible solutions. It frequently takes up valuable time and effort and may lead to deterioration in relationships. While this is happening the situation on-site may well be getting worse, and more expensive.

Additionally, traditional contracts have written notification procedures for the recording of claims, problems and issues, but these often lead to protracted letter-writing, rather than rapid and combined action to discuss and seek solutions, which is the intention of the early warning procedure.

The other problem in traditional contracts is that because 'claims' (similar to 'compensation events' in NEC3 ECC) often result in cost plus profit (rather than payment at tender rates), they can actually be to the *benefit* of the contractor. The temptation for some unscrupulous contractors was to be late in notification, or vague in the terms of the notification, so that events could actually get worse, thereby resulting in more potential work and extra profit.

NEC3 ECC, however, has two innovations which encourage early warnings and simplify agreement of costs. The first is the stipulation that if a contractor *does not give* an early warning within prescribed time scales, then its potential financial recovery is reduced. The second is assessment of compensation events based on the schedules of cost components (where applicable). Clause 63.5 says:

- If the project manager has notified the contractor of his decision that the contractor did not give an early warning of a compensation event which an experienced contractor could have given, the event is assessed as if the contractor had given an early warning.

Of course this clause can put the contractor in a difficult position. The project manager will not welcome being bombarded with numerous early warnings, because they divert everyone's management time and effort. However, if the contractor unfortunately decides that an event is trivial and does not notify it formally as a compensation event, then it could later be penalised under this clause if it later turns out to be a serious problem. By notification within eight weeks the contractor will satisfy the requirements of the contract and protect its position. The answer might be for the contractor to notify all potential early warning events, but use a 'traffic lights' system, with red representing the most serious (potentially) early warnings and green the least. In this way, everyone can concentrate on what are likely to be the most significant events to the contract. Of course, it would have to be agreed that the contractor assumed no liability for the choice of traffic lights' colour.

Early warning matters might be unexpected ground conditions, groundwater problems, problems with access, delays by service providers such as statutory undertakers, potential late supply of any items that the employer is to provide, problems with the employer's operational staff, intervention by a regulator such as the police or fire service and so on. Early warnings will also include the contractor's problems such as issues with labour or sub-contractors if they fall within the definition of clause 16.1 below.

Early warnings and alternative solutions

The idea of the early warning is to alert everyone to a problem so that alternative solutions can be considered. It is quite probable that the problem can be solved by a change in the contract requirements, such that the works information is changed. The contractor and the project manager thus try to find solutions, before becoming locked in a potential contractual wrangle about who pays – either the contractor or the employer. Clause 16.1 says:

- The contractor and the project manager give an early warning by notifying the other as soon as either becomes aware of any matter which could
  - increase the total of the prices,
  - delay completion,
  - delay meeting a key date or
  - impair the performance of the works in use.

These four items are clearly of prime significance if the project is to be a success. It is important to comply with the requirement for the contractor *or* the project manager to issue an early warning. The early warning process is one of the many ways that NEC seeks to make this contract a project management tool as well as a set of contractual rules and procedures. NEC use the phrase 'foresight applied collaboratively shrinks risk' and this is exactly the purpose of the early warning system. If the project manager and contractor are aware of a potential problem before it arises then they have the greatest opportunity to minimise or avoid it.

Let us take an example. As part of a factory project, a new electricity sub-station is to be constructed. It is on the critical path of the contractor's programme. On digging the foundation, the contractor discovers an unmarked power cable that may be live and may belong to the local electricity company. Rather than wait for the cable to be identified and possibly moved, an early warning meeting might simply result in a change to the works information to move the sub-station a few metres to avoid the problem. The change to the works information would enable the contractor to notify a compensation event. The cost here might only be a few hours of the excavation gang's time, rather than a delay of potentially many days.

Risk-reduction meeting

Following the early warning, the next step is set out in clause 16.2 which covers the 'risk-reduction' meeting. It is worth noting the title of this meeting. The intention at this stage is to try and solve problems and to reduce risks, not determine liability. In a traditional contract a similar meeting might be called a 'claims meeting', which has a very different intention. Clause 16.2 says:

- Either the project manager or the contractor may instruct the other to attend a risk reduction meeting. Each may instruct other people to attend if the other agrees.

'Other people' would be other persons or organisations associated with the problem or possible solutions, perhaps a sub-contractor, a supplier, the employer's operational staff or a statutory undertaker. At the risk-reduction meeting, the NEC3 ECC contract insists that people cooperate, in clause 16.3, thus:

- At a risk reduction meeting, those who attend cooperate in
  - making and considering proposals for how the effect of the registered risks can be avoided or reduced,
  - seeking solutions that will bring advantage to all those who will be affected,
  - deciding on the actions which will be taken and who, in accordance with this contract, will take them and
  - deciding which risks have now been avoided or have passed and can be removed from the risk register.

A less formal risk-reduction meeting may sometimes be appropriate; perhaps at the location of the site problem, or even over the telephone. However the project manager should still update the risk register as appropriate and issue any changes to the works information that may be required.

In many ways the early warning system is of particular benefit to the employer, because it allows changes in contract requirements, such as the works information, to be made at an early enough stage to prevent costs escalating where the event turns out to be an employer's risk. Additionally, since many NEC3 ECC contracts use the target price options C and D, the employer also has a vested interest in preventing costs from rising, because the employer is paying the contractor's 'costs', subject to the later operation of the pain/gain sharing mechanism. The lower the contractor's overall costs, the more will be the gain that can be shared between the contractor and the employer.

The contractor must also *notify* potential compensation events under clause 61.3. An early warning alone is not sufficient.

It is very important to remember that a contractor must also notify any potential compensation event as is required by clause 61.3, where the contractor has eight weeks after 'becoming aware' of the event. The contractor's early warning alone would not constitute the notice requirement of clause 61.3 because the early warning is not about compensation events.

Ambiguities and inconsistencies, clause 17

Unfortunately these can be very common. In preparing the many contract documents it is all too easy for the design organisation to say one thing in the specification, and something else on a drawing. It is also common to have inconsistencies between layout and general drawings and detailed drawings, particularly dimensional inconsistencies. Most construction contracts allow for the correction of such errors. Clause 17.1 obliges the project manager *or* the contractor to notify any ambiguities and inconsistencies. In practice it is usually contractors who identify such problems as they examine the detail in order to order materials and construct the works.

The project manager must give an instruction resolving the ambiguity or inconsistency. Since both ambiguous documents will normally be in the works information, the correction of one of them will usually be a change to the works information, which is a compensation event.

When assessing the compensation event in clause 60.1(1) resulting from a change to the works information, clause 63.8 comes into play. This says that the assessment assumes that the prices, completion date and key dates were based on the interpretation more favourable to the party who did not provide the works information. This would usually be the contractor, since most of the works information is usually provided by the employer. This is a restatement of the legal rule of *contra proferentem*, which means that documents which are ambiguous are read against the person that wrote them, which is only fair really.

In a similar way clause 60.3 states that where the ambiguity or inconsistency is within the site information (including the information referred to in it), the contractor is assumed to have taken into account the physical conditions more favourable to doing the work.

The final point on ambiguity is the second part of the compensation event exceptions set out in clause 60.1(1). This says that there is no compensation event if the ambiguity is between the works information provided by the employer and the works information provided by the contractor for the design. This simply means that where the contractor is asked to provide some design then the employer's requirements take precedence over anything the contractor puts forward, if an ambiguity is found later.

Illegal and impossible requirements, clause 18

This clause sensibly says that the contractor notifies the project manager as soon as it considers that anything in the works information is illegal or impossible and, if in agreement, the project manager changes the works information. Most contracts have a similar clause relating to the 'illegal and impossible' but the subject is not without legal problems. ICE contracts have a similar wording. The word 'impossible' has been ruled to mean absolutely impossible in *Yorkshire Water Authority* v. *Sir Alfred McAlpine* (1985). Yet impossible has also been interpreted in the ordinary commercial sense in *Turriff* v. *Welsh National Water Development Authority* (1980) where compliance with tolerance requirements would have been 'impossibly' expensive.

Prevention, clause 19

The Guidance Notes suggest that this is a *'force majeure'* clause. In other words, matters beyond the control of the parties. Clause 19 states that if an event occurs which:

- stops the contractor completing the works (at all) or by the date shown on the accepted programme, and which
- neither party could prevent and
- an experienced contractor would have judged at the contract date to have such

a small chance of occurring that it would have been unreasonable to allow for it

then the project manager gives an instruction to the contractor stating how it is to be dealt with. There would no doubt have been an early warning and a risk-reduction meeting. The event becomes a compensation event under clause 60.1(19) which means that the event becomes an *employer risk* both in terms of time and cost. Termination by the employer may be necessary under clause 91.7.

An example of prevention might be the foot and mouth outbreak of 2001, which was not foreseeable, but resulted in a ban on access to land containing cattle, sheep and other similar animals. Cross-country pipeline projects ground to a halt, and work could not be resumed for many months.

Many employers will be unhappy about clause 19, which a contractor may attempt to apply to strikes or perhaps insolvency of key suppliers for example. The other issue is that it relies on a judgment of 'forseeability' on the part of the contractor (by use of the words 'judged to have such a small chance of occurring'). Any judgment of 'forseeability', rather like the infamous ICE7 clause 12, opens up the potential for numerous arguments and possible unreasonable claims by a contractor so disposed. As a result many employers will consider deleting or modifying this clause.

Core clauses two: the contractor's main responsibilities

Providing the works, clause 20

This section begins with a clear statement of what may seem obvious, but which needs saying so that there is no doubt. Clause 20.1 says that the 'contractor provides the works in accordance with the works information'. However it is worth remembering that the 'works' may not simply be the physical construction works, but could also be design work as discussed below.

The contractor's design, clauses 21, 22 and 23

NEC3 ECC anticipates that the contractor can be given no design responsibility, some design responsibility or complete design responsibility such as in the design and build contract strategy. Unfortunately NEC3 ECC covers this very important subject in a very cursory way compared with many other forms of contract. Clause 21.1 simply says that 'the contractor designs the parts of the works which the works information states he is to design.' The rest of the clause briefly covers submission of the contractor's design and acceptance. Clause 22 deals with the employer's use of the contractor's design, and clause 23 with the design of equipment. The important subject of contractor design is covered in Chapter 20.

People, clause 24

Clause 24 requires the contractor to employ the people named in the contract data part two or a replacement which the project manager has accepted. The project manager will usually be looking for relevant qualifications and experience and a proposed replacement with less experience or qualifications is a reason for non-acceptance. In common with most forms of contract, this clause also gives the project manager the power to have a contractor's employee removed. There would have to be good reasons of course. Such reasons might include incompetence, lack of regard for health and safety or environmental pollution, or unacceptable or offensive behaviour.

Working with the employer and others, clause 25

The work of others

Clause 25.1 states that the contractor cooperates with others in obtaining and providing information which is needed in connection with the works, and the contractor shares the working areas with them as stated in the works information. This is not only intended to increase cooperation (in addition to clause 10.1) but to allow others to do work in connection with the project. Others could be statutory providers such as power or water companies, or it could be contractors working for the employer on a separate contract to do with the project, possibly an installer of electrical or mechanical plant.

Clearly the contractor must understand the extent of intervention by others and whether the contractor needs to interface with them when the contractor is pricing the tender. This is why this clause expects others and their work to be defined in the works information. Failure by the employer to adequately define the work of others in the works information is a compensation event as stated in clause 60.1(5). The works information should also be clear about whether the contractor provides any shared services or facilities such as access roads, storage areas, hard standings, welfare and security, cranes and lifting equipment, power supplies, water and so on. This is obviously a fair requirement to allow the tendering contractors to make adequate provision for such items.

When considering the work of others we must always think about health and safety. It is likely that the contractor will be the principal contractor under CDM 2007 and will therefore have duties to plan, manage and monitor the construction phase, and communicate and coordinate the work of other people and other contractors on the site.

The contractor must also show any interfaces with others on the programme. The programme should show the timing and duration of the work of others together with logic links and dependencies into the contractor's activities. Contractors can only discharge their responsibility to programme and provide the works safely if they are given full information as to the nature, activities and timing of the work of others. This should all be clearly detailed in the works information by the employer.

One of the most complicated health and safety problems is where many contractors are working on one site. This often happens in a building where the employer places separate contracts for perhaps the construction work, the fitting-out and the electrical and heating systems. The coordination of this work in terms of health and safety is very difficult. Wherever possible, it is good practice to have clear demarcation of areas of work, with marked access routes and agreements as to working times such as permits to work.

Clause 25.2 states that the employer and the contractor provide services and other things as stated in the works information. Here for example the contractor may be providing security and welfare facilities for other contractors of the employer. The clause goes on to say that any cost incurred by the employer as a result of the contractor not providing such 'services and other things' is assessed by the project manager and paid by the contractor.

Failure to meet a condition for a key date

Clause 25.3 is very important and a very useful inclusion in NEC3 ECC. It provides the *employer* with a right to compensation if the project manager decides that the contractor's work fails to meet the condition stated for a key date by the date stated. Since key dates are a way of tying the contractor's work into the work of 'others' it is logical to find the sanction on the contractor failing to meet a condition here in clause 25, which relates to working with the employer and others. Key dates and their relationship to the programme are discussed in Chapter 23.

So in constructing a lecture theatre we might have a key date for access to the installer of the AV projector, under a separate contract. The date and condition might be April 24, 'ceiling and support brackets complete, electrical supply available with safe access provided'. If the contractor does not meet these conditions by 24 April, the AV contractor will be delayed and the employer will probably be subject to a compensation event such as clause 60.1(2) for delayed access under the employer's contract with the AV installer.

Clause 25.3 states that the contractor pays the additional cost which the employer will incur or has paid on the same project either

- in carrying out the work, or
- by paying an additional amount to others in carrying out work.

The project manager must assess this cost within four weeks of when the condition is finally met.

Eggleston in his book *The NEC3 Engineering and Construction Contract: A Commentary* (Blackwell, 2006) suggests that clause 25.3 replaces the employer's right to sue the contractor for breach of contract. This would rarely be a viable proposition unless the losses are very large, so clause 25.3 gives the employer a practical alternative. The contractor's liability under this clause can be limited by the inclusion of option X18 which might be a wise move to avoid the contractor putting in a high price in its tender to cover the perceived risk.

It is important to differentiate the sanction on missing a key date from the sanction on failure to achieve a sectional completion which can be included in the contract as option X5. The sanction for being late with a sectional completion is delay damages, provided that delay damages are included as option X7. Key dates are intended to define the completion of a condition for *other contractors* to have access to do their work, where a sectional completion is a defined *geographical* part of the works to enable the *employer* to take it over and use it before completion of the whole of the works.

Sub-contracting, clause 26

This subject is important because sub-contracting is so prevalent in modern UK construction. The clause begins by making the statement that sub-contracted work and sub-contractor equipment are treated exactly as if they were the contractor's own. The contractor should submit the name of every sub-contractor to the project manager for acceptance.

Where the proposed sub-contract conditions are not from the NEC range, the contractor should submit them to the project manager for approval. These provisions are quite strict, probably because options C, D and E are cost-reimbursable and in simple terms the employer pays sub-contractors' invoices, whereas for the contractor's work the schedule of cost components is used. The latter is much more restrictive in terms of what can be included in the price for work done to date, which is used to determine regular payments to the contractor. Finally, if the contractor appoints a sub-contractor for substantial work before the project manager's acceptance and has not corrected this default within four weeks, it is grounds for termination by the employer under clause 91.2, which is a very serious matter.

NEC3 ECC does not have the facility to nominate sub-contractors, as some earlier forms of traditional contracts do. Such contracts normally had extensive clauses dealing with the process of nomination and possible refusal by the contractor. Nomination effectively forces a contractor to use a sub-contractor of the employer's choice, rather than the contractor's choice. Nomination (as a described contract clause) is now used rarely in forms of contract.

What happens in practice today is that the employer will expect the contractor to use named framework or other suppliers, who are listed in the works information. Any process to be used or payment terms are inserted in the works information by the employer. In many ways, this practice is nomination without any 'rights of refusal' by the contractor.

Other responsibilities, clause 27

Design

An important stipulation of clause 27.1 is that contractors must obtain approval of their design from others where necessary. Such approval might apply to permission under the Building Regulations or other statutes for example. The contractor remains liable for any contractor design even if it has been accepted by the project manager.

Access

Clause 27.2 ensures that the contractor provides access for the project manager, supervisor and *others*, to work being done and plant and materials being stored for this contract. However *others* must be *notified* by the project manager. There will therefore be a record in case others do not work within the times shown on the accepted programme or within conditions stated in the works information, which is compensation event 60.1(5).

Instructions

Clause 27.3 is the mirror of clause 14.3, discussed above. It simply states the fundamental point that the contractor obeys an instruction in accordance with this contract given by the project manager or supervisor.

Health and safety

Perhaps unnecessarily, clause 27.4 states that the contractor acts in accordance with the health and safety requirements stated in the works information. It is difficult to imagine that any competent contractor would not do this anyway.

Chapter summary

This chapter covers core clauses one and two of NEC3 ECC which deal with:

- General matters such as definitions, early warning, risk registers, instructions and working areas.
- The use of risk-reduction meetings.
- Dealing with ambiguities.
- Working with the employer and others.
- Sub-contracting and approval of sub-contractors.
- The contractor's main responsibilities for design where it is specified.

23

NEC3 ECC core clauses three and four

Aim

This chapter aims to explain the meaning and operation of the most significant clauses in core clauses three and four of the NEC3 ECC contract.

Learning outcomes

On completion of this chapter you will be able to:

>> Explain in detail the significant clauses of core clauses three and four of the NEC3 ECC form of contract.
>> Apply these clauses to typical contract issues.
>> Manage and administer an NEC3 ECC contract.

Core clauses three: time and important dates

Time is such an important issue in all contracts that this book devotes Chapter 16 to the general subject of times and programmes. This chapter will deal with how NEC3 ECC treats the contractual issues around this subject.

There are a number of important terms relating to dates in NEC3 ECC. These are listed below in *chronological order*, and assume that the contractor achieves completion before the completion date. These terms will be described in more detail in this chapter. Most are defined in clause 11.2.

Contract date	the date the legal contract comes into being
Starting date	identified in the contract data by the employer
Access date	identified in the contract data by the employer
Contractor's date for access	may be shown on the contractor's programme
Key dates	may be inserted in the contract data by the employer
Planned completion	must be shown on the contractor's programme
Date of completion	decided by PM and certified by PM
Takeover	certified by PM
Completion date	usually inserted in the contract data by the employer
Defects date	identified in the contract data as a number of weeks from completion

These dates are shown diagrammatically in Figure 23.1.

Contract date

The contract date is when the contract comes into being in a legal sense, possibly by the employer's letter of acceptance. On many larger contracts, however, acceptance will be formal signature of a form of agreement by both the employer and the contractor. NEC3 ECC contains no standard form for this but appendix three of the Guidance Notes includes forms of agreement.

After the contract date a legally binding contract has been formed and the key dates and completion date can only be changed under the terms of the NEC3 ECC contract itself (usually as a result of a compensation event but also by acceleration). Before the contract date, dates and contract data can be changed, if the parties agree, because there is no formal contract at that time. Here of course we are still in the tender period or the tender assessment period, before the contract is signed.

Starting date

The starting date is identified by the employer in the contract data part one. Before the starting date, the employer provides any necessary insurance. The starting date is also relevant in that a number of time intervals, especially the time interval for payment, begin not later than the assessment interval from this date as described in clause 50.1.

The benefit of having this provision for a starting date (if it is used) is to enable the contractor to be paid for any design work or possibly off-site fabrication work before access to the site is given. Hence in a contract with significant contractor design, the starting date could be the commencement of design, and the first access date (to site for construction work) could be many months later. Similarly in a contract with off-site manufacture or fabrication, this can also take place before the access date (to site). Advantageously, contractors can be paid for this work before they arrive on-site.

Access date

The access date is the date on which contractors can begin work on-site, although they do not have to begin on this date, and may start at a later date. Clause 30.1 says 'the contractor does not start work on site until the first access date.' On many projects access is not a problem, and can be given on the starting date. However, some projects are much more complicated than this, with multiple access dates, and access restrictions. Access dates (there may be more than one) are stated by the employer in the contract data part one. Where there are more than one access date, the areas of site affected by each one should be clearly stated.

An example where a contractor has general access to a university building for renovation work except for one lecture theatre might be: access date to the Cameron Building (except lecture theatre C4) shown on drawing 001 is the 1 December 2012, access date to lecture theatre C4 as shown on drawing 002 is 1 February 2013.

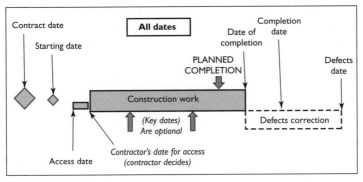

Figure 23.1 All dates under NEC3 ECC.

Hence 'access' in NEC3 ECC may not be sole access (possession), and others may also have access to all or part of the site. A compensation event as clause 60.1(5) may arise if this 'access for others' is not properly defined in the works information and it causes the contractor to incur cost or delay.

It is very important that the employer, usually through the design organisation, arranges all accesses to site and possession of the site itself, so that the contractor can *start* and *proceed* without being delayed. Any restrictions on access, possession or phasing of the works must be clearly set out in the contract data, by the employer, and ideally discussed with tenderers so that they really understand the implications when preparing their tenders.

NEC3 ECC thus facilitates much more flexible methods of procuring design, manufacture and construction services than more established forms of contract such as ICE7 and JCT. If we again take the example of a contract to paint a lecture theatre but with specially manufactured paint, we might see:

- Starting date 1 March 2012 (special paint manufacture may commence, and payments can be made to the contractor).
- Access date 1 May 2012 (painters may physically start work on-site).
- Completion date 1 June 2012 (if completion is later than this, delay damages may apply if option X7 is included in the contract).
- Planned completion 20 May 2012 (the date shown on the contractor's accepted programme).
- Completion achieved on 23 May 2012 (work completed to the definition in clause 11.2).
- Date of completion is therefore 23 May 2012.
- Project manager certifies that completion was achieved on 23 May within one week, that is, by 30 May 2012.

Key dates

The key date concept is a very useful addition to the NEC3 ECC contract. Clause 11(9) defines a key date as the 'date by which work is to meet the condition stated.' Key dates and the condition to be met are inserted by the employer, where required, in the contract data part one. Key dates are intended to tie other activities, or other contract work in other contracts, into the contract in question. Key dates do not have to be used. The sanction on a contractor who fails to meet the condition for a key date can be found in clause 25.3 which was discussed in the previous chapter.

> The concept of key dates is a useful addition to NEC3 ECC, but there is no obligation on the employer to include them in the contract.

One of the problem areas of construction contracts is deciding what constitutes completion. 'Completion' always refers to the whole of the construction work, but it also comes into play in NEC3 ECC where key dates are used, by the definition of the 'condition to be met'. This is rather like a 'mini-completion'. In the same way as completion itself (of the contract), there are likely to be disagreements over whether a *condition* for a key date has been met or not, particularly if the employer has incurred considerable expense as a result. The project manager and contractor are likely to have different views on the subject.

> The condition to be met by a key date requires very careful definition in the works information.

Additionally if key dates are used the programme must show key dates and they must be carefully monitored as the project proceeds. A contract with numerous key dates is much more likely to be problematic to administer than one with few, or none. Hence, useful as they are, key dates should only be used where they are essential.

> Useful as they are, key dates should only be used where they are essential.

An example of the use of key dates might be the early completion of an access road to an existing administration building (to allow access to a separate contractor engaged in internal repairs and refurbishment). Another example might be the completion of foundations for some plant in a building required before the completion of the building to allow 'others' to install the plant, such as an installation requiring a mobile crane before the roof can be constructed.

Completion date

This is usually inserted in the contract data part one by the employer. It is probably the most important of all the dates because this date is when the employer is expecting to be able to take over and use the new asset. Additionally, after this date, delay damages will be paid by the contractor if they have been included as option X7.

The completion date must be chosen carefully and have due regard to health and safety. It is a very important date.

The completion date must be determined with great care and must not result in too short a construction period. Regulation 9 of CDM 2007 makes it a requirement for the employer to allocate sufficient time for the construction work to be carried out, as far as reasonably practicable, without risk to the health and safety of any person. The employer will usually take advice from the designer or project manager on a suitable construction period for the project in question and hence set the completion date.

If the employer wishes, the contractor can be allowed to choose the completion date, in which case the contractor inserts its chosen date in the contract data part two. The advantage here is that the completion date can really suit the contractor's resources, workload and programme. The big drawbacks are that the contractor's proposed completion date may not really suit the employer, and different dates inserted by tenderers can make tender comparison difficult.

Planned completion

A contractor will usually produce a programme to show *completion* (the state), *before the completion date* (the contractual date), by indicating an earlier *planned completion* on its programme (clause 31.2). This is the *contractor's* planned completion and because of its importance is shown in capitals in Figure 23.4 below. The difference between the completion date and planned completion is *terminal float*. Terminal float results in a shortened programme for the construction work itself. Some shortening of the programme is normal but highly shortened programmes can have serious contractual implications. Terminal float and shortened programmes were discussed in Chapter 16.

Planned completion, as shown on the contractor's programme, is very significant in NEC3 ECC.

It is possible of course for the contractor to show planned completion at the completion date, thereby showing no terminal float, but very unlikely. Such a situation would allow contractors no leeway if delays occurred which were their 'fault', in other words not an employer risk or a compensation event.

On a project that is in delay, planned completion may have to be shown after the completion date on the contractor's programme (probably on revised programmes, later in the contract). It is likely in this event that the contractor will be pursuing compensation events to secure a revised completion date. However, the contractor's obligation under clause 31.2 is to show planned completion and the completion date. Where planned completion is after the completion date then delay damages will be due from the contractor if they have been included as option X7. Since late completion is usually in nobody's interests, it is to be hoped that the project manager and contractor are cooperating to try and address the problem.

Planned completion is particularly significant in NEC3 ECC because of the way that delays to the completion date are assessed by the operation of clause 63.3. This is a very important clause and is discussed further in Chapter 25 which deals with compensation events.

Date of completion

The date of completion must not be confused with the completion date. The *completion date* is written into the contract whereas the *date of completion* occurs when the contractor has 'finished', that is, achieved '*completion*'. As described in Chapter 22, completion is a state (as in physical state), not a date. Employers should avoid any uncertainty by ensuring that they clearly set out in the works information the exact nature of the works to be finished in order to achieve completion. In addition to defining the physical work to be completed, this description of completion might include the provision of as-constructed drawings, manuals, and health and safety information for example.

> In NEC3 ECC, completion, completion date and date of completion have very different meanings and must not be confused.

The project manager decides the date of completion and *certifies* completion (the state) within one week of completion (clause 30.2). This simple provision is to ensure that there is a clear record that completion (the state) has been achieved and on exactly what date (the date of completion). One assumes that the project manager will issue a completion certificate of some form but, unlike traditional contracts, NEC3 ECC does not give this certificate a name. JCT SBC05 for example calls it the 'practical completion certificate'.

The completion date can and must be changed by the project manager as a result of accepted compensation events which occur on most contracts owing to changes or unanticipated events. Thus on most contracts the completion date moves back in time as the contract proceeds. Compensation events are the subject of Chapter 25.

Take-over

Take-over is defined in clause 35. If there is nothing stated in the contract data, the employer takes over the works within two weeks of completion. The project manager should certify the date of take-over of any *part* of the works within one week of that date. Hence the project manager must certify both completion and take-over.

At take-over the asset becomes the employer's to use as it wishes, and from this date the contractor must request access for the correction of defects.

Defects date

The defects date is inserted by the employer in the contract data part one. It is usually set at 52 weeks from completion. It is the period of time for which the contractor has a responsibility to correct defects after completion (this is rather

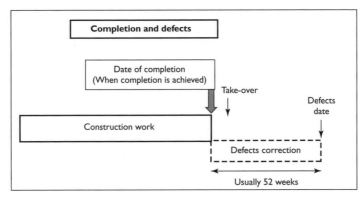

Figure 23.2 Defects date.

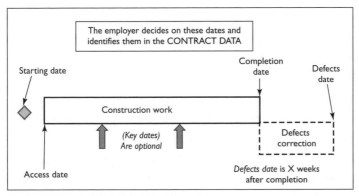

Figure 23.3 Dates identified by the employer.

like a twelve-month warranty on a domestic appliance). Figure 23.2 shows the defects arrangements. Because the defects date is set as a number of weeks after completion, the actual calendar date (25 July for example) cannot be determined until the contractor has finished the contract work sufficiently to be given 'completion' by the project manager. The date of completion is the date that completion is achieved.

The sequence in which these dates are developed

Dates that are usually decided by the employer

Subject to the provisos above, the employer determines the starting date, completion date, any key dates (if used) and by inserting a time period after completion (in the case of the defects date). These dates are shown in Figure 23.3.

Dates identified by the contractor

The contractor should submit a programme after the contract date and in accordance with clause 31.2. The contractor's programme should repeat the dates given by the employer and add the contractor's decisions as to the following dates, all as shown in Figure 23.4.

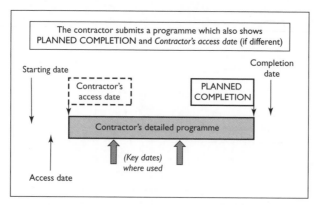

Figure 23.4 The dates identified by the contractor.

- The planned completion (discussed above)
- The dates when the contractor intends to meet the condition for a key date (if key dates are inserted by the employer)
- The contractor's access date (if different from the access date given by the employer)
- And where appropriate, dates for acceptances, items to be provided by the employer, and information needed from others (not shown in Figure 23.4).

Contractor's access date

It is possible that the contractor intends to start on the whole of the site or part of the site later than the access date. If this is the case, and the contractor wishes to state this, then clause 31.2 suggests that the contractor shows 'access to a part of the site if later than its access date' on the programme. It is difficult to see why a contractor would do this since, under clause 33.1, if a *later date for access* is shown on the contractor's accepted programme, then this becomes the access date, until subsequently changed.

So in the example above, if the contractor's accepted programme showed 8 May as the contractor's access date, then 8 May would become the access date and the employer would have one extra week in which to provide access. This is because compensation event 60.1(2) uses the *later* of the access date and the 'date shown on the accepted programme' as the base-line date.

Core clauses three: the clauses in detail

Starting, completion and key dates, clause 30

This clause confirms a number of fundamental points, which are that:

- The contractor does not start on-site until the first access date.
- The project manager decides the date of completion
  - and certifies it within a week of completion.
- The contractor does the work so that the condition stated for each key date is met by the key date.

Programme, clauses 31 and 32

Programme

NEC3 ECC uses the programme as a key management tool for running the contract effectively. The programme is covered in detail in clause 31. The requirements of the programme *itself* are covered more thoroughly in NEC3 ECC than in ICE7 and JCT forms of contract.

> The programme is important in all forms of contract, but NEC3 ECC gives it primary significance.

Items to be shown on the programme

Clause 31.2 lists a wide range of items that should be shown on the programme. The exhaustive list in this clause has been simplified below:

- The important dates, such as starting date, access date, key dates, completion date and so on.
- *Planned completion* (a very important date, referred to above).
- The order and timing of the contractor's operations.
- The order and timing of the work of the employer and others.
- The dates when the contractor intends to meet the conditions for any key dates.
- Provisions for float and **time-risk allowances**.
- Requirements for access to parts of the site if different from the access date.
- Date when the contractor needs acceptances, items provided by the employer and information from others.
- For each operation, a statement of how the contractor plans to do the work including principal equipment and resources.
- Other information which the works information requires the contractor to show on the programme.

Most contractor programmes contain the majority of these items, but the requirements to show float and time-risk allowances and statements of how the contractor plans to do the work and principal equipment and resources are more unusual, but very useful additions. On smaller projects, these additions could be expected on the first programme. However, on larger projects, much of the detail for later operations may have to be added on revised programmes as the work proceeds.

Time-risk allowance is 'spare time' within an activity. So if the estimated activity duration was actually 4.5 days, and it is shown on the programme as 5 days, then it would have half a day time-risk allowance.

Submission of a first programme

If a programme is not identified in the contract data (this is the usual case), then the contractor should submit a *first programme* within the period stated in the contract data. This period is counted from the contract date (the date when the contract is signed).

NEC3 ECC contains a very important sanction. If the contractor does not submit a programme, by the date calculated on the above basis, then 25 per cent of the *'price for work done to date'* is retained, until the contractor does submit a programme (this stipulation can be found in the payment section in clause 50.3). This sanction will only come into operation from the first time the contractor is paid of course. This first payment is made within the assessment interval (five-weeks maximum) counted from the starting date.

The other sanction is that if a programme has not been accepted, or if the programme has not been revised in accordance with the contract, then the project manager makes his or her own assessment of the programme when judging compensation events. In other words the project manager decides on future events and dependencies, not the contractor. This can be found in clause 64.2 (see Chapter 25) and is a situation that most contractors would wish to avoid.

Project manager's response to the programme

Clause 31.3 deals with the response of the project manager and says that the project manager should accept or reject the programme within two weeks of submission by the contractor. The accepted programme plays a key role in NEC3 ECC and is discussed below.

Contractors may submit programmes showing many critical activities and few activities with free float. It is quite easy to do this by adjusting the logic links in the programme. The reason contractor's may do this is that 'employer' delays to a critical activity produce a revised completion date, thereby reducing the contractor's exposure to delay damages and giving more time to achieve completion. The more 'critical activities' that are shown, then the more chance that one of them will be delayed by an employer risk, such as a change to the works information.

Such 'employer' delays would also justify a quotation for increased 'preliminary costs' for the extended time period on-site. Such costs might include working area overheads for example. Hence the more critical activities that a programme depicts, the more opportunity there will be for more time and more preliminaries, when a compensation event occurs. Consequently, the project manager would be well advised to consider all programmes very carefully before acceptance.

If the project manager rejects a programme, this rejection must be notified to the contractor and rejection must be for one of the reasons set out in clause 31.3, which are:

- The contractor's plans show that it is not practicable.
- It does not show the information which this contract requires.

- It does not represent the contractor's plans realistically.
- It does not comply with the works information.

Unfortunately even these 'objective reasons' are capable of some interpretation and disagreement, particularly the words 'not practicable' and 'realistically'. In traditional contracts it was not unknown for a project to be completed without an *accepted* programme, because the arguments about its validity had not been resolved. This is a very regrettable situation which would undermine the intentions of NEC3 ECC. In common with most forms of contract there is nothing to stop the contractor proceeding even in the absence of agreement to the programme by the project manager. This is correct of course since the contractor's prime obligation is to proceed to construct the works, in compliance with the works information, by the completion date.

Programmes as contract documents

The programme is not usually a contract document. It can be made so, by incorporating it into the original accepted tender but such a move would be very unwise because it locks the parties into fulfilling exactly what is at most an estimate of future events and activities. Hence programmes are usually not incorporated and pre-contract programmes should be given a clear status as documents that will be revised as the work proceeds.

NEC3 ECC seeks to do this by its descriptions of the programme. A programme is a best estimate at the time it is produced, by the contractor, of the phasing and timing of the works, and should remain flexible and capable of updating to reflect changing conditions on-site and incorrect programming assumptions. However, it is the best means we have of controlling the work and monitoring the effects of events.

Programmes and compensation events

The programme in NEC3 ECC is therefore used not only to plan and carry out the contractor's work, but also to ascertain the effects of compensation events. What helps in this assessment is the stipulation that the programme also includes the allocation of resources and equipment to relevant activities.

> In NEC3 ECC the programme is also used to assess compensation events.

Hence the programme and its associated resourcing are effectively the way that the contractor proves its losses in terms of time and the financial consequences of delays. For this reason, the project manager will wish to see a realistic programme and there can be some debate as to whether a submitted programme achieves this objective.

Yet another use of the programme is to consider 'what-if?' scenarios. This can be very valuable at risk-reduction meetings, where the contractor and project manager are seeking solutions to problems in order to minimise time and cost effects.

Revising the programme

NEC3 ECC is unusual in that it defines what a revised programme should show in clause 32.1, which states that this includes:

- Actual progress and the timing of remaining work.
- The effects of implemented compensation events.
- How the contractor plans to deal with any delays and to correct notified defects.
- Any other changes that the contractor proposes to make to the accepted programme.

The contract data also identifies the period for submission of *revised programmes*. This period is often a maximum of five weeks to allow for monthly revisions to coincide with site meetings, which are usually monthly. Revised programmes can also be issued when the contractor wishes, or within the period for reply, where instructed by the project manager. This provision might be used to see the effect of important changes or compensation events for example.

The submission procedure for revised programmes is covered in clause 32.2 which states that the contractor submits revised programmes for acceptance by the project manager:

- In the period for reply, following a project manager instruction.
- Whenever the contractor chooses to submit.
- But in any case, no later than the interval stated in the contract data from the starting date until completion of the whole of the works.

The accepted programme

The programme or any revision to it should eventually become the accepted programme, which we have seen to be an important concept in NEC3 ECC. The accepted programme is therefore the latest *agreed* depiction of the contractor's plans, timing and resources, together with the dates when information, materials and design information, as appropriate, are required from the employer.

Importantly, the accepted programme also shows the *latest* contractor's planned completion date. It is this planned completion date which is used to determine any new completion date for the contract itself as a result of compensation events.

Finally a comprehensive agreed programme is the best mechanism for the contractor to provide the works efficiently, and for the project manager and contractor to determine the effect of unanticipated events and investigate alternative actions at risk-reduction meetings.

Access to and use of the site, clause 33

This clause states the employer's obligation to allow access to the contractor to do the contract work. If the employer does not do this it is a compensation event as clause 60.1(2). However this access to the contractor can be later than the access date, if a later access date is shown by the contractor on its accepted programme. Hence if the access date stated in the contract data were 1 April, and the contractor showed 20 April on the accepted programme, then a later access date of 8 April, given by the employer, would not be a compensation event.

Instructions not to start or to stop work, clause 34

In NEC3 ECC this is a very short clause. Traditional contracts tend to go into more detail over what can be a very serious contractual matter. Only the project manager can issue these instructions, and they are a compensation event as clause 60.1(4). They may also be a reason for termination, where work to a substantial part of the works cannot commence for more than 13 weeks (clause 91.6). So in common with many other forms of contract, such instructions can have very serious consequences and should always be considered very carefully.

However, there may be times when some delaying event means that construction cannot proceed in whole or in part. Such delaying events are often external, such as a delay to planning permission, or delay to moving major services under the site, such as gas or water mains, or even possibly the discovery of extensive archaeological remains; all these mean that the contractor cannot proceed for some time, perhaps a few weeks. Sometimes these events will affect the whole site, and sometimes a part of the site.

The project manager is then faced with a difficult decision. Is it better to allow the contractor to continue working or not? If the contractor has restricted access, it may work inefficiently with a possible major reprogramming exercise, with financial consequences that may be difficult to agree and to determine and which would be compensation event 60.1(2). Alternatively is it better to give an instruction to stop or not to start work which would be compensation event 60.1(4)? If such an instruction is given then the employer is likely to have to pay the contractor's demobilisation and remobilisation costs. However, these costs may be much easier to determine and agree than costs arising from a major change to the programme's logic and dependencies, which are much more indeterminate and difficult to ascertain. The project manager would do well to bear this in mind when comparing the likely costs of the two actions.

The other consideration will be the effect on the completion date. Restricted working will have less effect on the completion date than an instruction to stop all work, and this may be important to an employer who has a crucial completion date in mind.

If the project manager decides to give an instruction not to start or to stop work, it will be a compensation event as clause 60.1(4), however the contractor is not entitled to any change to prices or dates if the event 'arises from a fault of the contractor', as clause 61.4. Such an instruction not to start or to stop work may arise from insufficient consideration to safety, or lack of quality or non-compliance with the works information. One would hope that such occurrences would not arise when a competent contractor is constructing the project.

At an appropriate time, the project manager may instruct the contractor to restart or start work.

Take-over, clause 35

Normally, the employer is waiting to take over its new asset, and will be happy to do so. Clause 35.1 states that the employer takes over the works not later than two weeks after completion.

Early completion

In common with most contracts the contractor may achieve completion before the completion date. Many employers will be pleased to have early completion. However, early completion can cause the employer to incur additional costs. These will certainly include insurance, but may include heating and security for example. To address this issue, NEC3 ECC allows the employer to state in the contract data that it is 'not willing to take over the works before the completion date'.

Taking over part of the works

Another possibility is that the employer wants to take over part of the works before they are complete and before the completion date. The contractor is unlikely to have a problem with this, but the contract has a compensation event under clause 60.1(15) if the contractor had not achieved completion of that part. Such an action on the part of the project manager might be necessary if the employer needed access through part of the works, or was so keen to put it into use that lack of completion was not an issue.

This could occur in a university extension contract where the employer needed a lecture theatre for the start of term when in some respects it was not quite finished. Under most circumstances the contractor would be more than happy to hand over an almost finished part of the works although, unlike traditional contracts, half the retention on the part is not paid to the contractor, but delay damages liability is reduced if they have been included as option X7. Clause 35.3 ensures that the date and extent of take-over of a part is certified by the project manager within one week of the date of take-over.

It can be assumed that take-over of the whole of the works must also be certified, but the contract is not very clear on the matter, unlike traditional contracts which give the process more formality. Certification of take-over is important because after this date the employer is generally responsible for 'loss of or wear or damage to parts of the works' under clause 80.1. There are some other provisos in the detail of the clause.

Acceleration, clause 36

Many contracts provide for the possibility of acceleration. NEC3 ECC defines acceleration as the achievement of completion before the completion date. This is not speeding up the work (although this might be the result), it is changing the existing completion date. The project manager may instruct the contractor to submit a quotation for acceleration and the project manager also states any changes to key dates which may be affected. The contractor submits a quotation for acceleration, or its reasons for not doing so. The contractor's quotation must include changes to the prices, and a revised programme and details of how the assessment has been calculated.

However the contractor cannot be forced to accelerate, which is of course contractually correct. It is not reasonable for one party to *force* a unilateral obligation on the other after the contract has been made.

Acceleration is usually contemplated when a series of compensation events have resulted in a revised completion date, which may be many weeks later than the original one. If the employer wants its asset on or near the original completion date, then acceleration may be the answer.

Acceleration is usually a serious and expensive business. The reason is that the original project was usually tendered in competition, but under acceleration the contractor is not subject to market forces when supplying the quotation. From the contractor's point of view, it will have put a lot of effort and time into planning and resourcing the project. Acceleration will probably mean that this planning exercise must be repeated and sub-contractors will also have to be contacted to agree revised terms. Additionally, acceleration will force a shortened programme (relative to the current completion date) onto the contactor and expose it to a greater risk of paying delay damages where these have been included as option X7. For these reasons, acceleration is quite rare.

> Acceleration is usually a serious and expensive matter and should never be undertaken lightly.

Another difficulty is what happens if the employer pays for acceleration and then the contractor does not achieve it. The situation is probably that the employer can recover delay damages, where included as option X7, but is unlikely to be able to recover the costs paid to the contractor to accelerate. Additionally, after acceleration has been agreed, the new agreed completion date will also be subject to potential change arising from any compensation events that arise. Thus the employer's wishes may not be realised after all.

If acceleration is contemplated, one assumes that the project manager will get the employer's approval. Oddly, NEC3 ECC does not actually state this, whereas contracts such as ICE7 only allow acceleration under clause 46.3 after the employer and contractor have agreed the terms.

Core clauses four: testing and defects

Before looking at the detail of this section it is very important to grasp a basic concept about the word 'defect' in NEC3 ECC. In this contract, a defect is usually something which has not been constructed in accordance with the works information (generally the drawings and specification given in the works information). There is a difficulty here, in that in normal language a 'defect' is something unsatisfactory or deficient. In contracts using NEC3 ECC, work can be unsatisfactory, but *not* be defined as a defect under the contract. This particularly applies to work that has been incorrectly or poorly specified or designed by the designer. On reflection, we can see that this is fair. The contractor should only be *obliged* to provide what it is contracted to provide; nothing more and nothing less. Contractors may be *willing* to provide more, often to please a valued client, but they are not *contractually obliged* to do so.

Items that are 'defective' may not be a defect under the contract.

Hence, in this contract, defects relate to things which are the 'contractor's fault' and for which the contractor is contractually responsible, and hence what the contractor must correct. Unsatisfactorily constructed work which is not the 'contractor's fault' is not a defect and would presumably be corrected by an instruction or change to the works information given by the project manager, which would produce a compensation event and the employer would have to pay the cost and accept any delay effects on the programme. Such unsatisfactory work would usually be due to deficient design or specification, by the design organisation, which is discussed below.

Unfortunately the issue of whether work or materials constitute a defect is one of the most contentious contract issues. NEC3 ECC brings some clarity to the discussion, but the arguments will no doubt continue.

Definition of a defect

NEC3 ECC defines defects in clause 11.2(5), which says that a defect is:

- a part of the works which is not in accordance with the works information or
- a part of the works designed by the contractor which is not in accordance with the applicable law or the contractor's design which the project manager has accepted.

Typical defects arise from setting-out errors or not meeting tolerances, such that the constructed work is dimensionally incorrect. Other defects will occur if the contractor does not follow the specification given in the works information, particularly in terms of the specified quality of work or materials. In order to establish the constructed quality of work, tests are usually specified. Hence NEC3 ECC relies on the works information giving comprehensive descriptions of tests.

Tests and inspections

Tests generally

It is normal for contracts to prescribe tests, to ensure that items in the constructed works comply with the specification given in the works information. Typical tests would be concrete cube tests, pipeline pressure tests, site testing of electrical panels and so on. It is also normal to carry out inspections to verify levels, dimensional accuracy, quality of workmanship and similar items.

The responsibility for carrying out such tests and inspections on quality matters lies with the supervisor. Apart from issues connected with defects, the supervisor has few other powers in the NEC3 ECC contract. Most powers are vested in the project manager. Hence the supervisor may discover a defect and must notify the contractor under clause 42.2 up until the defects date; but the project manager has to issue any relevant instructions where necessary.

Prescribing tests

The crux of the defects issue is prescribing the tests which must be carried out, in the works information. The NEC3 ECC contract itself gives little advice here, but the Guidance Notes devote many pages to the description of tests, and how they operate in practice. The Guidance Notes suggest that tests should be specified with respect to:

- The nature of the tests
- When they are to be done
- Where they are to be done
- Who does the test
- Who provides materials, facilities and samples
- Their objectives and procedures
- Whether or not payment or authorisation to proceed to the next stage of the work depends on the test results.

It is clearly better for tests to be thoroughly described as suggested. Where descriptions are insufficient or missing, the project manager can instruct tests but this becomes compensation event 60.1(1) because the works information would need to be changed. It is worth noting that the supervisor does not have the power to order tests or change specified tests, because this is not a power vested in the supervisor. Instead the supervisor would report the matter to the project manager, who would decide what action to take.

The timing of tests during the contract must also be stated clearly in the works information. Further details of tests might include:

- Tests between the starting date and access date
- Tests before delivery to site (often at the manufacturer's premises)
- Tests at certain stages of the work
- Tests required as part of the condition stated for a key date
- Tests which must be passed before further specified work takes place
- Tests needed to achieve completion

- Further tests after take-over (often part of the employer's commissioning or optimisation arrangements)
- Performance tests (usually for plant).

Tests and inspections and clause 40

Clause 40 deals with the procedure for conducting tests and inspections that are required by the contract or by 'the applicable law'. Applicable law would include tests on pressure vessels or lifting equipment, which are required under safety legislation for example. The contractor is obliged to do these even if they are not prescribed in the works information. The contractor is at liberty to carry out any tests that are required or considered beneficial, but these are not covered by this clause.

The contractor and the employer provide whatever is required of them as stated in the works information. This may include materials, testing equipment or facilities, samples or personnel. The contractor and supervisor should *notify* each other of tests and inspections, and the results. Notification effectively means in writing, which gives the procedure the required formality. It is usual for the contractor to carry out the tests, and the supervisor to observe them.

Where supervisors wish to carry out their own tests and inspections, they should do this without causing unnecessary delay to the work. If the supervisor causes unnecessary delay to the contractor's work, this is covered by compensation event 60.1(11).

Defining 'unnecessary delay' might be difficult unless the works information is explicit on the matter. It is not difficult to state that the supervisor will complete off-site inspections within two days, and inspections of excavations and concrete pours within two hours of notice being given. An outline programme of tests and inspections needed in the following week would also be a useful requirement of the contractor. In this way, supervisors can plan their time accordingly, and ensure their availability.

Defects

Where a test or inspection shows that any work has a defect

Clause 40.4 states the important point that where a test or inspection shows that any work has a defect, then the contractor corrects the defect and the test is repeated.

Whether the contractor is paid for correcting defects and repeating tests depends on the option A to F selected by the employer. In options A and B the contractor will bear the cost of correcting defects and of repeat tests and inspections, because payment is based on activity schedules or bills of quantities. An incentive to correct defects quickly in options A and B is that only completed work without defects (which would either delay or be covered by immediately following work) is paid for.

However in options C, D and E, the contractor will be paid for correcting defects *and* its own costs of repeating tests provided the defect is corrected before completion (defects corrected after completion are a disallowed cost). This provides some contractor incentive for ensuring that defects are corrected before completion. The payment will ultimately be subject to pain/gain share in options C and D of course. Many employers consider that any payment to the contractor to correct its own defects is unreasonable, and amend the relevant clauses.

However, clause 40.6 makes it clear that the contractor pays any costs incurred by the *employer* if a defect is found and a test or inspection needs repeating. Such employer costs are assessed by the project manager.

Searching for defects

The supervisor may instruct searches for defects up until the defects date but must give the reason along with the instruction. This issue is dealt with in clause 42 which says that *searching* may include:

- uncovering, dismantling, re-covering and re-erecting work,
- providing facilities, materials and samples for tests and inspections done by the supervisor
- doing tests and inspections which the works information does not require.

It may seem odd to have a clause with this title. It may be better to think of it as a search to ascertain whether there is a defect under the contract, after some 'defective item' has been discovered. What we are searching for here is the root cause of the defective item, and ascertaining whether the contractor complied with the works information or not. Thereby we establish contractual responsibility, and hopefully a solution.

> 'Searching for a defect' is really searching for the root cause of a defect, in order to establish the problem, and liability under the contract.

This is actually quite a difficult concept. Let us take an example. It is normal for house window frames to be primed before delivery and final painting. Suppose the specification (contained in the works information) did not specify primer. This could be an oversight by the designer. Nine months after installation, the paintwork on the windows begins to peel off; they only had a coat of undercoat and a coat of topcoat as specified in the works information. In normal speech, peeling paint is a 'defect'. However, it is not a defect in this case, since the window frames complied with the works information. The window frames have *defective* paintwork, but it is not a *defect*. To remedy this situation, the project manager would have to issue a change to the works information, specifying suitable remedial action (possibly sanding down followed by primer and two coats of paint); this would be a compensation event, and the employer would have to pay for it. The employer may want to talk to the design organisation that wrote the specification!

Let us take another example. Suppose a reservoir wall exhibits severe cracking three days after pouring the concrete. In normal language this is clearly a defect, because the reservoir might leak and the reinforcement will become corroded. The reservoir wall is 'defective'. However this is not necessarily a defect under the contract in NEC3 ECC terminology. The question is whether the construction of the wall complied with the works information. This is sometimes called the 'design/workmanship problem'.

> In dealing with defective items, the first action is to determine the root cause and then ask whether the contractor's work complied with the works information.

In order to find the root cause and determine liability, a 'search for a defect' might include checking the cement content and strength of the concrete by means of core samples, tests to determine the reinforcement actually fixed and checks on dimensions and joints. If these all prove that the contractor has complied with the works information, then the cracks are not 'defects' under the contract; although the wall is certainly defective. The cracks are probably due to early thermal shrinkage caused by an incorrect design. If this is the case, correction will be at the employer's cost, and the project manager will have to instruct the contractor what to do. This could be repair or, in extreme cases, reconstruction. The next step would be for the employer to consider whether it has a claim for negligence against the design organisation. For the contractor, however, this would not be a defect for which it has any responsibility under the contract.

Where the supervisor instructs a search for a defect and no defect is found it becomes compensation event 60.1(10), unless the search is needed because the contractor gave insufficient notice of doing work which obstructs a required test or inspection.

Notifying defects

The final provision of clause 42 is that, until the defects date, the supervisor notifies the contractor, and the contractor notifies the supervisor, of any defect as soon as it finds the defect. However the contractor must correct defects whether or not the supervisor has notified them, which is the first part of clause 43.

> The contractor is obliged to correct defects whether or not the supervisor has notified them.

Correcting defects 'whether notified or not' is just the simple contractual responsibility of the contractor to comply with the employer's works information and construct the works in return for payment. However, the notification obligation has some difficulty for the contractor. The contractor may see a 'defect', such as the cracking in the reservoir wall example above, but may not know as yet whether it is a defect under the contract or not. In theory, the contractor may also be obliged to give an early warning under clause 16, since this issue could cause delay, extra cost or impair the performance of the works in use. Perhaps the sensible course of action for the contractor is to notify a 'potential defect' whilst avoiding accepting liability at this stage, and also give an early warning. There is no doubt that a matter of this importance should be brought to the attention of the project manager and a risk-reduction meeting should consider actions and implications, which will no doubt include searching to determine cause and liability, together with possible actions such as repair or reconstruction.

Defect correction period

This period is identified in the contract data by the employer, and is often two or three weeks. In traditional contracts, phrases similar to this are used to denote the total rectification period, which is often 52 weeks. NEC3 ECC does not work like this, and the defect correction period is a novel idea. In NEC3 ECC, the total rectification period is still usually 52 weeks, but it has no specific name. It is set by the defects date.

In contrast, the defect correction period is the period of time within which the contractor must correct a defect. The period of time runs from notification, either by the contractor or the supervisor, *but operates in a different way before and after completion.*

> The defect correction period is the period of time within which the contractor must correct a defect. It is usually specified as two or three weeks.

The defect correction period really comes into prominence *after* completion, and clearly it is only likely to be the supervisor who notifies since the contractor will have left site. It is more likely that the employer will remain in contact with the design organisation after completion for general contract administration requirements, and the supervisor is usually employed by them. Where employers employ their own staff, they will frequently also employ site supervision staff and supervisors. It is also quite likely that an employer will require regular meetings or inspections before the defects date, and will pay for the supervisor to carry these out. After completion the contractor must correct a defect within the defect correction period.

Before completion the contractor must correct a defect before the end of the defect correction period, *added to the date of completion.* This can be found in clause 43.2 but the clause is not easy to understand and requires careful reading. Hence before completion the contractor usually has many weeks in which to correct a defect, although in practice the contractor must usually act quickly so that subsequent work can proceed. This is because, during construction, the contractor is likely to need to correct defects quickly, because they are likely to affect ongoing work. Defective columns on the second floor of a tower block, for example, will prevent the contractor constructing the third and higher floors. However, before completion the contractor can choose to correct notified defects at any time subject to the above constraint. Additionally, the contractor will need to remember that it cannot achieve completion if there are notified defects remaining which 'would prevent the employer from using the works or others from doing their work' as stated in the definition of completion in clause 11.2(2).

The contract data has the provision to have different defect correction periods for different stated defects. The employer would presumably take into account the importance of rapid defect correction for defects that could significantly affect the use of the works (perhaps defects with mechanical and electrical plant), and longer periods for minor defects (perhaps problems with landscaping such as plants dying). As always, the contractor will price the risk; a short defect correction period will increase the contractor's risk because it has to determine much more quickly how to correct the defect and mobilise people and equipment to do so.

Defects after take-over

The employer also has an obligation under clause 43.4 to give the contractor access to correct defects, after take-over. After take-over of course the employer is likely to be using or operating the asset, and any access for the contractor must be agreed with due regard to the health and safety of both the contractor and any of the employer's personnel or possibly the public by carrying out thorough risk assessment. In this case, the defect correction period begins when the employer gives access to the contractor.

Defects date

The defects date is identified by the employer in the contract data part one. It is usually 52 weeks from completion and is the period of time for which the contractor has a responsibility to correct defects after completion. The defects date has four features:

- It is the last date when either the supervisor or contractor can notify defects, as clause 42.2.
- It is the last date when the supervisor can instruct a search for a defect, as clause 42.1.
- It is the date on which the supervisor issues the defects certificate, as clause 43.3, although if necessary the defects correction period can be added to the defects date for defects notified just before the defects date.
- It sets the final date for notification of compensation events, as clause 61.7.

Accepting defects

This clause is a useful addition to NEC3 ECC and gives the project manager the power to accept defects. Since the clause does not restrict this power, the power presumably runs until the defects date, not just up to completion. Quite often defects may be minor, and the cost of correcting them may be out of all proportion to the benefit to the employer. Additionally, in options C to E the employer is paying for the cost of correcting defects (up to the completion date), but eventually this payment will be subject to pain/gain in options C and D. Hence there is a real incentive to accept minor defects, but of course no obligation on the project manager.

Suppose we have a contract for the construction of a new house, a domestic project, with oak skirting boards. The contractor has nearly completed the house but has installed pine skirting boards and has sanded and varnished them. All the plastering and painting of the walls are complete. On a site inspection the supervisor sees the problem and notifies a defect. To replace the skirting boards and repaint the walls as necessary will be very expensive and disruptive, especially as oak skirting boards are on a four-week delivery. The contractor will no doubt hope that the project manager will accept the defect. Replacement will cost £5,000 including delay costs, and the contractor may well offer a price reduction of £4,000. The project manager can accept the defect if the contractor and project manager agree, for a reduction of £4,000. Of course a sensible project manager will consult the 'employer' (or possibly the employer's wife!) to see if oak skirting boards are really important, and whether the employer is happy with a £4,000 reduction in price for the construction of the house.

In traditional contracts the contract administrator would often accept defects but does not strictly have the contractual power to do so. In contrast, in NEC3 ECC, the project manager may be willing to accept specific defects, often in exchange for a reduction in price or an earlier completion date, *or both*; and has the power to do so. This is covered in clause 44. In these circumstances, the project manager would be wise to consult the employer first as in the example above. This is to ensure that the employer is happy to accept the 'defect'. The project manager will also no doubt consult the supervisor for an opinion, although the supervisor has no power to accept a defect. It is important to note that accepting a defect can bring the completion date *forward to* an earlier date, which is very unusual in contracts, except for acceleration clauses. This could be a key item for the contractor to offer in exchange for acceptance of a defect; but the project manager is not obliged to accept a defect.

The project manager is not obliged to accept a defect.

If the project manager is willing to accept a defect under clause 44.2, he or she does so by instructing a change to the works information. This acceptance of a defect is not a compensation event as explained in clause 60.1(1). It is necessary to change the works information to ensure that the completed project is in full compliance with the works information in case there are any future legal issues or questions from auditors. It is also necessary to ensure that any as-constructed drawings and the health and safety file required by CDM 2007 are correct.

Whilst the project manager is not obliged to accept a defect, unreasonable behaviour by the employer or project manager would run counter to the spirit of cooperation required by clause 10.1. Many defects are quite minor, but the cost to the contractor of putting them right can be out of all proportion to the value gained by the employer.

Let us take another example. Suppose the concrete finish on walls to a basement car park is poor. The walls have some undulations, and hence do not comply with the tolerances in the specification given in the works information. So they are a defect. Remedies might be to take down and replace the walls (very expensive), or to plaster over the walls (less expensive). However, the employer may be happy to live with this defect (in a basement car park) for a reduction in price and possibly an earlier completion date. If this were the case, the project manager would accept the defect.

A final important consideration is that in options C to E the contractor is paid its costs, and these costs include correcting defects before completion. Hence in the examples above, the contractor would be paid (by the employer) for correcting the defects. The result may well be to exceed the target price in which the pain/gain share mechanism would apportion the costs (in options C and D) between the contractor and the employer. However the employer would almost certainly pay more, and is therefore disposed to agree to the project manager accepting defects.

Uncorrected defects

Contracts usually deal with the case where the contractor does not correct a defect, and here clause 45 deals with this situation. This issue often occurs after completion, when it can be very difficult to get a contractor to return to a 'completed' project to correct defects. Firstly, the contractor is not paid to do this, even under options C, D and E, because the correction after completion is now a disallowed cost. Secondly, correction may be very expensive for a contractor who has to travel many miles to do so and also incurs the costs of arranging the work with the employer and in compliance with CDM 2007, where the work is of significant duration, which will require the availability of suitable welfare facilities for example. Thirdly, the contractor will probably be concentrating on current contract work, and will not wish to divert people or resources away from this. These reasons are all very understandable, but equally the employer may be very inconvenienced by the defect.

The sanction on contractors in clause 45.1 applies if they have been given access to correct a notified defect and have not corrected the defect within its defect correction period. We must remember that the defect correction period begins when the defect is notified, for defects notified after completion, or when the employer gives access for correction of the defect. Then the project manager assesses the cost to the *employer* of having the defect corrected by other people, and the contractor pays this amount.

In practice some compromise will be necessary if the contractor is doing its best to correct a defect, and it is still not resolved within the defect correction period. This is quite likely to happen with serious defects or ones that are complicated to correct. It is also worth remembering that correction of a defect within the defect correction period can be quite difficult for the contractor after completion, when the contractor is no longer present on-site, especially as the defect correction period is usually only two or three weeks.

Clause 45.2 deals with the issue of the contractor not being given access to correct a defect. Again, the parties should act in a spirit of cooperation, but there may be good reasons why the employer cannot give the contractor access. This could happen on a classified defence contract, where the employer has confidential papers or equipment or in a hospital or prison for example. In this event the project manager assesses the cost to the *contractor* of correcting the defect, and the contractor pays this cost. No doubt the assessment would be subject to discussion, since it is difficult for the project manager to assess the contractor's costs. In some ways the operation of clause 45.2 is a little similar to 'accepting a defect'. The employer receives some payment from the contractor and can choose when, how and if to correct the defect, and who should do it.

Defects certificate

It is worth remembering that we have the construction period, ending with completion, and then a further period in which defects which arise must be corrected. This is rather like a 'twelve-month warranty' on a domestic appliance. The defects date marks the end of this period (usually 52 weeks) during which the contractor is responsible for correcting defects which arise after completion.

In traditional contracts the defects certificate confirms the rectification of all defects. In JCT contracts it is called a 'certificate of making good'. NEC3 ECC does not work like this. In NEC the defects certificate *records* that there are no defects left to be corrected *or records* which defects still remain uncorrected. Clause 11.2(6) says:

'The defects certificate is either a list of defects that the supervisor has notified before the defects date which the contractor has not corrected or, if there are no such defects, a statement that there are none.'

Hence the defects certificate places on record any defects still outstanding at the defects date (or the end of the last defects correction period).

Clause 43.3 says that 'the supervisor issues the defects certificate at the later of the defects date and the end of the last defect correction period. The employer's rights in respect of a defect which the supervisor has not found or notified are not affected by the issue of the defects certificate.'

Hence the certificate is normally issued at the defects date or within three weeks of that date for a late notified defect and a defect correction period of three weeks. One would hope that the certificate will usually contain a statement that 'there are none'.

Where there are still outstanding defects, clause 45.1 comes into play and the project manager assesses the cost of having the defect corrected by other people and the *contractor pays this amount*. If option X16 (retention) has been included, the employer will have half of the total retention from which to deduct this assessed cost. This is another incentive for the contractor to correct defects, for it is the project manager who assesses the cost, and correction by 'other people' may be more expensive than the contractor, who has to pay this cost.

The latter part of clause 43.3 confirms that the issue of the defects certificate does not affect the employer's rights with regard to **latent defects**. Latent defects are those which are not yet known about but become apparent at some time in the future. If they are serious enough to constitute breaches of contract, the contractor is obliged to put them right within the limitation period. This is a complicated legal subject and is discussed briefly in Chapter 17.

Chapter summary

This chapter covers core clauses three and four of NEC3 ECC which deal with:

- Time and programmes
- Instructions to stop work
- Take-over by the employer
- Acceleration
- Defects, tests and inspections and their associated procedures
- Accepting defects
- Defect correction period
- The defects certificate.

24

NEC3 ECC core clauses five

Aim

This chapter aims to explain the meaning and operation of the most significant clauses in core clauses five (payment) of the NEC3 ECC contract.

Learning outcomes

On completion of this chapter you will be able to:

>> Explain in detail the significant clauses of core clauses five of the NEC3 ECC form of contract.

>> Apply these clauses to typical contract issues.

>> Manage and administer an NEC3 ECC contract.

Core clauses five: payment

Payment under NEC3

Pricing mechanisms and payment mechanisms have been covered in general terms in Chapter 14. In NEC3 ECC, options C and D use a different pricing mechanism from the payment mechanism. This is why it is important to understand the difference.

> In options C and D the pricing mechanism is different from the payment mechanism.

The payment mechanisms in NEC3 ECC may seem complicated at first, but it must be remembered that they cover payment under the six options, A to F. Thus the payment section five in the core clauses (which apply to all options) is just over a page long and some of the options A to F add as much again in the sections relating to each option. The other important point about NEC3 ECC is that regular payments are usually determined in a different way from payments for compensation events depending on the option. Compensation events are those that are the employer's risk. If a compensation event occurs and it is not caused by a fault of the contractor, then the contractor is compensated, all as explained in Chapter 25.

> Regular payments and payments for compensation events are usually determined in a different manner, but this depends on the option used.

In order to understand the payment provisions of NEC3 ECC we have to understand a number of new terms. The main ones are the **prices**, the **price for work done to date (PWDD)** and **defined cost**. These terms are defined *separately in clause 11.2 of each option*. Before looking at these terms and the options in detail, it is worth examining the general provisions of section five, which apply to all options.

Assessing the amount due, clause 50

The role of the project manager in assessment

In NEC3 ECC, the project manager assesses the amount due at each **assessment date**. In traditional contracts the contractor generally 'applied' for payment, with a valuation which the contractor considered was representative of the work done, and hence the amount to be paid. NEC3 ECC is completely different and puts the responsibility clearly on the *project manager* who 'assesses the amount due' as clause 50.1.

However, NEC recognises that in practice the contractor will almost certainly need to submit an application for payment, which the project manager 'considers' as clause 50.4. It would in fact be very difficult for the project manager alone to determine a payment in options C to F, which are based on contractor costs, rather than on an activity schedule (option A), or a bill of quantities (option B) which the project manager could determine if necessary without the contractor's assistance.

Many employers will amplify these very brief provisions in options C to F in order to specify more precisely what is required of the contractor, because these options are cost-reimbursable in one form or another. Employers may wish to see invoices, plant records, and labour and personnel details such as timesheets. The employer might specify these requirements as a Z clause or in the works information. It is hoped that in a cooperative contract, such as NEC3 ECC, the contractor and project manager will arrange an efficient and thorough way of agreeing regular payments, based on these employer requirements, especially because the turnaround time for agreeing the monthly amount due is very short.

Regular payments on forecast costs

Traditional contracts usually pay the contractor in arrears. In other words, monthly (regular) payments are made on the amount of work done in the previous month. Options A and B of NEC3 ECC work in the same way.

However, options C to F of NEC3 ECC are completely different. Here the price for work done to date is 'the total defined cost which the project manager *forecasts* will have been paid by the contractor before the next assessment date, plus the fee.' This important definition can be found in the identical clause 11.2(29) addition to each of the options C to F. This payment in advance aids the contractor's cash flow and should result in lower tender prices. Because of the short timescale for assessment and payment, the project manager is likely to need the contractor's assistance in determining these future estimated costs. Additionally the contractor is the only organisation in a position to predict them.

This quick turnaround of regular payments is likely to be followed by a thorough check on costs or an audit procedure taking place later. This may well result in corrections being made in later payments. Hence, in common with traditional contracts, the project manager can correct any 'wrongly assessed' amount due, in a later payment certificate, as clause 50.5. This is particularly important in NEC3 ECC since payment timescales are very short and regular payments in options C to F are made on a forecast basis. Hence it is likely that the project manager will be obliged to make a best-estimate of payment, particularly on options C to F, which may be subject to later audit or financial scrutiny. However, regardless of the audit procedure or perhaps the involvement of cost consultants, the *project manager* must make any necessary correction.

Assessment intervals

The assessment interval is identified in the contract data part one, and must not exceed five weeks. It is usual to make this interval either four weeks or a calendar month. Hence regular payments are usually made on a monthly basis.

Regular payments to the contractor are usually monthly.

The first assessment date is determined by the project manager to suit the procedures of the parties (employer and contractor) but must not exceed the assessment interval after the starting date. The starting date is identified in the contract data part one, and may be many weeks or months before the site start (access date), which can allow the contractor to be paid for such items as design work, or off-site fabrication or manufacture.

After the first assessment, further assessments of the amount to be paid to the contractor occur at the assessment interval until four weeks after the defects certificate is issued, with an additional assessment at completion of the whole of the works (as clause 50.1). Traditional contracts do not provide for as many assessments during the period between completion and the defects date, and many of the assessments during this period may well be zero.

The amount due to the contractor

The amount due is stated in clause 50.2 as:

- The price for work done to date (PWDD)
- Plus other amounts to be paid to the contractor
- Less amounts to be paid by or retained by the contractor.

However it is very important to remember that the definitions of these phrases vary with the option A to F, and it is necessary to refer to the specific definition in each option.

Deduction if no programme is identified, clause 50.3

NEC3 ECC relies heavily on an up-to-date programme and hence has a 'penalty' provision. A programme may be identified in the contract data, as one submitted at tender (Chapter 23 deals with programmes). Where no programme is identified, then one-quarter of the price for work done to date (PWDD) is retained until the contractor has submitted a first programme for acceptance, which shows the information that the contract requires.

Payment, clause 51

The project manager certifies payment within one week of each assessment date and each certified payment is made (by the employer) within three weeks of the assessment date. These are short timescales relative to traditional contracts. This means that either a large number of quantity surveyors will be needed or the project manager will certify a sensible payment, subject to later correction, as discussed above. The employer makes payment within three weeks of the assessment and hence has just two weeks after certification to pay the contractor. If a payment is late, interest will be due, in common with traditional contracts.

Defined cost, clause 52.1

This clause is supplemented in the clause 11(2) addition to each of the options A to F where defined cost is defined separately for each option. However, the general clause, 52.1, states that all the contractor's costs that are not included (as thus defined) are assumed to be included in the fee. Such costs would generally be for headquarters staff and expenses, together with profit.

Defined cost is important not just for regular payment purposes in options C to F but also for the assessment of compensation events in all options. Compensation events are explained in Chapter 25.

Defined cost is not used for *payment* purposes in options A and B.
Defined cost is used for the assessment of compensation events in all options.

The activity schedule and bill of quantities

These are used in options A to D, always as a pricing mechanism (to explain the tender price), and also as a payment mechanism in two of the options, options A and B. However, NEC3 ECC takes pains to ensure the status of the activity schedule and the bill of quantities is not in doubt. They are not intended to redefine the contractor's responsibilities, which are set out in the works information. This is different from many traditional contracts such as JCT SBC05, where ambiguities between the BoQ and works information (in JCT the drawings and specification) have to be corrected and usually result in compensation for the contractor. NEC3 ECC says specifically:

- Information in the activity schedule is not works information or site information (clause 54.1 of options A and C).
- Information in the bill of quantities is not works information or site information (clause 55.1 of options B and D).

The schedules of cost components (SCC and SSCC)

These two schedules are an important feature of NEC3 ECC. They are used as part of the payment calculation in options C, D and E, and for the assessment of compensation events in options A to E, with some exceptions as explained below. They are not used in option F.

- The schedule of cost components (SCC) is used in options C, D and E, both as part of the calculation used for regular payment and also for valuing compensation events.
- The shorter schedule of cost components (SSCC) is used in options A and B for valuing compensation events but *not for payment*. Payment calculations in options A and B use the activity schedules or the bill of quantities respectively.

However, the SSCC (rather than the SCC) may be used to value compensation events in options C, D and E *by agreement*. Use of the SSCC is more straightforward as explained below, because it uses **published lists**.

The SCC and SSCC are defined in separate sections of NEC3 ECC which follow the sections on the W, X, Y and Z options. They describe how cost components for people, equipment, plant and materials, charges, manufacture and fabrication, design and insurance will be valued. In order for the SCC and SSCC to be used, the *contractor* must insert rates and percentages against identified items in the contract data part two.

> The SCC and SSCC are not schedules of rates; they are lists of items, the actual cost of which will be paid together with various percentage additions inserted by the contractor in the contract data.

It should be remembered that the SCC and SSCC are not schedules of costs or rates as such, because the contract data part two as tendered contains little detail. What the contractor enters in the contract data part two is lists of percentage additions together with rates for special plant, or hourly rates for designers for example. These percentages will later be applied to actual costs as the contract work proceeds.

Using the SCC and SSCC

The introduction to the SCC and SSCC say that an amount is included only in one cost component; and only if it is incurred in order *to provide the works*.

The introduction to the SCC section makes it clear that the 'contractor' means the contractor and not their sub-contractors.

The introduction to the SSCC is different. In the SSCC the 'contractor' means the contractor and not the sub-contractors in options C, D and E. In options A and B clause 11.2(22) says 'defined cost is the cost of the components in the shorter schedule of cost components whether work is subcontracted or not'. Hence in options A and B sub-contractor costs are processed through the SSCC and the rates and percentages inserted by the contractor in the contract data are applied. Additionally, this wording restricts the contractor's costs to items listed in the SSCC so the contractor cannot just put forward sub-contractors' invoices as evidence of defined cost.

Components of the schedule of cost components (SCC)

The SCC contains the following headings of components of cost:

People includes staff or labour who are directly employed by the contractor and whose normal place of working is within the working areas either permanently or temporarily. This component includes all the normal salary on-costs such as bonus, national insurance, overtime, absence, travel and so on. 'People' also includes the amounts paid by the contractor to people who are not direct employees, for the time they spend working in the working areas under item 14. Contractor's staff outside the working areas, such as headquarters staff, should be included in the fee.

Equipment used within the working areas includes equipment hired or owned by the contractor, *including accommodation* such as offices, canteens, welfare facilities, telephones and faxes and surveying equipment. Where contractors obtain equipment or special equipment specifically for the contract, they insert details, rates and charges in the contract data part two.

Plant and materials deals with purchasing plant and materials, delivery and tests. It should be remembered that NEC3 ECC uses the term 'plant' to mean items to be included in the works, such as panels, boilers, gantry cranes and so on.

Charges covers miscellaneous costs incurred by the contractor such as charges for gas, electricity, water and also payments to public authorities, and for inspection certificates, royalties and cancellation charges resulting from a compensation event.

Importantly 'charges' also includes a *charge for working area overheads*. This is a percentage inserted by the contractor in the contract data part two. This percentage is applied to the people items and includes provision and use of equipment, supplies and services. However the working area overhead does not include the cost of accommodation for a list of specific purposes which can be found under item 44, which should be included in the component for equipment.

Manufacture and fabrication items are paid for by the employer, partly as the direct costs paid by the contractor for these activities which are usually off-site in a factory or fabrication yard and partly by the application of overheads where applicable. The contractor inserts a percentage overhead in the contract data part two and hourly rates for different categories of staff for such manufacture and fabrication, provided it is outside the working areas. Presumably this would be used where the contractor itself is providing the staff and facilities, rather than directly paying the cost of manufactured items.

Design is used where the contractor is given design responsibility further to clause 21.1 and as detailed in the works information. The contractor inserts hourly rates for categories of employees and a design overhead percentage in the contract data part two at tender. These are applied to designers working outside the working areas.

Insurance is included where an event occurs for which the contract requires the contractor to insure; and other costs paid to the contractor by insurers. The insurance category is a deduction.

For construction professionals used to traditional contracts, the contractor priced the first section of the bill of quantities at tender which was called 'preliminaries' (part A in CESSM3). These often amounted to about 30 per cent of the total of the work items. NEC3 ECC splits this item between the sections on people and equipment in the SCC. As an added complication, this split was not the case in NEC2 ECC, the earlier version of this contract.

Components of the shorter schedule of cost components (SSCC)

The SSCC is simpler to use in practice than the SCC because it uses a specified 'published list' for prices for equipment. The one most frequently specified is the *CECA Schedule of Dayworks carried out Incidental to Contract Works*, published in the UK by the Civil Engineering Contractors Association, assumed to be used here, and called the CECA schedule. The CECA schedule lists hourly and daily rates for most of the commonly used plant and equipment, and covers not only cranes, lorries and excavators but also shuttering, scaffolding, piling equipment and pumps.

The CECA schedule also has items for simple office accommodation (with the usual fittings), mess rooms, stores, toilets and so on. The rates quoted in the CECA schedule are adjusted by a percentage tendered by the contractor, and inserted in the contract data part two (so again visible to the employer at tender). This inserted percentage represents the contractor's own particular addition or reduction applied to the listed rates in the CECA schedule. Clause 27 covers equipment which is not listed in the CECA schedule. Payment for such equipment is based on competitively tendered or market rates.

Most typical accommodation costs will be covered by the CECA schedule and are hence included in the heading of 'equipment' as in the SCC.

The other main difference between the SSCC and the SCC is that the SSCC uses the term 'people overheads', rather than working area overheads as a percentage applied to the people items.

The significance of the SCC and SSCC at tender

As we have seen, the SCC and SSCC are used for assessing compensation events in options A to E, and the SCC is used as part of regular payment in options C to E. The rates and percentages inserted in them, plus the fee (divided into direct fee and sub-contract fee), can have a significant effect on the contract outturn costs, particularly where there are a lot of compensation events, or where the contractor decides not to sub-contract much of the work. However, these rates, percentages and the fee do not appear in the tender total itself because this is made up of activity schedules in options A and C or BoQs in options B and D.

Employers and their advisors really need to scrutinise these elements of the tender very carefully. It is not unusual for employers themselves (or their design organisation) to price up a number of different contract scenarios to test the significance of these items, before deciding which contractor is most likely to have provided the lowest tender overall.

Regular payments under the options A to F

Regular payments under option A

Option A uses the activity schedule as both a pricing mechanism and a payment mechanism for determining regular payments to the contractor. The detail can be found in the 11.2 additions to the option A clauses, which say:

- The **Price for Work Done to Date (PWDD)** is the total of the prices for *completed* activities; and where activities are in a group, completion of the whole group of activities. A completed activity is one without defects which would either delay or be covered by immediately following work.
- The **Prices** are the lump sum prices for each of the activities on the activity schedule unless changed later in accordance with this contract (this change would normally be the result of a compensation event).

Hence in option A, the contractor is only paid for *completed* activities (without defects as defined above) which then form the PWDD. This means that to aid cash flow, the contractor is likely to break the contract down into many activities. This can lead to administrative problems and also issues with the programme, which should be linked to the activity schedule as clause 31.4. In option A it is important for the contractor to describe activities in a way that makes them and their completion clearly recognisable for payment purposes.

Since the activity schedule is used for payment, it has to change when events dictate change. Thus clause 54.2 gives the contractor the right to change the activity schedule if the planned method of working changes or for the intervention of compensation events as clause 63.12 (found in the option A clauses). The project manager can reject a revised activity schedule under clause 54.3 if:

- It does not comply with the accepted programme
- Any changed prices are not reasonably distributed between the activities
- The total of the prices is changed.

As we can imagine, the phrase 'are not reasonably distributed' could lead to some heated discussions between the contractor and the project manager.

Regular payments under option B

Option B uses a bill of quantities (BoQ) as both a pricing mechanism and a payment mechanism for determining regular payments to the contractor. Definitions are found in the clause 11(2) additions to option B and are summarised below:

- The **Price for Work Done to Date (PWDD)** is the result of adding together the total quantities of work completed multiplied by the relevant rates in the BoQ, together with any amounts against lump sum items (calculated by the proportion of work completed).
- The **Prices** are the lump sums and the amounts obtained by multiplying the rates by the quantities for the items in the BoQ.

The definition of the PWDD makes option B very similar to a traditional contract, and so it is relatively easy to administer on-site. Every month, the project manager or staff measure the amount of work done (preferably in conjunction with the contractor), as so many square metres of formwork, tonnes of reinforcement and so on, and multiply each item by the relevant rate from the BoQ. Since a standard method of measurement such as CESMM3 is used to compile the BoQ, the method of measurement can also be used to determine what each item covers for payment purposes. Chapter 15 covers BoQs in detail.

At the start of the contract, the *prices* are effectively the tender total. When the contract is complete, the prices become the total of the final account. As site work proceeds and quantities are remeasured and corrected, and the effect of compensation events are added in, the prices usually increase to ultimately give the final account total. The difference between the original tender total and the final account total is typically a few per cent on a well-defined and properly managed contract.

Since the BoQ is used for payment, it has to change when events dictate change, usually by adding in additional items. Thus clause 63.13 states that the BoQ is changed to reflect agreed compensation events. Such changes for work that is yet to be done will be new priced items, or changes to the rates and quantities. For work already done, the change as a result of a compensation event is a new lump sum item. Hence in option B the BoQ is a dynamic document, produced at tender but adjusted to include all changes and compensation events.

Of course, since there is no pain/gain adjustment in options A and B the difference between the tender price and the final total of the prices does not affect payment. It will however provide a record of what items (such as compensation events) produced the difference.

Regular payments under options C and D

The pricing mechanism in options C and D

Options C and D are target price contracts. NEC3 ECC calls these target contracts. The operation of target price contracts was described in Chapter 14. In option C the activity schedule and its total price are used as a pricing mechanism to compare tenders and to set the initial target price, *but they are not used for payment.* Option D works in a similar way, except that the BoQ (rather than the activity schedule) is used as a pricing mechanism. NEC3 ECC defines the prices as follows:

- In option C the **Prices** are the lump sum prices for each of the activities on the activity schedule unless later changed in accordance with this contract.
- In option D the **Prices** are the lump sums and the amounts obtained by multiplying the rates by the quantities in the BoQ.

It is important to remember that in target contracts both the pricing mechanism and the payment mechanism remain important throughout the contract. The tender price is the initial total of the prices and is usually called the target price. Compensation events and other matters as the construction work proceeds move this initial total of the prices (the 'target'), usually upwards. As the prices move upwards, the contractor can ultimately be paid more because of the movement of the pain/gain bands. This is why the prices are also important when determining payments to the contractor in options C and D.

The final *total of the prices* is ultimately used in options C and D as part of the pain/gain mechanism, by comparing it with the contractor's *total PWDD*, and proportioning the difference between the employer and contractor.

The target price at tender moves with compensation events, and is ultimately compared with the contractor's **total** PWDD to assess pain/gain share.

The payment mechanism in options C and D

The payment mechanism is used for assessing regular payments. Payment is assessed by estimating the **Price for Work Done to Date (PWDD)**. The price for work done to date (PWDD) is the total defined cost which the project manager assesses will have been paid by the contractor *before* the next assessment date, less **disallowed cost plus the** *fee*. The fee has two components, identified by the contractor in the contract data part two, completed at tender stage. These are the 'direct fee percentage' and the 'sub-contractor fee percentage'. These percentages are typically around ten per cent.

- **Defined cost** is basically payments to sub-contractors plus the cost of components in the **Schedule of Cost Components** (SCC) for other work, less disallowed cost.
- The **Fee** is the sum of the amounts calculated by applying the sub-contracted fee percentage to the defined cost of sub-contracted work and the direct fee percentage to the defined cost of other work.
  - The fee will therefore represent the contractor's profit plus any overhead items (such as headquarters costs and expenses) that are not recovered as part of the defined cost calculation. Site overheads will be recovered as part of the SCC.

> PWDD = total defined cost – disallowed cost + fee
>
> Total defined cost = payments due to sub-contractors + costs from SCC for other work

Options C and D cater for the fact that much work these days is sub-contracted. Hence defined cost separates payments to sub-contractors (paid at invoiced cost) from 'other work' which is paid using the SCC. Other work would generally be work carried out by the contractor's own staff or using the contractor's own equipment.

Disallowed costs apply to payments under options C to F and are described below.

Disallowed costs

The employer is paying contractor costs in options C to F. Clearly there will be some contractor costs which are not acceptable or cannot be verified. Appropriate costs are usually called 'allowable' costs, and those which are not appropriate are usually called **'disallowed costs'**. NEC3 ECC defines disallowed costs in the clause 11.2(25) addition to options C to E (the option F version has some small differences).

The project manager decides what constitutes disallowed costs. There are a number of reasons listed which include:

- Costs that are not justified by the contractor's accounts and records.
- Costs that should not have been paid to a sub-contractor or supplier in accordance with the sub-contractor's or supplier's contract.
- Costs which the contractor incurred only because the contractor did not
  - follow a procedure stated in the works information for acceptance or procurement or
  - give an early warning which he should have done according to the contract terms.
- Costs of
  - Correcting defects after completion (note here that correcting defects *before* completion is an allowed cost in options C to E). However correcting defects caused by the contractor not complying with a constraint on how he is to provide the works stated in the works information is disallowed.
  - Plant and materials not used to provide the works (after allowing for reasonable wastage). We need to remember the wide definition of plant and materials in NEC. However this does not apply to those resulting from a change to the works information.
  - Resources that were not used to provide the works (after allowing for reasonable availability and utilisation).
  - Resources that were not taken away from the working areas when the project manager requested.
  - Preparation for and conduct of an adjudication or proceedings of the tribunal.

The words 'not complying with a constraint' may cause some difficulty. They reflect the words used in clause 11.2(19) which defines works information. NEC2 used the word 'requirement' which is rather different. A requirement is a very general word, but constraint means something that limits the contractor's freedom of action. Such constraints should be set out in the works information. Examples might be 'access to area D shall only be between 10am and 4pm on weekdays' or 'the contractor shall not allow more than four workers to be in area B at any one time', or 'in the area of the gas holder, only explosion proof equipment shall be used'. If a defect arises as a result of the contractor ignoring such constraints, then its rectification becomes a disallowed cost.

Clearly costs must be auditable and hence the phrase 'not justified by the contractor's accounts and records' should provoke little disagreement, especially if the employer is clear in the works information as to what records are required. Clause 52.2 sets out the minimum required of the contractor as:

- Accounts of payments of defined cost
- Proof that the payments have been made
- Communications about and assessments of compensation events for sub-contractors and
- Other records as stated in the works information.

The phrase 'should not have been paid to a sub-contractor or supplier in accordance with the sub-contractor's or supplier's contract' ensures that the contractor is only paid by the employer for *appropriate* payments that are made to sub-contractors and suppliers.

It is worth noting that correcting defects *before* completion is not disallowed (unless the contractor does not comply with a stated constraint), although some employers amend this clause in order to *disallow* such costs. It is also clear that there could be some disagreement between the project manager and contractor over phrases such as 'reasonable availability and utilisation' and 'reasonable wastage'.

The contractor is not paid for any costs associated with disputes procedures except in option F.

Finally, the contractor is not paid for costs associated with disputes procedures except in option F. This will clearly include costs associated with adjudication.

Pain/gain (the contractor's share) in options C and D

Payment under options C and D is complicated by the operation of the pain/gain mechanism. NEC3 ECC calls this the **contractor's share**. The contractor's share percentage and share ranges are identified by the employer in the contract data part one.

The share calculation is based on the difference between the adjusted target price (the total of the prices) and the contractor's costs (the price for work done to date). NEC3 ECC describes this process in clauses 53.1 to 53.4 of option C and clauses 53.5 to 53.8 of option D. Despite the difference in numbering, the clauses are identical. Clauses 53.1 and 53.5 say:

- The project manager assesses the contractor's share of the difference between the total of the prices and the price for work done to date. The difference is divided into increments falling within each of the share ranges. The limits of a share range are the price for work done to date divided by the total of the prices, expressed as a percentage. The contractor's share equals the sum of the products of the increment within each share range and the corresponding contractor's share percentage.

The wording of this clause is not particularly easy to understand. Hence a more general discussion of the operation of pain/gain can be found in Chapter 14.

Forms of contract vary as to how often the share is calculated and paid. Clearly it is something of an administrative burden. NEC3 ECC prescribes share calculation and payment at completion of the whole of the works (which will be an estimate) and as part of the **final amount due**. Calculation at completion will usually be an estimate because it is likely that the cost of compensation events and similar items will not have been agreed and finalised at this stage.

However a real difficulty is where the calculation demonstrates that the contractor owes the employer money, particularly if this is at or after completion, where few future payments are likely to be made to the contractor. If retention has been identified under option X16 then at least the employer will benefit from half of the retention total until the final payment is determined.

NEC3 ECC is unspecific about procedure or certification of the 'final amount due'. Traditional contracts speak of final certificates and final accounts and set down time limits and procedures. One assumes that the final amount due is calculated and paid within four weeks of the date (clause 50.1) of the defects certificate under clause 43.3. It is worth noting that the supervisor issues the defects certificate, but the project manager assesses payment. Clearly some coordination will be required.

In summary, the total of the prices determines the target price but payment is made on the basis of 'costs' or the PWDD to be correct. On completion and 'final account', a pain/gain share is applied on the difference between the total of the prices and the total PWDD. This pain/gain share is adjusted as necessary when the final costs and prices are known. A compensation event in options C and D adjusts the prices, in other words it changes the target price (usually upwards), and therefore potentially increases the contractor's overall payment after pain/gain share is applied.

Regular payments under option E

Option E is a cost-reimbursable option. Here there is no target, and the contractor is simply reimbursed reasonable and justified costs, again on a forecast basis. The option is quite risky for the employer but has its uses for ill-defined or emergency contracts for example. This was discussed in Chapter 14. Payment in option E is similar to that in options C and D, but there is no pain/gain adjustment.

Regular payments under option F

Option F is dissimilar to the other options in that it is fundamentally a contract strategy rather than a pricing/payment mechanism. Payment under option F is similar to options C and D, but without the operation of pain/gain. However this option makes some small modifications to the definitions clauses. The option F equivalent of clause 11.2(22), defined cost, is clause 11.2(24). The option F equivalent of clause 11.2(25), disallowed cost, is 11.2(26).

Option F adds to defined cost 'the prices for work done by the contractor himself', in clause 11.2(24). This is an important modification, since in options C and D the schedule of cost components (SCC) is applied to such work done by contractors themselves, rather than *actual* contractor costs as option F. Presumably this is because option F is a management contract and it is assumed that contractors will sub-contract all, or most of the construction work, and will therefore have few of 'their own prices' to be paid by the employer.

Option F reduces the list of disallowed costs in clause 11.2(26) to remove reference to defects correction, wastage of plant, materials and resources. This is slightly curious. It no doubt assumes that most sub-contract work will be awarded under lump sum contracts (which is often the case), in which event the sub-contractors will be responsible for wastage and the correction of defects under their contracts with the contractor. Hence they need not concern the project manager who administers the contract with the contractor under option F.

The other interesting difference is that option F does not disallow costs relating to dispute procedures.

Chapter summary

- In order to understand the payment section of NEC3 ECC it is necessary to understand the difference between pricing mechanisms and payment mechanisms, which were described in Chapter 14.
- Regular payments to the contractor are determined at each assessment date. These dates are usually monthly.
- Regular payments in option A are based on work done against activity schedules.
- Regular payments in option B are based on work done against bills of quantities.
- Regular payments in options C to F are based on defined cost plus fee.
- Defined cost in options C to F is basically payments to sub-contractors plus the cost of components in the Schedule of Cost Components (SCC) for other work less disallowed cost.
- Fee is the total of the direct fee plus the sub-contracted fee applied to the relevant defined cost.
- The operation of pain/gain to target contracts applies to options C and D.

25

NEC3 ECC core clauses six

Aim

This chapter aims to explain the meaning and operation of the most significant clauses in core clauses six (compensation events) of the NEC3 ECC contract.

Learning outcomes

On completion of this chapter you will be able to:

>> Explain in detail the significant clauses of core clauses six of the NEC3 ECC form of contract.
>> Apply these clauses to typical contract issues.
>> Manage and administer an NEC3 ECC contract.

Core clauses six: compensation events

What are compensation events?

Compensation events are those that are the employer's risk. If a compensation event occurs and it is not caused by a fault of the contractor, then the contractor is compensated. In NEC3 ECC, contractors are compensated for any effect that the event has on their *prices*, the completion date and any key dates. Thus the contractor gets 'time and money' where there is a justification.

In NEC3 ECC a compensation event gives the contractor time and money.

However the way in which the 'money' (the financial effect of the compensation event) is calculated is often different from the calculation of regular (monthly) payments, which were discussed in the core clauses five in Chapter 24.

The financial effect of compensation events may be calculated in a different way from regular payments, depending on the option A to F.

The 'money' or, to be more precise, the financial effect of compensation events is complicated by the choice of option:

- In options A and B, the change to the prices (including the fee) is incorporated in the activity schedules/BoQ respectively and hence the contractor will receive the 'money' in regular (monthly) payments as the work covered by the compensation event proceeds.
- In options C and D the contractor receives regular payments based on 'costs' (defined cost plus fee to be precise). The costs of course are likely to increase because of the compensation event. However, contractually, the compensation event moves the *prices* and hence the bands of pain/gain. This gives the contractor more 'headroom' below the target price (since it has moved upwards) before the contractor begins to suffer from reduced payments arising from the operation of pain/gain (this depends on the way the pain/gain operates in any given contract). If the estimate of cost is correct, the prices move by the same amount, and so the contractor is fully compensated, since the extra cost is balanced by the change in the prices.
- In options E and F contractors are receiving their costs anyway, so a compensation event has little effect except to act as a record of increased prices.

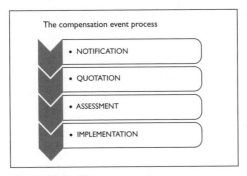

Figure 25.1   The compensation event process.

ICE7 and JCT contracts are fundamentally different in that time and money are treated separately and some events, such as exceptionally adverse weather, only give time. So in this respect, NEC3 ECC is a simplification.

Additionally NEC3 ECC has a much more carefully prescribed process for agreeing compensation events than do traditional contracts, which comprise notification, quotation, assessment and implementation, as shown in Figure 25.1, and will be described in detail below.

Avoiding compensation events

Before we look at the detail of this process, it is worth thinking about avoiding compensation events in the first place. Compensation events are not rare occurrences. They happen frequently and divert management time and effort from the construction work itself to agree the 'time and money'. It could be argued that a compensation event represents a failure by the design organisation to:

- Produce clear and accurate contract documents
- Ensure that contract documents are not ambiguous
- Ensure that site information is relevant, referred to properly and is comprehensive
- Identify all programme constraints in the works information
- Clearly specify work by others
- And so on.

Or a failure of the employer to:

- Provide access to site on the agreed date
- Provide specified items by the date shown on the accepted programme
- Provide materials, facilities and samples at the agreed times, if specified
- And so on.

Or a failure of the project manager and supervisor to:

- Allocate sufficient time to fulfil their roles diligently
- Reply to a communication within the period for reply
- Carry out inspections promptly
- Minimise change wherever possible
- And so on.

Of course this is a rather simplistic argument in that unusually adverse weather cannot be predicted, and there will often be ground conditions that nobody can foresee despite a good ground investigation. However, poor site investigation, poor contract documents and poor administration will always lead to more compensation events than will good ones.

Before the compensation event

As we have seen, NEC3 ECC relies on the contractor and project manager attempting to foresee risks and then working together to remove the risk or minimise the impact on the project. Before a compensation event is *notified*, we should normally have:

- An accepted, up-to-date programme
- An early warning
- Resulting in a risk-reduction meeting
- An up-to-date risk register, containing all the risks that have previously been notified and actions taken together with the problem in hand.

The idea of the risk-reduction meeting is to consider actions that may reduce or avoid the risk, seek solutions and decide on actions. The result may be to avoid an issue that would have become a compensation event, or to reduce the impact of a compensation event when it is assessed.

There is a sanction on the contractor for not giving an early warning. As stated in clause 61.5, the project manager can notify the contractor that it did not give an early warning which an experienced contractor should have given; at the time when the project manager instructs the contractor to submit quotations. This is followed by the provisions of clause 63.5, where the event is assessed as if the contractor did give an early warning. These clauses are not without difficulty, because the assessment will be based on what might have happened, rather than what has actually happened.

A risk-reduction meeting will often be followed by the contractor's formal notification of a compensation event under clause 61.3, in which there is an eight-week time bar after the contractor became aware of the event.

The effect of compensation events (CE)

Compensation events have a number of effects, depending on the option both in terms of time and money.

'Money effects'

- In options A and B the value of the compensation event changes the prices and is 'converted' into new or amended items in the activity schedule or bill of quantities under clauses 63.12 (option A) and 63.13 (option B). The CE therefore affects the price for work done to date (PWDD) in regular payments and thus the amount the contractor is paid as the work related to the CE proceeds.
- In options C and D the value of the compensation event changes the prices and so moves the target price. It does not *directly* affect the regular payments to the contractor which are based on defined cost (which is basically a forecast of the 'cost of work to be done' that month plus fee). A higher target price will of course give the contractor more 'headroom' at full payment, before payments are reduced by pain/gain share. By its effect on the target price, the CE will also affect the contractor's final share payment; and will usually increase it.
- It is possible in options C and D for the prices to be reduced, as clause 63.11. This would usually arise from the project manager changing the works information by deleting work, or substituting work that has a lower cost. Whilst the target price will be reduced, so will the cost of the work itself.
- In options E and F a compensation event has no effect on what the contractor is actually paid.

Time effects

In all options compensation events potentially change the completion date and also key dates, where they have been included, and the contractor's planned completion. Another effect is on the contractor's programme where, according to clause 32.1, revised programmes should show the effects of implemented compensation events.

Administrative effects

Compensation events also affect the procedures and resources of the contractor and project manager. The contract has fixed timescales for the various stages of the CE process, and assumes that assessment and implementation will be achieved promptly, and that CE discussions will not be left to accumulate to produce a pile of unfinished administrative work at completion. Many project managers decide to have regular meetings with the contractor (possibly weekly) in order to progress and agree compensation events. There is no doubt that the administration of compensation events takes up a lot of staff time, both for the contractor and the project manager. On a large contract it is not unusual to need a number of quantity surveyors just working on the administration and valuing of CEs.

It should not be assumed that compensation events are few and far between. A change of dimension on one drawing from 2.000 metres to 2.100 metres is strictly speaking a compensation event, since it is a change in the works information. May all CAD technicians bear this in mind!

On a typical project there may be dozens of compensation events. On larger projects it is not unusual to have hundreds. The CE process is very thorough and is not really designed to have hundreds of CEs to deal with, particularly with their requirements for quotations and alterations to the accepted programme. In particular a reprogramming exercise may take days of work for the contractor. As a consequence many project managers and contractors attempt to simplify the procedures or group compensation events together, especially the ones of less significance.

The 19 compensation events (CEs) of clause 60.1

There are 19 compensation events listed in section six, plus a number in options B and D relating to measurement issues, and yet more in some of the X options. Of the 19 compensation events, some should be notified by the project manager and are shown with (PM) in the title below. The others can be notified by either the project manager or the contractor. However, they would usually be notified by the contractor and are shown with a (C) in the title below. The 19 compensation events are as follows:

(1) Changes to the works information (PM)

This is one of the most frequent compensation events. The contract assumes that all details, dimensions and specifications are given on drawings and in documents that are listed in the works information. Unfortunately construction contracts, being complex, are rarely this complete. Hence additions, subtractions or changes to the works information are all CEs *unless the change* is:

- To accept a defect.
- A change in the works information provided by the contractor for his design which is made either at the contractor's request or to comply with other works information provided by the employer.

As we saw in section four of the core clauses, the PM can accept a defect in NEC3 ECC by following the provisions of clause 44. Since this is usually done by agreement it would not be logical for it to be a CE. Again it would not be reasonable for a compensation event to apply if the contractor changes its own design, or if it has to be amended because it does not comply with works information provided by the employer. This is the second bullet point above. The contractor must ensure that any works information given in the contract data part two complies with the employer's works information in the contract data part one. This is because the employer's works information in part one effectively takes precedence over the contractor's works information in the contract data part two.

If the project manager needs to resolve an ambiguity (clause 17) or to correct an illegality or impossibility (clause 18) then these would usually generate compensation events under clause 60.1 because the works information would normally need changing.

It used to be common for the employer to issue further design details some weeks after tender acceptance, such as reinforcement drawings. In NEC3 ECC such actions would be CEs, possibly with zero cost and time implications, but CEs nonetheless. Similarly finalisation of specification requirements, such as specific types of bricks (often to comply with planning requirements) or colours or types of finishes, would also be CEs in NEC3 ECC, because they change (modify or amplify) the works information. So for example changing the colour of the walls of a lecture theatre from blue to green would be a compensation event. It may attract extra payment as well if for example the contractor already had a stock of blue paint, and had to wait for delivery of green.

(2) Access to site (C)

In common with all contracts, the contractor is entitled to be given access to site to proceed with the contract work on the due date for access (the access date) but is not obliged to start on this date. If the employer does not allow access to a part of the site (or all of it) on the access date, or a later date shown on the accepted programme, then it is a compensation event.

Clause 33 is relevant here and deals with the situation where the contractor shows a later access date. If the contractor shows a later access date on its accepted programme, this effectively gives the employer more time to provide access without it becoming a compensation event.

(3) Late provision by the employer (C)

If employers do not provide something which they are to provide by the date shown on the accepted programme this is a CE. However contractors must show this information on their programme as clause 31.2. This emphasises again the need for the project manager to check the programme carefully, and to inform the employer of any requirements of the employer. Such employer provision may be 'free-issue' materials, manufactured items or possibly services.

Traditional contracts often have clauses relating to the late supply of design information or drawings. Presumably this clause could be used for this purpose provided the contractor was clear about the requirements for further details or drawings on the programme. When issued, such further drawings would fall under the heading of 'changes to the works information'.

(4) Instructions to stop or not to start work (PM)

The PM is empowered to issue such instructions under clause 34 as discussed in Chapter 23. Since the contractor is entitled to proceed with the contract work and complete it, such instructions are clearly a compensation event unless they arise from a fault of the contractor.

(5) The employer or others and their work (C)

This compensation event quite rightly compensates the contractor if the employer or others do not act as agreed. Specifically if the *employer or others* do not work:

- Within the times shown on the accepted programme
- Within the conditions stated in the works information.

Again, the contractor must show this information on its programme as clause 31.2 requires. Once again this clause emphasises the need for the employer (or the design organisation acting for the employer) to be clear in the works information as to who 'others' are, what they are to do, and when and how they might interface with the contractor's work. In this context, others will often be contractors carrying out work for the employer on separate contracts, or possibly service providers such as gas, electricity or water companies.

Finally, the clause also applies if the employer or others carry out work on the site that is not stated in the works information at all.

(6) Late reply (C)

The compensation event applies if:

- The project manager or the supervisor does not reply to a communication from the contractor within the period required by this contract.

The period for reply is referred to in clause 13.3 and is identified in the contract data part one. It is often two or three weeks. Hopefully, project managers will always be competent and prompt in their dealings with the contractor. This clause is of help to a contractor who does not receive a reply in the specified time, which will often have financial consequences because the contractor's work may well be delayed or carried out less efficiently.

(7) Objects of value or historical interest (PM)

There are laws to protect some items of historical interest, and work should be stopped until the relevant authorities have been consulted. The PM is likely to direct the contractor to stop work under clause 34 and then order further action such as fencing and guarding under clause 73.1. Any PM instruction for dealing with an object of value or of historical or other interest found within the site is a compensation event. The contractor has no 'title' to such finds.

(8) Changing a decision (PM)

It is a compensation event if the PM or the supervisor changes a *decision* which has previously been communicated to the contractor. This is not the same as changing an assumption, which is CE (17).

(9) Withholding an acceptance (C)

As we have seen, the PM has to give reasons for rejection, all as set out in clause 13.4. So it is a compensation event if the PM withholds an acceptance (other than acceptance of a quotation for acceleration or for not correcting a defect) for a reason not stated in this contract.

(10) Searching for defects (PM/supervisor)

Quality control is the supervisor's responsibility. A problem may occur on-site, such as severe cracking of a wall. However, this may or may not be a defect *under the contract*. It depends on whether the contractor has complied with the works information or not. Therefore the phrase 'searching for a defect' really means ascertaining whether or not the contractor has so complied. A search may discover that the wall complied with the works information but was designed incorrectly by the designer, in which case it is not a defect under the contract. In other words it is not the 'contractor's fault'. Such a search would be a compensation event if:

- The supervisor instructs the contractor to search for a defect and no defect is found, unless the search is needed only because the contractor gave insufficient notice of doing work and so obstructed a required test or inspection.

(11) Delayed test or inspection (C)

It is a compensation event if a test or inspection done by the supervisor causes unnecessary delay. It would be wise to state in the works information the notice requirement on the contractor, and the period of time which the supervisor has to carry out the inspection or test. This will avoid future argument as to what 'unnecessary' means in practice.

(12) Physical conditions (C)

This is the NEC3 ECC equivalent of the infamous ICE7 clause 12, and interestingly the compensation event has the same number. Perhaps the writers of NEC3 ECC have a sense of humour. This is actually a very difficult area. Civil engineering contracts in particular are usually in the ground and conditions are not always as indicated in any ground investigation reports. ICE7 and also NEC3 ECC therefore have a 'foreseeability test' to determine whether the conditions should be the employer's risk or the contractor's. Firstly we need to look at clause 60.1(12) which describes the criteria for a compensation event under this clause:

> The contractor encounters physical conditions which:
> - are within the site
> - are not weather conditions and
> - an experienced contractor would have judged at the contract date to have such a small chance of occurring that it would have been unreasonable for him to have allowed for them.

Clause 60.1(12) is one of the most common compensation events, and needs to be administered correctly. Unfortunately clauses such as this create some of the greatest problems of agreement, but are a fair reflection of the difficulties that ground conditions can cause, however comprehensive the ground investigation information is. Some employers delete this clause, in which case the contractor bears the whole risk and will almost certainly increase its tender prices as a result.

The first thing to note is that the clause only refers to the site, and not to working areas. Weather conditions are excluded because they appear in the next compensation event. However 'conditions due to weather conditions' are not excluded as they were in the ICE7 version of this clause. This could be significant in a site adjacent to a river for example, where heavy rain (weather conditions) may be excluded, but subsequent damage caused by flooding of the river (conditions due to weather conditions) is not potentially excluded. The employer could reduce the uncertainty of such situations by being clear in the contract data part one which risks the employer will take, such as 'river levels above 52 metres AOD' for example, following the use of clause 80.1 (the final bullet point relating to additional employer risks).

A practical difficulty is the phrase 'judged at the contract date to have such a small chance of occurring'. It must be assumed that the contractor is competent (CDM 2007), but should it be assumed that the contractor is *expert* in this particular type of contract when making this 'foreseeability' judgment? This issue has been a problem on some specialist contracts such as tunnelling and drilling water supply boreholes for example.

In a similar way to the **'weather clause'**, the contractor is paid for the 'excess effects'. This is because the clause says that 'only the difference between the physical conditions encountered and those for which it would have been reasonable to have allowed is taken into account in assessing a compensation event.'

The effect of clause 60.2

Provided the first part of clause 60.1(12) is satisfied, we then need to apply the 'foreseeability test'. The criteria for judging 'foreseeability' are set out in clause 60.2, which says that in judging the physical conditions for the purpose of assessing a compensation event, the contractor is assumed to have taken into account:

- the site information,
- publicly available information referred to in the site information,
- information obtainable from a visual inspection of the site and
- other information which an experienced contractor could reasonably be expected to have or to obtain.

There are three further important points here. Firstly, in common with ICE7 contracts, contractors are *assumed* to have made a visual inspection of the site, *whether they actually have or not*. Most contractors will inspect a site when tendering, and some potential issues will be quite evident from such an inspection. The contractor is expected to take into account publically available information (this could be flood levels, geological drift deposits, previous ground investigations and so on), but only if it is referred to in the works information. Hence, quite rightly the project manager cannot later, in a compensation event discussion, refer to an obscure document that the contractor had no knowledge of.

Secondly, the contractor is entitled to rely on the site information provided by the employer in the tender documents. Hence it should be as correct and comprehensive as possible. It is also important for the employer to ensure the site information is properly separated from works information, because of the different contractual treatment of the two sets of information.

Thirdly, the phrase 'other information which an experienced contractor could reasonably be expected to have or to obtain' could cause difficulty. It should be remembered that the baseline for this clause is the contract date; the date at which the contract is signed. So information which comes to light after the contract commences cannot be taken into account in making this judgment. This may be important where the start date is a long time before the access date for example, where conditions may change, or more information may become available. However, the real problem is the judgment of what a contractor could reasonably be expected to obtain, particularly if it is not included in the tender documents or referred to in other information supplied by the design organisation.

Another potential area of disagreement is the phrase at the end of clause 60.1(12) which says 'only the difference between the physical conditions encountered and those for which it would have been reasonable to have allowed is taken into account in assessing a compensation event.' What this phrase is trying to do is only to compensate the contractor for the 'extra' problems caused by physical conditions which would have been unreasonable to allow for. The difficulty will be the likely argument over what would have been a reasonable allowance at tender stage.

(13) Weather (C)

Most contracts compensate the contractor for weather which is exceptionally adverse and could not reasonably have been expected. However traditional contracts such as ICE7 and JCT only give the contractor more time for completion. NEC3 ECC gives 'time and money'.

However NEC3 ECC brings more certainty to the process of determining 'exceptional adverse weather', by referring to weather measurements and data taken at a weather station (place) identified in the contract data part one. The clause relates everything to a weather measurement which is recorded:

- within a calendar month,
- before the completion date for the whole of the works, and
- at the place stated in the contract data

the value of which, by comparison with the weather data, is shown to occur on average less frequently than once in ten years.

This means that the weather station referred to in the contract data must not only record all 'weather' so listed, but must have records going back ten years.

The contractor is only compensated for the *difference* between the effects of the weather occurring less frequently than once in ten years with the weather data showing conditions likely within a ten-year period. This is because the clause goes on to say that 'only the difference between the weather measurement and the weather which the weather data show to occur on average less frequently than once in ten years is taken into account in assessing a compensation event.'

In other words the 'excess' exceptional weather is considered (in a similar way to the 'excess' argument in clause 60.1(12)). Weather which the weather data shows as likely to occur within a ten-year period is the contractor's risk. Whilst this mechanism goes a long way towards a more certain outcome, one can imagine that the 'excess' effects of the weather on a construction project are not easy to determine.

> Short spells of adverse weather are excluded because the recorded weather has to be over a calendar month.

Another point is that the clause relates to weather data at the 'place' stated, which will often be the nearest meteorological station, not at the site itself; because it is important to have records going back over ten years. In most of the UK this should not present a problem, but the difference between site weather and that at the 'place' is the contractor's risk or benefit. There will be contracts on high or low ground, or near rivers where this difference may be significant.

The contract data sets out the weather data that is monitored and it mainly relates to rainfall, temperature and snow. However the employer may wish to add additional items for monitoring (such as wind speed for the construction of high buildings which could put the tower cranes out of use) and must list them in the contract data part one.

(14) Employer's risks (PM)

This clause refers to employer risks listed in clause 80.1 and also if identified specifically by the employer under the heading 'these are additional employer risks' in the contract data part one. If such a risk occurs, it is by definition a compensation event. This subject is discussed in Chapter 26.

(15) Employer takes over a part of the works (PM)

For this clause to apply, the project manager must have certified take-over of a *part* of the works before both completion and the completion date. In other words the employer is willing to take over a part of the works before it is complete. However the trigger for this compensation event is the project manager's certification. Whilst the contractor may be quite pleased with this state of affairs, the compensation event ensures that the contractor is reimbursed for any adverse cost or time effects, because contractually the contractor is entitled to *complete* the works.

(16) Materials and other items for test (C)

The employer does not provide materials, facilities and samples for tests and inspections as stated in the works information. Clause 40.2 refers to this.

(17) The project manager notifies a correction to an assumption (PM)

Clause 61.6 says that 'if the project manager decides that the effects of a compensation event are too uncertain to be forecast reasonably, he states assumptions about the event in his instruction to the contractor to submit quotations. Assessment of the event is based on these assumptions.'

Hence compensation event (17) applies if the project manager notifies a correction to such an assumption which he or she has stated about a compensation event. This should not be confused with any forecasts or estimates made by the contractor in connection with a compensation event.

(18) Breach of contract by the employer (C)

A breach of contract by the employer which is not one of the other compensation events in this contract is a compensation event in itself. This clause allows the compensation event procedure to be operated, before the parties consider use of the law.

(19) Prevention (C)
This compensation event deals with something which neither party could prevent and which stops the contractor completing the works. The clause refers to an event which:

- Stops the contractor completing the works.
- Stops the contractor completing the works by the date shown on the accepted programme.

This part of the clause should be fairly straightforward to agree in any particular case. However the clause goes on to require the exercise of considerable judgment, about which the contractor and project manager are likely to disagree. This is because the event has to be something that:

- Neither party could prevent.
- An experienced contractor would have judged at the contract date to have such a small chance of occurring that it would have been unreasonable for him to have allowed for it.

Finally, the event must not be one of the other compensation events stated in the contract. Despite the 'would have judged' provision this is a wide-ranging clause and was discussed in Chapter 22 with reference to clause 19.

Further compensation events in options B and D
There are additional compensation events in options B and D relating to measurement issues and bills of quantities. These would normally be notified by the contractor. The clauses are identical in both options B and D.

The first is found in clause 60.4 and compensates the contractor for a difference between the final total quantity of work done and the quantity stated for an item in the BoQ. There are three provisos however. The first and most important is that the difference in quantity 'does not result from a change to the works information.' Changes to the works information will be the most frequent causes of changes in quantity and under clause 60.4 are not a compensation event.

The other two provisos restrict the application of a compensation event to significant changes in quantity only, thus:

- The difference causes the defined cost per unit of quantity to change.
- The rate in the bill of quantities for the item multiplied by the final total quantity of work done is more than 0.5 per cent of the total of the prices at the contract date.

In other words, the correction to the quantity has to cause the defined cost per unit to change and is has to be significant. An example would be a rate multiplied by a quantity which is £5,000 different from that originally shown in the BoQ on a contract with a tender total of £1,000,000 (which is 0.5 per cent). This compensation event is meant to compensate the contractor for incorrectly billed or measured items. Changes to the works information will also frequently change the quantities, but this is not a compensation event under this clause 60.4 which deals with quantities that are *initially* billed incorrectly in the tender documents. The compensation event 60.1(1) covers the effect of changes in the works information.

A second compensation event in options B and D is detailed in clause 60.5 and is a difference between the final total quantity of work done and the quantity for an item stated in the bill of quantities which delays completion or the meeting of the condition stated for a key date. This is a sensible provision. An increase in the quantity of an item on the critical path is very likely to delay completion unless extra resources can be brought in by the contractor and if they are available in the time frame required. Again clause 60.5 is intended to address issues arising from incorrect billing of quantities in the tender bill of quantities.

A third compensation event in clause 60.6 is based on the correct use of a standard method of measurement. Hence a compensation event arises when the project manager corrects mistakes in the bill of quantities which are departures from the rules for item descriptions and for division of the work into items in the method of measurement or are due to ambiguities or inconsistencies. Each such correction is a compensation event; which may potentially lead to reduced prices, although this is unusual.

Clause 60.7 states that in assessing a compensation event which results from a correction of an inconsistency between the bill of quantities and another document, the contractor is assumed to have taken the bill of quantities as correct. The inconsistency is often between a drawing and the BoQ, and this clause means that the contractor simply prices the BoQ and is not obliged to abstract the quantities from the drawings and check that the two are compatible. This is a repetition of the legal rule of '*contra proferentem*' which means that an ambiguous document is read against the person that wrote it.

Notification, clause 61

There are four parts to the compensation event process, as shown in the box below. They all have to be carried out to strict timescales. Capital letters highlight the part under consideration in this chapter. We begin with 'notification'.

NOTIFICATION > Quotation > Assessment > Implementation

Project manager notification

Clause 61.1 says that where the compensation event arises from the project manager or supervisor giving an instruction or changing an earlier decision, the project manager notifies the contractor at the time of giving the instruction or changing the earlier decision. The relevant compensation events are 1, 4, 7, 8, 10, 15 and 17 and are shown above with a (PM) after the title.

The PM also instructs the contractor to submit a quotation unless the CE arises from a fault of the contractor or quotations have already been submitted. In this case it would be wise to make this clear in the notification such as 'notification under clause 60.1(4) but from an event arising from a fault of the contractor so no quotation is required.' Clearly, the project manager should give reasons for considering the event to be a fault of the contractor.

Assessing the effect of possible changes

Clause 61.2 is intended to allow the project manager to ask for a quotation for a *proposed* instruction or *proposed* changed decision. This may well apply to a proposal to change the works information. Changes to the works information are usually revised drawings or specifications. It is not an instruction to proceed, but a request for a quotation which will include the cost and time implications so the PM can assess the potential effects. The project manager should remember that under options A and B the contractor is not paid for producing quotations. However, under these more defined options, compensation events and changes are less likely.

Contractor notification

Where the project manager does not notify a compensation event as above, the contractor should notify the project manager of an event which has happened or which the contractor expects to happen as a compensation event if:

- The contractor believes that the event is a compensation event.
- And the project manager has not notified the event to the contractor.

It is necessary for the contractor to '*notify*', in other words to comply with clause 13.1, which basically means giving notice in writing. A telephone call or conversation will not suffice. An 'event' is something that has already happened or that is expected to happen that the contractor believes is a compensation event, so some foresight may be needed. What is also needed is the contractor's 'belief'. Hence some events will be notified which do not turn out to be compensation events upon scrutiny, or in the way things unfold. The project manager should expect this, and the contractor should not be criticised for notifying events that do not materialise as compensation events with the passage of time.

Importantly the clause goes on to provide a 'time-bar', because it says 'if the contractor does not notify a compensation event within **eight weeks** of becoming aware of the event, he is not entitled to a change in the prices, the completion date or a key date unless the project manager should have notified the event to the contractor but did not.'

The sanction for a contractor not notifying is this eight-week time-bar which runs from the contractor 'becoming aware of the event.' This is presumably intended to avoid the late submission of 'rolled-up' claims towards the end of site work, or even afterwards, which dogged traditional contracts. A positive intention will be to attempt to make decisions, allocate liability and assess compensation events near to the time when the event occurs, when it is still in everyone's mind and records and relevant personnel are available. However, legal opinion is divided over whether this well-intended provision is a **condition precedent** (a legal phrase meaning something that must be done before further actions can be taken) thereby destroying any further rights the contractor may have in law if the eight weeks are exceeded.

Project manager's response

The project manager's response is set out in clause 61.4 (headings in *italics* are not part of the clause).There are three possibilities:

1 The 'project manager's disagreement'

The project manager may disagree in a number of ways. The most difficult objection is when the project manager decides that an event notified by the contractor is not one of the compensation events stated in the contract. In reality many events arise that do not easily fit into a compensation event category, but still cost the contractor time and money. So this particular decision by the PM may take some time and discussion to settle. Similarly the project manager may disagree if the PM decides that the notified event arises from a fault of the contractor, or has not happened and or is not expected to happen. Finally the PM may agree that the event is a compensation event but reject it because the PM considers that it has no effect upon defined cost, completion or meeting a key date.

If the project manager disagrees in any of the above ways, the PM notifies the contractor of this decision that the prices, the completion date and the key dates are not to be changed.

The contract implies that this is the end of the matter. If dissatisfied the contractor then has to notify a dispute, and proceed to adjudication. In practice this is most unlikely, because such action will just upset everyone including the employer. A likely contractor response will be to ask for a meeting to understand why the project manager has 'disagreed' and then seek to provide further evidence or information to convince the project manager of the validity of the application for a compensation event. In traditional contracts, such discussion over a major event could go on for many months and involve technical experts, legal experts, further investigations and extensive correspondence.

2 The 'project manager's agreement'

If the project manager 'decides otherwise' (*in other words, agrees*), the contractor is notified accordingly and instructed to submit quotations. Of course, this is the response the contractor is hoping for.

3 'No response from the project manager'

If the project manager does not notify the decision to the contractor within either

- one week of the contractor's notification or
- a longer period to which the contractor has agreed

then the contractor may notify the project manager instead. Here the contractor *notifies* the PM that a decision is awaited and the PM has two weeks in which to reply, or the PM is treated as having given acceptance. This PM acceptance is not only acceptance that the event is a compensation event, but it is also treated as an instruction for the contractor to submit quotations.

Once the contractor has submitted a quotation, the project manager is obliged to consider it. Hence this clause places pressure on the project manager to reply promptly within the periods stated. For straightforward compensation events this should not represent a problem, but for difficult situations the periods are tight.

Hence it would sometimes be wise to explore the prospect of agreeing a longer period with the contractor, as above.

Failure to give an early warning

Clause 61.5 deals with the contractor's failure to give an early warning. Here, if the project manager decides that the contractor did not give an early warning of the event which an experienced contractor could have given, this decision is notified to the contractor when the PM instructs the contractor to submit quotations.

The result of this is that the compensation event is assessed as if the contractor had given an early warning as provided by clause 63.5. This is logical but very difficult in practice because it involves assessing what *might* have happened. The arguments over this are likely to run for some time.

Uncertain effects

If the project manager decides that the effects of a compensation event are too uncertain to be forecast reasonably, all assumptions made about the event in the instruction to the contractor to submit quotations must be stated.

Assessment of the event is then based on these project manager assumptions. If any of them is later found to have been wrong, the project manager notifies a correction. Correction of such assumptions is compensation event (17).

Late notification of a compensation event

Clause 61.7 prevents very late notification by saying 'a compensation event is not notified after the defects date.' This is a good attempt by NEC to prevent the unfortunate practice of 'late rolled-up claims' by contractors. These were often made 'from nowhere' sometime after completion when memories and records were sparse and were consequently very hard to agree in many cases.

Quotations, clause 62

Clause 62.1 describes the quotation process. It is worth remembering that quotations are for *changes to the prices* and will result in amendments to the activity schedules or BoQs as clause 63.11 or clause 63.12 depending on the option A to D. This change to the prices is based on 'defined cost' which is described differently in the clause 11(2) addition to each option A to F.

All definitions of defined cost, except option F, include the use of either the schedule of cost components or the shorter schedule of cost components. Therefore, as stated previously, the assessment of compensation events is often different from the calculation of regular (monthly) payments to the contractor.

> **Notification > QUOTATION > Assessment > Implementation**

Alternative quotations

The project manager may ask the contractor to submit alternative quotations after they have discussed different ways of dealing with the compensation event. The contractor is also at liberty to submit alternative quotations for different solutions, on its own account.

Format of quotations

Quotations for compensation events should comprise not just money but time effects as well, both to the completion date and any key dates as described in clause 62.2. Clearly the contractor needs to submit sufficient detail with the quotation for the project manager to assess it properly. Clause 62.2 says that the details include:

- Proposed changes to the prices
- Any delay to the completion date
- Any delay to key dates.

Quotations include time and money implications.

One aspect of this clause which is very logical but can give administrative problems is the requirement for the contractor to show alterations to the accepted programme with the quotation. This is only a requirement if the remaining work is affected by the compensation event. However producing new or revised programmes does put the contractor to considerable extra work.

Timescales for quotations

The contractor submits quotations within three weeks of the instruction to do so, and the project manager replies within two weeks of the submission as set out in clause 62.3. The project manager's reply is one of four options, which are:

1. The project manager accepts the quotation.
2. Or the project manager gives an instruction to the contractor to submit a revised quotation. The project manager must give reasons for this and the contractor then has a further three weeks to resubmit (as clause 62.4).
3. The project manager may decide not to proceed. In which case the PM notifies the contractor that a proposed instruction will not be given or a proposed changed decision will not be made. This might occur over a proposed change to the works information for example, where the project manager has second thoughts upon seeing the likely cost and effect on the programme.
4. The project manager can issue a notification that the PM will be making his or her own assessment, in which case the project manager has three weeks from the time of notification to respond.

Preparing quotations can be costly to the contractor both in money terms and the diversion of personnel from other activities. It is to be hoped that the project manager will bear this in mind when seeking revised quotations or deciding not to proceed. The contractor is paid for preparing quotations for compensation events in options C to E which are cost-reimbursable, but *not* in options A and B as set out clearly under clause 11.2(22) at the beginning of the specific option clauses.

Presumably the main reason for these timescales is to keep the contract moving with proactive management of events and agreement of the way forward and its cost. The other idea in options C and D is to move the target price (the total of the prices) on a quotation basis (an agreed best-estimate) to allow the contractor to try and better this target price move with actual costs, thereby reintroducing efficiency by the operation of pain/gain.

Unfortunately, many contractors and project managers prefer to wait and see what the actual costs are before agreeing, which rather defeats this objective. However it does avoid the potential embarrassment of one or the other where a predictive assessment proves to be wide of the mark when actual costs are known.

The project manager can extend the time allowed for contractor submission or project manager reply, if the contractor and project manager agree. In this case the project manager notifies the agreed extension. This is likely to be necessary for complex and expensive events.

No reply from project manager within the timescale

It is to be hoped that the project manager always replies within the prescribed timescale. However the contract again seeks to allow the contractor to keep the process moving by the provisions of clause 62.6 regarding quotations (similar to clause 61.4 where the project manager does not reply to a notification) which contains the following stipulations:

- if the project manager does not reply to a quotation within the time allowed
- then the contractor may notify the project manager to this effect.

If the contractor submitted more than one quotation for the compensation event, the contractor should state in the notification which quotation it is proposed should be accepted.

If the project manager does not reply to the contractor's notification within two weeks, then the contractor's notification is treated as the acceptance of the quotation by the project manager. However this assumption of project manager acceptance does not apply to a quotation for a *proposed* instruction or a *proposed* changed decision as discussed in clause 61.2 above.

Assessment, clause 63

As we have seen, when a compensation event occurs it changes the prices and usually also the completion date and any key dates. Assessment will cover both 'costs' and time. Clause 63.1 deals with 'costs' and clause 63.3 with time.

> **Notification > Quotation > ASSESSMENT > Implementation**

Assessment of changes to the prices – dividing the work

In all options, a compensation event *changes the prices*. This is very important particularly with options C and D. However the definition of 'the prices' varies with the option and hence has different effects on the ultimate payment that the contractor receives.

> Compensation events change *the prices.*

Compensation events are often related to previous work but they can affect future work as well. For example a 'weather-related' compensation event may have delayed excavation work. When the compensation event is agreed this cost will be in the past. However, the same compensation event may delay future access to the work and hence have a 'future cost' in addition. Clause 63.1 divides the work into work that has been done, and work that is yet to be done. 'The date when the project manager instructed or should have instructed the contractor to submit quotations divides the work already done from the work not yet done' as stated in clause 63.1.

This division into past and future concentrates on the changes to the prices. These changes are assessed as the effect of the compensation event upon:

- The actual defined cost of the work already done
- The forecast defined cost of the work not yet done
- The resulting fee (note that this is added in *all* options).

In many cases there will be no effect on the work already done, for example for many changes in works information such as drawings showing details of items yet to be constructed.

Assessment of changes to the prices in each option

The first and most important thing to note about clause 63.1 is that the change is to the *prices*. The *costs* of the work will be assessed separately as work proceeds in options C to E. In options C and D, the difference between costs and prices will be subject to pain/gain.

The second important point is the use of *defined cost* (plus fee) for this assessment of the changes to the prices. However, defined cost has a different meaning depending on the option.

Defined cost in each option

- Options A and B uses the *shorter* schedule of cost components (SSCC) whether the work is sub-contracted or not. In other words only the SSCC is used, and the contractor cannot use sub-contractor quotations as part of the assessment.
- Options C to E use the schedule of cost components (SCC) as part of the defined cost definition; which does not include payments to sub-contractors. Payments to sub-contractors in options C to E are a separate part of the defined cost calculation.
- Option F uses the schedule of cost components (SCC) as part of the defined cost definition which includes the prices for work done by the contractor themselves.

In traditional contracts, rates and prices are used as far as possible for valuing changes (called variations in traditional contracts), which is the NEC3 ECC equivalent of a change to the works information. NEC3 ECC does not do this as the default option which is based on defined cost. However NEC3 ECC does have a number of other possibilities, but only where the contractor and project manager agree.

In options B and D, quotations for compensation events *may* be valued using rates and lump sums from the BoQ but only if the PM and contractor agree to do so as set out in clause 63.13; which is a specific clause in these two options. This makes sense with these options B and D which use a BoQ containing rates and lump sums.

Curiously this option is also available in option A which uses activity schedules, as clause 63.14 which has identical wording to clause 63.13 in options B and D.

Options C to E use the schedule of cost components (SCC) as part of the defined cost definition. However another possibility is offered for options C to E, in clause 63.15 which allows the use of the shorter schedule of cost components (SSCC) rather than the schedule of cost components (SCC), the SSCC being simpler to use. This is subject to the project manager and contractor agreeing to proceed in this way.

The final important point about clause 63 is that the defined cost of work not yet done will often include 'prolongation costs'. As we shall see, the contractor may be entitled to a revised completion date. This will mean the contractor is on-site longer and hence will incur site establishment costs for running the site for this extended period. These costs will be working area overheads or people overheads and include the cost of providing offices, welfare facilities, fencing and security for example, together with overall site supervision costs such as the contractor's site management.

Delay to completion date

Traditional contracts often have specific whole clauses dealing with this issue, such as ICE7 clause 44 and JCT SBC05 clauses relating to relevant events. Curiously, NEC3 ECC covers this very important matter in a sub-clause, which can be found as clause 63.3. *It is fundamental nonetheless.*

> Determination of any delay to the completion date is very important and is found in clause 63.3.

This clause amends the completion date, based on the contractor's planned completion, which clearly means the contractor owns 'terminal float'. Thus NEC3 ECC brings clarity to this vexed question which traditional contracts do not make clear. This was discussed in Chapter 16. Clause 63.3 says:

- A delay to the completion date is assessed as the length of time that, due to the compensation event, planned completion is later than planned completion as shown on the accepted programme.

The clause goes on to apply the same logic to key dates, but this time using the 'planned date for meeting the condition.' The definition of a condition would follow a similar pattern to the definition of completion, but on a smaller scale:

- A delay to a key date is assessed as the length of time that, due to the compensation event, the planned date when the condition stated for a key date will be met is later than the date shown on the accepted programme.

The really important point is that the clause always focuses on 'planned completion'. There is no reference to the original completion date which the employer identified in the contract data.

> The additional time given is the amount of time that the new planned completion is later than the previous planned completion as shown on the accepted programme.

Hence the completion date may remain the same, or it may move backwards in time, giving the contractor longer to achieve completion. A compensation event cannot bring the completion date forward, even where work is removed or reduced. Only acceleration under clause 36 or acceptance of a defect under clause 44.2 can bring the completion date forward.

Project manager assessments

The project manager may make his or her own assessment under certain circumstances which relate to the time and money effects as set out in cause 64.1. This is not really what the contract envisages, and such actions by the project manager are unlikely to please the contractor.

The money effects are:

- If the contractor has not submitted a quotation with details of its assessment within the time allowed.
- If the project manager decides that the contractor has not assessed the compensation event correctly in a quotation and the project manager does not instruct the contractor to submit a revised quotation.

The time effects relate to programme and are:

- If, when the contractor submits quotations for a compensation event, the contractor has not submitted a programme or alterations to a programme which the contract requires the contractor to submit.
- If, when the contractor submits quotations for a compensation event, the project manager has not accepted the contractor's latest programme for one of the reasons stated in the contract.

We see that the clause once again emphasises the programme. If the contractor has not submitted a programme (where the programme itself is affected) with the quotation for the compensation event as required by clause 62.2, then the

PM assesses the compensation event. Contractors are unlikely to want this to happen, because they will no longer be in control of the quotation process and its assessment. Again the difficulty for the contractor is the administrative and technical time needed to keep the programme both up to date, and showing the latest situation with regard to compensation events. On contracts with many compensation events this can be an almost impossible task. One could argue that in circumstances like these, if the project manager assesses compensation events then it would be counter to the intentions of cooperation in clause 10.1.

The project manager notifies the contractor of the assessment of a compensation event under clause 64.3 and gives details of it within the period allowed for the contractor's submission of the quotation for the same event (three weeks).

Project manager does not assess a compensation event within the time allowed

This situation is covered in clause 64.4 whereby if the project manager does not assess a compensation event within the time allowed, the contractor may notify the project manager to this effect. In order to avoid any doubt where the contractor has submitted more than one quotation for the compensation event then the contractor should state which quotation it is proposed will be accepted by the project manager. The project manager has two weeks in which to respond to this notification. If the PM does not respond within this time the PM is assumed to have accepted the contractor's quotation. Hence in this way, the contractor can effectively force a decision despite the default or delay of the project manager.

Implementation, clause 65.1

Implementation is the NEC3 ECC word for confirmation. The changes to the prices, completion date and any key dates are agreed and confirmed. This is 'implementation'.

Notification > Quotation > Assessment > IMPLEMENTATION

Clause 65.1 says that a compensation event is implemented when the project manager notifies:

- Acceptance of the contractor's quotation
- The contractor of the project manager's own assessment.

Implementation also takes place in the clause 64.4 situation, discussed above, where 'a contractor's quotation is treated as having been accepted by the project manager.'

Clause 65.4 found in options A to D points out that the changes to the prices, the completion date and the key dates are included in the notification which implements the compensation event. Presumably this is done for record purposes and administrative accuracy.

Options E and F refer to the *forecast* amount of the prices in their version of the clause, which is 65.3.

Incorrect forecasts

Clause 65.2 states that the assessment of a compensation event is not revised if a *forecast* upon which it is based is shown by later recorded information to have been wrong. This clause is saying that the accepted quotation and changes to dates are final and are not revised if subsequent events prove that forecast to be incorrect.

This is not the same issue as described above under clause 61.6, where the project manager may state *assumptions* which may be subject to later correction. Such corrections would be a compensation event in themselves as clause 60.1(17).

Let us take an example. A contractor has received a change in the works information (additional work) to replace a pump in an existing pumping station but the project manager is having difficulty getting a decision from the employer with regard to lifting equipment. An assumption might be 'the contractor can utilise the employer's overhead crane for this purpose'. Clearly this would be quick, easy and safe for the contractor, rather than bringing in mobile lifting equipment. If this assumption proved incorrect, then the contractor's costs would rise, and should be compensated under clause 60.1(17).

A forecast might be the contractor's forecast that the work may take six hours, and this is the forecast on which the quotation is based. If the work takes ten hours, then the quotation (and hence the prices) is not revised by the operation of clause 65.2. We have to remember of course that the forecast moves the prices and so, on options C and D, the contractor would receive the extra *cost* of ten hours, but all costs would eventually be subject to pain/gain relative to the change in prices.

Chapter summary

- Compensation events are matters which are the employer's risk, and not the contractor's fault.
- The contractor is compensated for any effect on prices, completion date and any key dates.
- The compensation event process has four parts, which are Notification > Quotation > Assessment > Implementation.
- Compensation events change the *prices*.
- The calculation of the change in prices resulting from compensation events is based on defined cost and may be different from the assessment of costs for regular payments depending on the option.
- The schedule of cost components (SCC) is used as part of this compensation event price calculation for compensation events in options C to E. The shorter schedule of cost components (SSCC) is used to assess compensation events in options A and B.
- The fee is added to defined cost in all options (including options A and B).
- The effect on prices is based on quotations.
- The completion date is revised based on delay to the contractor's planned completion as shown on the accepted programme under clause 63.3.
- Thus the contractor owns terminal float.

26

NEC3 ECC core clauses seven to nine

Aim

This chapter aims to explain the meaning and operation of the most significant clauses in core clauses seven to nine of the NEC3 ECC contract.

Learning outcomes

On completion of this chapter you will be able to:

>> Explain in detail the significant clauses of core clauses seven to nine of the NEC3 ECC form of contract.
>> Apply these clauses to typical contract issues.
>> Manage and administer an NEC3 ECC contract.

Core clauses seven: title

This short section deals with ownership of plant and materials. We need to remember again that NEC3 ECC does not use the word 'plant' in the traditional sense of contractor's plant such as excavators and dumpers. Plant and materials are defined in clause 11.2(12) as items 'intended to be included in the works.' Hence plant could include boilers, electrical panels, an overhead crane and so on.

There can be project situations where it is advisable for the employer to secure ownership of plant which is off-site. Such plant could for example be expensive pieces of equipment before they are brought to site, or perhaps a large quantity of pipes when there is a long delivery of such pipes because of a national shortage of production relative to demand. This situation with pipes occurred in the 1990s. Similarly, many years ago a house-building boom led to a national shortage of bricks.

In traditional contracts, such as ICE7, the process of securing ownership off-site was called 'vesting' and it only applied to plant that had been specifically identified in the contract. However paying for plant off-site has many issues from which NEC3 ECC is not immune.

Some employer issues with paying for plant off-site

Before employers pay for plant, they wish to know that the contractor 'owns' it and that 'title' passes to the employer. This is a complicated legal subject and beyond the scope of this book. However as a minimum the plant must be properly marked (and NEC3 ECC deals with this as we shall see below).

Plant is often stored at the manufacturer's premises, possibly in a storage yard. Let us take the example of ductile iron pipes for use on a water main. Some difficult issues arising from transferring ownership to the employer, while the pipes are stored in the manufacturer's yard along with many other pipes for different customers, are as follows:

- If the pipe manufacturer's yard contains many thousands of pipes, for example, it is easy for the 'employer's pipes' to be confused or misplaced.
- If the manufacturer becomes insolvent the manufacturer's goods will be seized pending administration. Can employers rely on obtaining 'their' pipes when they need them, or could their pipes be seized in error? At best there could be a delay in delivery when the employer needs them.
- Protection, fencing and guarding are serious considerations. The pipes could be many miles from site or the employer's offices.
- Then we have the issue of insurance where presumably the employer insures the pipes but has little control over their storage and security arrangements.
- Protection from the weather and possible deterioration can be a real problem with some items of plant, less so with pipes, but it can still be an issue over a period of weeks or months. If an item of plant rusts prematurely after installation by the contractor, this would normally be the contractor's liability. However, after prolonged storage, the contractor will presumably blame the storage conditions for the deterioration.
- Then we have the possibility of theft or vandalism to consider.

Here we have only looked at ductile iron pipes, which are a simple, robust item of plant with coatings and protection. Imagine the difficulties with complex items of plant, such as electrical panels, which once installed have thermostatically controlled internal heaters, to prevent deterioration.

In some ways NEC3 ECC gets round this problem, at least in options C, D and E with the concept of working areas and the schedule of cost components (SCC), as discussed in Chapter 24. As we have seen, working areas can be much larger than the site itself (although they are determined by the contractor), and plant and materials could potentially be stored therein. Working areas can also be located at some defined place away from the site itself, and the contractor is responsible for the working areas. Plant and materials are included in the SCC and so payment can be made for them at the assessment interval. Finally, payment is made in advance in options C, D and E because it is assessed as the forecast of the amount paid by the contractor before the next assessment date.

The employer's title, clause 70

This clause deals with ownership (title) passing to the employer, thus:

- Whatever title the contractor has to plant and materials which is outside the working areas passes to the employer if the supervisor has marked it as for this contract.
- Whatever title the contractor has to plant and materials passes to the employer if it has been brought within the working areas. The title to plant and materials passes back to the contractor if it is removed from the working areas with the project manager's permission.

Hence whatever title the contractor has passes to the employer when plant or materials are brought into the working areas. So from the employer's point of view, valid title is only obtained if the contractor *has* title in the first place, which may not be the case. Looking at it from the contractor's point of view, in options C, D and E the contractor will presumably claim payment as forecast, before the arrival of the plant in the working areas, and hence before the employer gains whatever title the contractor has. However in options A and B, the contractor is paid against completed work either from the activity schedule or the bill of quantities, and hence will lose title to the employer, once items are brought into the working areas. In options A and B the contractor is only paid for these items once the relevant activities or BoQ items are completed. These may have items for 'delivery to site' or they may only be paid for once they are incorporated in the works.

Marking, clause 71.1

This clause says that the supervisor marks equipment, plant and materials which are outside the working areas if:

- this contract identifies them for payment and
- the contractor has prepared them for marking as the works information requires.

So rather like the process of vesting in traditional contracts, we would expect to see items for marking and hence payment when they are outside the working areas, carefully listed and described. A description of the marking procedure would also be a wise inclusion, together with details of storage and protection.

Objects and materials within the site, clause 73

This clause has two parts. The first states the contractor has no title to an object of value or historical or other interest within the site and that the contractor must notify the project manager if such an object is found. The project manager then gives instructions as to what to do with the object, which is a compensation event under clause 60.1(7).

This is the NEC3 ECC equivalent of the 'antiquities' clause 3.22 in JCT SBC05, and 'fossils' as ICE7 clause 32. In fact title to many such items is not the employer's but the State's, and the relevant authorities will need consulting by the project manager in case archaeological investigation is required. Archaeological remains of historical importance can cause serious delays to a project.

The second part states that the contractor has title to materials from excavation and demolition *only* if so stated in the works information. Hence for example the employer has title to lead from roofs and other similar materials from demolition work, unless otherwise stated in the works information. This sounds good for the employer but could cause problems for unwary employers (or their design organisations).

> The works information should be clear about 'title' to excavated materials taken off-site.

Material removed from site, such as surplus sub-soil, often has 'negative value'. In particular, waste materials or contaminated material can be very expensive to dispose of. A poorly drafted section in the works information will give title to contaminated material to the employer (by default) – not a result many employers would welcome. A safer alternative would be to include a standard clause in the works information, giving title to all material removed from site to the contractor. The contractor will then make allowance in its tender for suitable disposal and the resulting cost. This is a much safer option for the employer because, except in rare instances, there is nothing of value on most sites.

Core clauses eight: risks and insurance

Risk

Risk is inherent in all contracts. Risk is the probability that an event may occur, multiplied by the consequence if it does occur.

RISK = probability × consequence

Predicting what these events might be is the first difficulty. Many projects today have a risk register which is started by the employer or the design organisation. These registers tend to look at some of the larger risks such as not securing the best access to site, landowner difficulties, delays with planning and other consents and so on. Later, during the tender period, the tendering contractors will look very carefully at construction risks to assess as best they can the likelihood of the event occurring and who is likely to pay for the consequence.

Who is taking the risk (and paying for the consequence if it occurs) is supposed to be clear, whether it is the employer or the contractor. However, the organisation that actually takes the risk often depends on the *interpretation* of the words in the contract, which may not always be clear or may be subject to argument.

Risk assessment is far from easy or certain.

Let us take an example of the construction of a new highway bridge on the new bypass to the village of Barford in Warwickshire. The new bridge over the River Avon is in the flood plain. Construction will take two years, and the northern abutment is in an area that floods once every five years. The central piers are either side of the river in an area that floods once a year. The southern abutment is in an area that floods once every ten years. Flooding of the pier area can last for some weeks in a wet winter. The piers and abutments are on piled foundations, and work will commence in late summer. River Avon levels are measured officially at Stratford Weir, some five miles downstream. How likely is flooding to disrupt the construction work and who takes the risk (pays the cost and any time delay costs as well)? Firstly we can see that the piers are likely to be disrupted during construction, the northern abutment may be and the southern abutment should not be; but it is all a matter of estimation. Here are some possible contract terms:

- The contractor takes all risks arising from flooding of the River Avon.
- The employer accepts all costs of flooding arising from water levels at Stratford Weir above 52 metres AOD.

Questions in the tenderers' minds will be the likelihood and costs of possible flooding, which could be hundreds of thousands of pounds because of potential damage, but more importantly long-term disruption of the construction work. They will also contemplate perhaps whether the words 'all costs' include consequential costs away from the site, and what the actual water level at Barford might be when the Weir at Stratford is at 52 metres AOD, because this depends on hydraulic and storage effects in the River Avon over a stretch of five miles.

They will also look at the level of delay damages, and whether their resources could be diverted to other areas of work, perhaps the other new bridge on the bypass, which is not subject to flooding. Tenderers will also look at the ground investigation data to see how quickly the sub-soil will drain after flooding, because this will affect the routes and passage of construction plant and the method and cost of dewatering excavations. Finally they will scrutinise the insurance provisions and consider what their tender competitors may do. We begin to see how difficult this issue of risk assessment really is.

Construction risks

Let us now look at some typical construction risks and who takes them. The general theory is that risks should be allocated to the party most able to control them. Another consideration is that many larger employers have much greater financial stability than contractors and so can take the very unlikely but high consequence risks such as war, rebellion and revolution. These risks are often called 'force majeure' and most forms of contract allocate them to the employer.

A contractor's finances are only as good as margins (profit plus overhead contribution) on current work, and the expectation of winning future tenders. By contrast, employers such as government departments, local authorities and utilities have predictable and constant income streams from charges and taxes, often for many years ahead. Some risks are clearly with the contractor, such as supplier or sub-contractor problems or general lack of efficiency. Some risks are allocated depending on their likelihood, such as 'weather' as we have seen.

The lists below are an indication of typical risks, but they are not exact, the allocation of the risk often depends on the precise circumstances and their relationship with the words of the relevant clauses in the contract.

Typical employer risks (some are particular to NEC3 ECC)
- Giving access to site which is later than specified
- Failing to provide free issue materials on time
- Incorrect or late design information
- Archaeological remains found on the site
- Physical conditions (usually ground conditions) which an experienced contractor could not have foreseen
- Weather that occurs less frequently than once in ten years
- Correction of ambiguities in the contract documents
- Other contractors (working under other contracts with the employer) on the site who disrupt the contractor
- Mistakes in bill of quantities contracts, or measurement errors in options B and D.

Many of these and more are covered by the compensation event clauses of NEC3 ECC.

Employer risks as clause 80.1

Clause 80.1 tends to deal with the more serious and general risks that employers take under most forms of contract, and these are summarised below:

- Costs which are the unavoidable result of the works
- Negligence, breach of statutory duty or interference with any legal right by the employer
- A fault of the employer
- A fault in the employer's design
- Loss of or damage to plant and materials supplied to the contractor by the employer (or by others on the employer's behalf) until the contractor has received and accepted them
- Loss of or damage to the works, plant and materials due to
  - war, civil war, rebellion, revolution, insurrection, military or usurped power
  - strikes, riots and civil commotion not confined to the contractor's employees
  - radioactive contamination
- Loss of or wear or damage to the parts of the works taken over by the employer, except loss, wear or damage occurring before the issue of the defects certificate which is due to
  - a defect which existed at take-over
  - an event occurring before take-over which was not itself an employer's risk
  - the activities of the contractor on the site after take-over
- Loss of or wear or damage to the works and any equipment, plant and materials retained on the site by the employer after a termination, except loss, wear or damage due to the activities of the contractor on the site after termination
- *Additional employer's risks* stated in the contract data (see below).

Employer risks and additional employer's risks (as clause 80.1)

Employer risks are listed in clause 80.1 as above and cover the usual events such as war, riots, a fault of the employer or its design and so on. They tend to be connected with insurance matters in construction contracts. These clause 80.1 employer risks are included in the NEC3 ECC contract itself and the employer will have to pay if they arise. The employer will also have to pay for any valid and justified compensation events under clause 60.1.

The last section of clause 80.1 adds 'additional employer's risks stated in the contract data' to cover any very specific risks that employers wish to take onto themselves in addition to the normal ones listed in clause 80.1. These risks can be found in the contract data part one under the heading *'these are additional employer's risks'*.

If the employer did not take on such particular risks, then tendering contractors would have to assess them and might increase their prices as a result. When assessing the many risks that were not specifically stated as 'these are additional employer risks', the tendering contractors will have an eye on the possibility of recovering costs through the compensation event process if a risk event arises; but of course compensation events have to be agreed by the project manager, and this has a degree of uncertainty. In contract matters, as always, it is an issue of judging a sensible commercial balance and following the precept that 'risks should be allocated to the party best able to control them'.

Many employers will not wish to add items under the heading of '*these are additional employer's risks*' in the contract data. However, an employer may do so to cover a risk with a small likelihood but with significant effects, or a risk over which the employer rather than the contractor has some control, such as obtaining planning consent or some other statutory approval or relocating buried services belonging to a statutory service provider.

As we saw in Chapter 21, there are also risks included in the risk register, which is defined in clause 11.2(14). However, *these are different risks*. They are not risks which employers are taking on themselves (as clause 80.1 risks are), but risks that employers wish to alert the contractor and project manager to, so that they can be monitored and controlled. These risks will be added to as the contract proceeds and the risk register is developed to show the result of risk-reduction meetings, as clause 16.4.

The similarity of the wording between these two completely different types of risk can be a potentially confusing area of NEC3 ECC. In summary:

A. '*The following matters will be included in the risk register*' are risks inserted in the risk register to alert the parties to their significance. If one arises the compensation event procedure will have to be applied to determine liability.

B. '*These are additional employer's risks*' are risks inserted in the contract data by the employer for which the employer takes 'guaranteed' liability if they occur under clause 80.1 and the related compensation event as clause 60.1(14).

As an example, a risk of 'flooding from the River Avon' inserted into the contract data part one as part of the tender documents (case A above), would be an intention for the contractor and project manager to monitor this risk carefully and take steps to minimise its effect. If flooding occurred, it would have to be assessed as a compensation event if one of the lists of compensation events applied to the problem. This would probably hinge on the results of 'weather measurements' and the potential for unusually heavy rainfall to cause a river to flood.

Whereas, by contrast, if 'all consequences arising from the water level in the River Avon at Stratford Weir (where measurements are taken) exceeding 52 metres AOD' were inserted in the section 'these are additional employer's risks' in the contract data (case B above), then flooding arising from water levels higher than 52 metres AOD at Stratford Weir would *automatically* be a compensation event by the operation of clause 60.1(14), which is a compensation event covering 'an event which is an employer's risk stated in the contract' (with reference to clause 80.1).

Contractor risks

Contractor risks can be very difficult to predict for the contractor, and will be based on the works information and the *interpretation* of the site information, together with an inspection of the site, which the tenderer is assumed to have made (clause 60.2). The tenderer will also use experience of similar contracts or similar conditions in the past and the lessons learned from them. If the tenderer cannot persuade the employer to modify the contract documents, or cannot get away with a qualified tender (both unlikely), then the tenderer will have to price the risk. This again is difficult for tenderers because they are in competition to win the tender. A realistic price for the risk assessment may mean they do not win the tender. Hence tendering contractors have to take a view.

> Contractor risks begin when they price up a tender and continue throughout construction, and even extend to the correction of defects after completion.

Clause 81.1 simply says that 'from the starting date until the defects certificate has been issued, the risks which are not carried by the employer are carried by the contractor.' Typical contractor risks would be:

- Underpricing of items in the activity schedules or bills of quantities
- Omission of the cost of items from activity schedules
- Undercalculation of the fee, working area overheads or people overheads
- Misunderstanding items in the contract documents which are not ambiguous
- Poor programming, possibly by missing items or assessing activity durations incorrectly
- The effect of ground conditions which an experienced contractor could have foreseen
- Problems with suppliers and sub-contractors
- Problems with own labour force
- Problems with choice of plant or plant break-down
- Errors by the contractor's own management or supervision
- Non-compliance with health and safety regulations
- Inappropriate or inefficient construction techniques or construction plant
- Rises in the price of labour and materials (unless option X1 is included)
- Weather that occurs more frequently than once in ten years
- Faulty materials
- Poor workmanship
- Security problems such as theft and vandalism
- Design errors, where the contractor has been given responsibility for design of part or all of the works.

The 'failure to convince' risk

There is another risk that the contractor carries that may not seem obvious at first, and that is the 'failure to convince' risk. This risk is a very real one, but is not written into the words of contracts. Compensation events compensate the contractor for a list of events in terms of time and money. However, matching *actual* events and physical conditions to the words in the contract may not be easy, and persuading the project manager to accept the connection may be harder still. The contractor has to convince the project manager that the words of the compensation event clauses apply to the event in question.

One example is compensation event 60.1(12) covering the often encountered physical conditions which 'an experienced contractor would have judged at the contract date to have such a small chance of occurring that it would have been unreasonable for him to have allowed for them.' One can imagine the discussions and arguments arising from those words. There may be times when the contractor thinks it has a valid case, but fails to persuade the project manager.

In theory at least, the 'failure to convince' risk is covered by the dispute procedure, but as we shall see in Chapter 28 a contractor may decide not to pursue a dispute to adjudication. This could be because the contractor does not wish to alienate the employer, or on the basis of a risk assessment of potential costs against the probability of success.

Insurance

Insurance of course is a way of passing some risks to the insurer, for a premium. Under the insurance table in clause 84.2 the contractor provides the insurances listed below in the joint names of the parties, unless the contract data states otherwise. These items are at the contractor's risk:

- Loss of or damage to the works, plant and materials
- Loss of or damage to equipment
- Liability for loss of or damage to property (except the works, plant and materials and equipment) and liability for bodily injury to or death of a person (not an employee of the contractor) caused by activity in connection with this contract
- Liability for death of or bodily injury to employees of the contractor arising out of and in the course of their employment in connection with this contract.

As is normal in contracts, the contractor is required to send proof of such insurances to the employer.

Core clauses nine: termination

Termination means ending a contract prematurely and it is a very serious affair. A construction professional who thinks that termination is a real possibility would be wise to consult a legal expert or advise the employer to do so. The process and consequences are beyond the scope of this book.

Contracts can usually be ended by mutual agreement or by some gross default of one party, by frustration (which is an external event outside the control of the parties) or perhaps most frequently for the financial insolvency of one party (usually the contractor). As we saw in Chapter 17, some serious breaches of contract give a right of termination to the innocent party, in common law, regardless of any specific terms in the contract.

If either party wishes to terminate (this usually means the employer or the contractor) it must *notify* the other party and the project manager giving reasons for terminating. The termination table sets out reasons for termination, the procedure to be used and the amount due.

Clause 91.1 sets out a number of legal terms for the various types of financial insolvency. Clause 91.2 gives the employer rights of termination for serious contractor default such as substantially failing to comply with obligations, or appointing a sub-contractor for substantial work before acceptance by the project manager.

The contractor may terminate under clause 91.4 if the employer has not paid an amount certified within thirteen weeks of the certificate being issued.

In common with traditional contracts, an instruction to stop or not to start work, made by the project manager, can give the contractor a right to terminate, if an instruction allowing work to restart is not given within thirteen weeks.

Finally there are reasons for termination connected with prevention, which was briefly discussed under clause 19.

However, if termination for any reason becomes a real possibility, consult a legal expert in the field.

If termination becomes a real possibility, consult a legal expert in this area.

Chapter summary

- Title to plant and materials is a complex legal subject and advice should be sought.
- Risks are inevitable in construction but difficult to assess.
- A risk is the probability multiplied by the consequence of an event occurring.
- Risks are accepted and paid for by the employer or the contractor.
- Clauses in the contract such as compensation events allocate risks.
- Some other serious risks, such as *'force majeure'*, are allocated in section eight of NEC3 ECC.
- Some risks are insured against.
- Termination means cancelling the contract prematurely. Legal advice should be sought.

27

JCT Standard Building Contract SBC05

Aim

This chapter aims to explain the meaning and operation of the most significant clauses in the Joint Contracts Tribunal, Standard Building Contract (JCT SBC05) revision 2 (2009).

Learning outcomes

On completion of this chapter you will be able to:

>> Explain in detail the significant clauses of the JCT SBC05 form of contract.
>> Apply its clauses to typical contract issues.
>> Manage and administer a JCT SBC05 contract.

The traditional contract

JCT SBC05 is a traditional contract in common with ICE7. Such contracts have been with us for many decades, and the term 'traditional' should not be thought of as meaning out-of-date or old-fashioned. It simply reflects the traditional contract strategy where the employer has separate contracts with the design organisation and the contractor. JCT SBC05 (with quantities) is based on the assumptions that:

- Most or all of the design will be carried out by a design organisation (called the 'designer' for short) under a design contract with the employer.
- The pricing and payment mechanism is a bill of quantities.
- The construction is carried out by a contractor (sometimes called a builder) under a construction contract with the employer. This construction contract is JCT SBC05.
- The contract is administered on behalf of the employer by the 'architect/contract administrator' who is named in the contract and will often be a senior member of the design organisation.
- In carrying out the work, the contractor will often employ sub-contractors.

The structure of JCT SBC05 revision 2 (2009) with quantities

Like other traditional contracts, JCT contracts were drafted by lawyers and used 'legalistic' language that is intended to be legally clear, but that can be hard to understand at first. As we saw in Chapter 20 there are many versions of JCT contracts. This chapter will concentrate on the 'with quantities' version, denoted by the '/Q' in the name JCT SBC05/Q.

JCT SBC05/Q has its main clauses set out in nine sections. These are preceded by the articles of agreement, recitals, articles and contract particulars, and followed by the schedules. This chapter will refer to the 'with quantities' version simply as JCT SBC05.

> This chapter uses the abbreviation JCT SBC05 to denote JCT SBC05 revision 2 (2009) with quantities.

Articles of Agreement

State the employer's and contractor's name and address.

Recitals

- Recital first: states the nature and extent of the works.
- Recital second: confirms that a priced bill of quantities has been provided by the contractor (and also a schedule of activities if required) which should be signed and initialled by both parties. This is the significance of the '/Q' in JCT SBC05/Q.
- Recital third: lists the drawings, which should be signed and initialled by both parties.
- Recital fourth: status of employer for CIS (taxation) purposes.
- Recital fifth: refers to the information release schedule (delete if none provided).
- Recital sixth: explains the sections of the works (delete if not applicable).
- Recital seventh: allows for the use of a framework agreement.
- Recital eighth: is used if the supplemental provisions are to apply (these are found in schedule eight).
- Recital ninth to twelfth: used if a CDP (**contractor's design portion**) is included in the contract, and refer to the supply and acceptance of employer's requirements and contractor's proposals.

Articles

- State the contract sum.
- State the names of the architect/contract administrator, quantity surveyor, CDM coordinator and principal contractor.
- Article seven names the disputes procedure as adjudication, and then article eight or nine is chosen in case further reference is made to either arbitration or legal proceedings.

Contract Particulars Part 1

These contain a number of very important items which will be explained later in this chapter; a selection is given below:

- Gives further detail relating to recitals sixth to eleventh where they are used.
- Date for completion of the works (and the sections if used).
- Date for possession of the site (and the sections if used).
- Deferment of possession if under 6 weeks.
- Whether critical paths should be shown on the master programme.
- Liquidated damages.
- Rectification period (six months if none stated).
- Listed items, if used.
- Advance payment and advance payment bond, if used.
- Dates for interim certificates.
- Retention percentage.
- Insurance details.
- CDP insurance details.
- Joint fire code information.
- Period of suspension (two months if none stated).
- Name of adjudicator/arbitrator.

Contract Particulars Part 2

- Third party rights and collateral warranties (if used).

Execution as a Deed

- Relevant definitions and spaces for signatures (if used).

Conditions

1. Definitions and Interpretation
2. Carrying out the Works
3. Control of the Works
4. Payment
5. Variations
6. Injury, Damage and Insurance
7. Assignment, third party rights and collateral warranties
8. Termination
9. Settlement of disputes.

Schedules

1. CDP procedure (clause 2.9.5)
2. Variation and acceleration procedures (previously called 'Schedule 2 Quotation' as clause 5.3)
3. Insurance options (clause 6.7)
4. Code of Practice (refers to clause 3.18.4 'work not in accordance with the contract')
5. Third Party rights (clause 7A and 7B)
6. Forms of Bonds (clauses 4.8, 4.17 and 4.19)
7. Fluctuations options (taxes and inflation) (clauses 4.21 and 4.22)
8. Supplemental provisions (eighth recital).

Some important terms used in JCT SBC05

The architect/contract administrator

Under JCT SBC05, the employer *must* appoint an Architect/Contract Administrator (A/CA), who is named in article 3. The role of the A/CA carries clearly specified duties and obligations. In common with NEC3 ECC and ICE7, the A/CA is not a party to the construction contract (JCT SBC05); the parties are the employer and the contractor. The A/CA is appointed to administer the construction contract, certify payment and completions and all the other essential matters that must be attended to in order to ensure a well-administered contract.

The employer's rights to act on its own behalf are very limited under the contract. The most important roles of the employer are to give possession of site, to make payments on time and to take over the works when completed. On most day-to-day items the A/CA takes action under the terms of the contract.

Issues can sometimes arise between the employer and the contractor (such as late possession of site), and the A/CA is required and empowered to try to resolve these, but *only within the terms* of the construction contract, JCT SBC05.

Other named people in the contract

The employer may also appoint a **Clerk of Works** whose sole duty is as an inspector (of quality issues) under the A/CA's directions. Any directions given to the contractor by the clerk of works have to be confirmed in writing within two days by the A/CA if they are to have any contractual effect. The employer also appoints a **Quantity Surveyor**, whose duties relate to measurement and valuation (money).

The employer may also appoint an **Employer's Representative** under clause 3.3, to carry out any of the few employer functions under the contract. The employer's representative should not be confused with the A/CA, and the Guidance Notes to JCT SBC05 suggest that they are never the same person. It follows of course that the employer's representative (if one is appointed) is *not* empowered to carry out the duties of the A/CA.

Contractor's Design Portion (CDP)

Where the employer wishes to use a contract with complete design, then JCT DB is a better choice since it is drafted with this in mind. However JCT SBC05 can be used for contractor design. In fact it has very clear direction and terms relating to 'contractor design', unlike NEC3 ECC which is rather sparse in comparison. In any contracts where the contractor is to design some of the permanent works, the scope and extent of that design must be defined carefully and comprehensively, and the integration of the contractor's design with the rest of the design must also be considered.

Where it is used, CDP or Contractor's Design Portion is defined in the ninth to twelfth recitals. It is the part of the works where the contractor is responsible for *design* as well as construction. More detail is added in many of the other clauses of the contract, where the CDP is relevant, such as clauses relating to 'the documents', design and valuations (payment). It will often be the design organisation (usually an architect or building surveyor) that decides to use CDP, but it would be wise to obtain the employer's approval. This is because the employer has awarded a design contract to the design organisation, and quite rightly assumes that it will carry out the design unless another arrangement to 'subcontract' some of the design to the contractor is agreed.

The employer's requirements will be set out in various documents, and these are referred to in the tenth recital of the contract particulars. These requirements could range from a simple performance specification describing what the design and construction are to achieve, to a comprehensive set of detailed documents. It is worth remembering that the more detail the employer gives, the more risk the employer takes. This occurs by the operation of clause 2.13.2 which says that the contractor is not responsible for the content of the employer's requirements or for verifying the adequacy of any design contained within them.

The contractor's proposals are the contractor's response to the employer's requirements. These will be referenced in the eleventh recital in the contractor's returned tender documents. The contractor will typically include drawings, specifications and possibly a list of qualifications or assumptions. These should be considered carefully when examining tenders because they can affect the balance of risk between the employer and contractor. The breakdown of the contractor's price for the CDP work is called the CDP analysis and is set out under the contractor's proposals in the recitals.

In the recitals (which summarise all the main aspects of the agreed contract), the twelfth recital states that the employer has examined the contractor's proposals and is satisfied that they meet the employer's requirements.

Clearly it would not be reasonable to expect a number of tenderers to complete a complex design, since only one of them will become the successful contractor. Hence JCT SBC05 sets out a very comprehensive submission procedure for the detailed documents forming the contractor's design in schedule one which is used for detailed design submissions after the contract is awarded. This procedure is referenced in the contract clauses in clause 2.9.5.

In law, where a contractor designs *and* constructs works, the contractor's design liability would be to a standard of 'fitness for purpose', which is very onerous. In common with most contracts, JCT SBC5 reduces this standard to the normal designer's standard of 'reasonable skill and care' by the operation of clause 2.19.1 which limits liability to 'that of an architect or other professional designer.' This protection also extends to any further contractor design resulting from a variation.

Relevant Matters and Relevant Events

Relevant Matters and Relevant Events are very important, because they allocate named events as employer risk and the contractor is compensated if one occurs. However, unlike NEC3 ECC, JCT contracts generally separate time from money. Some events are common to both categories, and some such as exceptionally adverse weather give time only. This gives JCT SBC05 the ability to share some risks between the employer and the contractor – 'weather' being one of them. A number of other events give 'time only' as can be seen by comparing the two lists below.

Relevant Matters

Relevant Matters as defined in clause 4.24 are those for which the contractor gets loss and expense (money). They are summarised below:

- Variations.
- Instructions of the architect/contract administrator (A/CA) for certain specified matters.
- Finding antiquities.
- Suspension by the contractor under clause 4.14 of the performance of his obligations under this contract.
- The execution of work for which an approximate quantity is not a reasonably accurate forecast of the quantity of work required.
- Any impediment, prevention or default, whether by act or omission, by the employer, the A/CA, the quantity surveyor or any of the employer's persons (except where caused by the contractor's default).

Relevant Events

Relevant Events as defined in clause 2.29 give extensions of time by the A/CA fixing a later completion date under clause 2.28. Some are very similar to the Relevant Matters but they cover a more extensive list of events. They are summarised below:

- Variations.
- Instructions of the A/CA (with some exceptions).
- Deferring giving possession of the site or any section under clause 2.5.
- Executing work for which an approximate quantity is not a reasonably accurate forecast of the quantity of work required.
- Suspension by the *contractor* of the performance of his obligations under the contract under clause 4.14.
- Any impediment, prevention or default, by the employer, the A/CA, the quantity surveyor or any of the employer's persons (except where caused by the contractor's default).
- The carrying out of work (or failure to carry out) by a statutory undertaker in relation to the works.
- Exceptionally adverse weather conditions.
- Loss or damage caused by any of the specified perils (see insurance clause 6.8).
- Civil commotion or the use or threat of terrorism. This also includes the activities of the relevant authorities in dealing with such events or threats.
- Strike, lock-out or local combination of workmen affecting the works or the trades engaged in preparation, manufacture or transportation of goods or materials for the works. This also applies to the preparation of the design for the contractor's design portion. This clause gives very wide protection to the contractor.
- The exercise after the base date by the United Kingdom Government of any statutory power which directly affects the execution of the works.
- *Force majeure.*

Section I Definitions and interpretation

People tend to skip over definitions, where they are actually very important, and there are many of relevance in this first section.

Clause 1.7 Notices and other communications

This clause affirms that all notices or other communications referred to in the contract between the parties (employer and contractor) or by the A/CA or quantity surveyor *shall be in writing*.

Clause 1.8 Issue of A/CA's certificates

This provides for all *certificates* issued by the A/CA to be copied to the employer and the contractor. In other words, certificates are copied to the actual parties to the contract.

Clause 1.9 Effect of Final Certificate

The Final Certificate is issued by the A/CA under clause 4.15 at the end of the rectification period. There are other provisions of clause 1.9 and what follows is a simplification. There have been a number of legal cases as to the status of the Final Certificate, and particularly whether it reduces the employer's rights to pursue defects or latent defects. The Final Certificate is very important and, provided that no adjudication, arbitration or other proceedings have been commenced, the Final Certificate provides:

- That all amounts and adjustments have been made to the contract sum (money).
- That all extensions of time have been given under clause 2.28 (time).
- That direct loss and expense properly justified under clause 4.23 have been reimbursed to the contractor (claims).

Clause 1.9 also confirms that where the approval of the A/CA was required in respect of qualities of materials, goods or workmanship then they are to the reasonable satisfaction of the A/CA. Hence in cases where work has been stated to be 'to the approval of the A/CA', the A/CA should take particular care, and should not issue the certificate unless entirely satisfied with the contractor's work.

However the Final Certificate is not *conclusive* evidence that such materials or any other materials comply with any *other requirements* of the contract. In other words, the Final Certificate confirms the satisfaction of the A/CA where this was needed, but that is all.

Both parties have the right to challenge the issue of the Final Certificate by commencing proceedings within 28 days. The certificate is then only conclusive with respect to matters that are not challenged in the proceedings.

Section 2 Carrying out the works

Clause 2.1 General obligations

This clause sets out the contractor's basic obligation to carry out and complete the works in a proper and workmanlike manner in compliance with the contract documents.

Clause 2.2 Contractor's designed portion

This clause states the basic obligation to complete any contractor designed portion and to select suitable materials, goods and workmanship where these are not described in the employer's requirements or contractor's proposals. CDP work has been described above.

Clause 2.3 Materials goods and workmanship

This clause states that all materials and goods for the works, excluding any CDP work, shall so far as procurable be of the kinds and standards specified. Similarly, workmanship shall be as specified. CDP work must comply with the employer's requirements or, if not described therein, with the contractor's proposals. Where approval of materials, goods or workmanship is required, they shall be to the A/CA's reasonable satisfaction.

Clause 2.4 Date of possession, progress

The date of possession is fundamental in all contracts. It is stated in the contract particulars and is the date on which the contractor is given access to the site and from which it becomes responsible for the control and safety of the site in accordance with CDM 2007. Unlike NEC3 ECC this clause also states that the contractor will begin construction from this date and then 'regularly and diligently proceed with and complete the same on or before the relevant completion date.'

Clause 2.5 Deferment of possession

JCT is unusual in the inclusion of this clause which allows the employer to defer possession for up to six weeks, provided the contract particulars state that clause 2.5 applies. A period less than six weeks can also be specified. Such deferment is a relevant event, but not a relevant matter. Hence the contractor is given more time but receives nothing for any associated loss and expense.

Clause 2.6 Early use by the employer

This clause allows the employer to take over and use the site, or the works or part of them, before practical completion provided the contractor agrees.

Clause 2.9 Construction information and contractor's master programme

Programmes are very important in construction contracts and JCT SBC05 is rather light in its treatment of this issue compared with NEC3 ECC. However, revision 2 (2009) does allow critical paths to be specified as a programme requirement in the contract particulars. Clause 2.9 is basically concerned with the provision of the contractor's master programme but does not define what it should include, as NEC3 ECC does so well. It also stipulates that the contractor shall provide updated programmes within 14 days of a decision by the A/CA under clause 2.8.1 (extension of time).

Of course the programme is not a contract document and does not have to be followed rigidly. However it is the best indication of the contractor's plans and so is very important to the A/CA in supervising the works. The programme is also essential to determine any delay effects when assessing claims for loss and expense.

Clause 2.10 Levels and setting out of the works

The A/CA should provide whatever drawings and details are needed for the contractor to set out the works. If the contractor makes errors in setting out, then these errors are corrected at the contractor's cost. However, the A/CA may accept any errors and an appropriate deduction is made from the contract sum.

Clauses 2.11 and 2.12

As is usual, these clauses provide for the A/CA to issue further drawings or details to the contractor. They should be provided at a reasonable time, allowing for the contractor's progress. Where the contractor needs further information but thinks the A/CA may not be aware of the timescale, it should notify the A/CA under clause 2.12.3. Unlike ICE7 there is no *direct* clause giving the contractor more time or money if such drawings are late.

There is also the option to provide an information release schedule, specifying what and when the A/CA will release to the contractor.

Clauses 2.13 to 2.15 Errors, discrepancies and divergences

Clause 2.13 is important in that it confirms that, unless otherwise stated, the contract bills have been prepared using a specified standard method of measurement, which will usually be SMM7. Clause 2.14 goes on to give the contractor the right to more time and loss and expense if there is a departure from such method of measurement or instructions of the A/CA to correct any discrepancy between the drawings, contract bills and instructions of the A/CA. However the contractor has to give notice of such discrepancies under clause 2.15.

Clauses 2.26 to 2.29 Adjustment of completion date

These are very important clauses. The employer sets the 'date for completion of the works' in the contract particulars. The date when the contractor completes the works to a sufficient standard in the A/CA's opinion is called 'practical completion'. This completion date for the works can be revised by the A/CA for a number of reasons set out in the contract. During construction, unforeseen problems often arise and the work may need to be varied by the A/CA. Hence most contracts have provisions to move the completion date to account for such matters, unless they are due to the contractor's errors or default.

If it becomes apparent that the progress of the works is being or is likely to be delayed, the contractor should promptly give notice to the A/CA including particulars of likely effects and an estimate of the delay beyond the relevant completion date.

When the A/CA receives such notice he or she should consider it carefully and if justified give an extension of time (EOT) to the completion date for the works, by fixing a revised date. However, the contractor must meet two conditions:

1. The event must be one of the listed relevant events in clause 2.29 (summarised above).
2. Completion of the works is likely to be delayed beyond the relevant completion date.

The second provision effectively means that the event has to be on the contractor's critical path as demonstrated by the programme. A further difficulty that is clarified in NEC3 ECC is the ownership of terminal float (see Chapter 16). In NEC3 ECC reference is always made to the *contractor's planned* completion, and hence the contractor owns terminal float. However in JCT SBC05 reference to delay to the completion date effectively allows terminal float to be used up by relevant events. This is an area that generally provokes heated debate between contractors and A/CAs.

The A/CA must notify the contractor of the decision under clause 2.28.3. This clause goes on to allow the A/CA to bring forward the revised completion date *but only* for any relevant omission defined under clause 2.26.3. The A/CA cannot bring forward the *original* completion date (as stated in the contract particulars) and there has to be a relevant omission to justify movement of any revised completion date. Hence this is a mechanism for balancing EOTs with the time saved by relevant omissions, but only when fixing a revised completion date.

Clauses 2.30 to 2.32 Practical completion, lateness and liquidated damages

After the date of possession, probably the next most important date is the date when the works are completed, because this marks an end to the contractor's main obligation – to complete the works. The law typically envisages completion as being an absolute state. This is really not possible on most construction contracts, which are usually very complex, and JCT SBC05 recognises this with the concept of 'practical completion'. This is not defined in JCT SBC05 but is usually taken to mean, complete enough for the employer's safe occupation and use, or for other contractors to come onto the site and carry out their work under separate contracts, such as finishing works or landscaping contracts.

The A/CA decides whether practical completion has been achieved and, when satisfied, issues a certificate. The certificate has a number of effects:

• Half of the retention will be released for that section or the whole works.
• The relevant rectification period begins.
• The contractor's liability for liquidated damages ends.
• Regular interim certificates cease to be issued.

Where the A/CA does not consider that the works have reached the state of practical completion by the completion date, the A/CA should issue a 'non-completion certificate' under clause 2.31. Hopefully this will rarely be required, because it is normal for the contractor and A/CA to meet regularly in the period just before potential completion, and for them to agree a 'snagging list' of work still to be completed or corrected, before the A/CA is happy to issue a practical completion certificate.

A non-completion certificate is an essential prerequisite to the deduction of liquidated damages by the employer under clause 2.32. The employer must also notify the contractor that the employer intends to deduct liquidated damages. Following these notices, the employer may deduct liquidated damages at the rate stated in the contract particulars. If an EOT is subsequently given, the employer repays liquidated damages for the appropriate period of extended time.

Clauses 2.33 to 2.37 Partial possession by the employer

These clauses give the employer the right to take over a part of the works, called the 'relevant part', before practical completion of the works. The contractor has to give consent. The A/CA confirms the date of take-over in a notice.

The clauses go on to describe issues of defects, insurance and liquidated damages applying to the relevant part.

Clauses 2.38 and 2.39 Defects

Most contracts provide for a period after completion, often a year, in which the contractor must return to correct defects. JCT SBC05 calls this period the rectification period which is stated in the contract particulars by the employer. If no period is given, the default is six months. However, for larger contracts a year is normal. The A/CA delivers a schedule of defects to the contractor, no later than 14 days after the end of the rectification period. Before issue of this schedule, the A/CA may issue instructions for the correction of defects at any time during the rectification period. The contractor corrects defects within a reasonable time at no cost to the employer.

Clause 2.39 confirms that all defects have been completed to the satisfaction of the A/CA by requiring the A/CA to issue a 'certificate of making good' to the contractor.

Section 3 Control of the works

Clauses 3.1 to 3.6

These clauses begin by giving the A/CA rights of access to the site of the works and workshops and other premises that are being used to prepare work for the contract. Technically the contractor has possession and hence control of the site and this clause is necessary to give the A/CA rights of access.

By clause 3.2 the contractor is obliged to ensure that there is always a competent person in charge on the site. Instructions given to this person are deemed to have been given to the contractor.

Under clause 3.3 the employer may appoint an individual to act as representative by giving notice to the contractor. It should be remembered that the A/CA has the role of administering the contract, and the employer's representative (if appointed) only has a few actions under the contract.

Clause 3.4 entitles the employer to appoint a clerk of works (CoW) who acts as an inspector (largely of quality) under the direction of the A/CA. However the CoW has no authority to approve the works, as this is part of the role of the A/CA. Any instructions given by the A/CA only have contractual effect if confirmed in writing by the A/CA within two working days of the instruction.

Both the A/CA and the quantity surveyor may be replaced by the employer under clause 3.5 by the employer notifying the contractor. It is worth noting that under clause 3.5.2 the replacement A/CA cannot overrule any previous decision, certificate and approval of the previous A/CA unless the predecessor would have had such power.

Clause 3.6 is very important in that it states the contractor's basic contractual responsibility to carry out and complete the works in accordance with the conditions of contract, regardless of any obligation of the A/CA to the employer and whether a clerk of works is appointed or not. This contractual responsibility is not changed by inspections carried out by the A/CA or by the A/CA certifying payment for work or issuing a practical completion certificate or certificate of making good.

Clauses 3.7 to 3.9 Sub-contracting

Most contractors today sub-contract much of their contract work. In common with most forms of contract, clause 3.7 requires the contractor to notify the A/CA of sub-contracting and the A/CA to consent to it. However, sensibly, the contractor remains totally responsible for the work of any sub-contractor. Where there is CDP work, the contractor must obtain the *employer's* consent to sub-contract the design work.

Clause 3.8 deals with the situation where certain work in the contract bills is to be carried out by persons named in a list annexed to the bills and selected from the list by the contractor. This list shall contain at least three persons and the contractor may add to it. A person selected from the list becomes a sub-contractor. This is a little like the process of nominating sub-contractors but without all the formal clauses previously used for this practice, and allows the employer to name selected firms for certain work. Clause 3.9 sets out the requirements of the sub-contract.

Clauses 3.10 to 3.22 Architect/Contract Administrator's instructions

Events arise on most contracts which were not foreseen when the contract was signed and provision has to made to deal with them and also to make payment to the contractor. Clause 3.10 gives the A/CA the power to issue instructions to the contractor and the contractor is obliged to comply. There are two provisos:

1. The instruction falls under clause 5.1.2 (discussed in the section on that clause) and the contractor notifies a reasonable objection.
2. Or where a variation quotation under schedule two is required, in which case the contractor must not proceed until the A/CA gives a confirmed acceptance.

In the unlikely event that the contractor does not comply with an instruction, clause 3.11 gives the employer power (after an A/CA notice has been given and seven days have elapsed) to use other persons to do the work and deduct the cost from the sum due to the contractor. It is important to note the seven days. Thus if an instruction were issued on 1 May by the A/CA and, on the 6 May, the A/CA becomes aware that it has not been followed and issues a written notice asking the contractor to comply, then after 13 May the employer may use other contractors to carry out the work.

On most contracts it will be necessary for the A/CA to give instructions to the contractor. Clause 1.7.1 above required the use of writing for notices or other communications. However, verbal instructions are frequently given by the A/CA and later *confirmed* in writing. This situation is covered by clause 3.12.2 which says that if a verbal instruction is confirmed by the A/CA within seven days, it takes effect from the date of confirmation. Clause 3.12.1 allows the contractor to confirm an instruction of the A/CA in writing, and if the A/CA does not dissent within seven days the instruction takes effect after the latter seven-day period. Under clause 3.13 the contractor may ask the A/CA to confirm which provision of the conditions of contract gives the A/CA power to issue such instruction.

Clause 3.14 gives the A/CA power to issue instructions which require a variation (change). Variations are frequently necessary in construction work and are defined and discussed further in clause 5.1.

Clause 3.15 gives the A/CA the power to postpone any work to be executed under the contract. The contractor is compensated in terms of time and money because clause 3.15 is a relevant event and a relevant matter. The measures in this clause are serious, and should only be used in rare circumstances. Long postponements (several months) can be treated as abandoning the contract – subject to time and notice provisions. However, there may be times when some delaying event means that construction cannot proceed for a few weeks. Such delaying events are often external, like delays to planning permission, or delays to moving major services under the site, such as gas or water mains, or even possibly extensive archaeological remains, which mean the contractor cannot proceed for some weeks.

The power to issue instructions regarding the expenditure of provisional sums is given to the A/CA in clause 3.16. Provisional sums are inserted by the design organisation in the contract bills as sums of money. This is usually done for work that cannot be fully described or defined at the contract preparation stage. Once the extent and nature of the work are clear, the A/CA can issue directions as to how the work in the provisional sum shall be implemented or whether it shall be omitted.

The power to order opening-up for inspection or further tests is given to the A/CA in clause 3.17. As is usual in contracts, the cost of such work or tests is added to the contract sum, unless the inspection or test shows that the workmanship or materials are not in accordance with the contract, in which case the contractor pays for the tests and for the making-good. Clause 3.18 goes on to discuss this matter in more detail. Where any work, materials or goods are not in accordance with the contract, the A/CA may have them removed from the site, or may allow them to remain subject to a reduction in the contract sum.

Clause 3.22 deals with fossils, antiquities and other objects of value. These become the employer's property and the contractor is obliged not to disturb them, take any steps necessary to preserve them in the exact position they are found and inform the A/CA. The A/CA issues instructions and again the contractor is compensated for any additional time or cost.

Clauses 3.23 and 3.24 CDM regulations

These clauses repeat some of the duties set out in CDM 2007 and confirm that both the employer and contractor will comply with the regulations.

Section 4 Payment

Clauses 4.1 to 4.5

The contract sum is the contractor's price for carrying out all the work set out in the contract documents and is inserted at tender stage, by the contractor, in Article Two. When employers sign the contract, they undertake to pay the contract sum at the times and in the manner specified in the contract. Hence JCT SBC05/Q (called JCT SBC05 for short in this chapter) is a lump sum contract but regular payments are made, and the contract sum can be adjusted for variations and a number of other reasons.

In contrast, 'JCT SBCO5 with Approximate Quantities' is a remeasurement contract. Approximate quantities (but the most accurate that can be ascertained from the drawings) are given for all of the work, and the contractor submits a fully priced copy of the bills of approximate quantities at tender stage, which forms the basis of the contract. There is no 'contract sum' entered in the Articles. All the work is remeasured prior to certification and the contractor is paid for the actual quantities of work carried out.

Returning to JCT SBC/Q, clause 4.1 sets out the contractor's basic obligation and says 'the quality and quantity of work included in the contract sum shall be as set out in the contract bills, and where there is a contractor's designed portion, in the CDP documents.'

Under clause 4.2, the contract sum is only adjusted in accordance 'with the express provisions of the contract.' This simply reiterates the fact that the A/CA must comply with the contract and only has power to operate within the terms of the contract.

Clause 4.3 goes on to list ways in which the contract sum may be adjusted. They include adjustments for variations, **variation quotations** (see clause 5.3 below), acceleration quotations and variation in insurance premiums as paragraph A.5.1. Deductions from the contract sum can be made for provisional sums that are not used, omissions and so on. Additions to the contract sum include fees, certain tests, variations, loss and expense claims, and so on. Hence whilst JCT SBC05 is nominally a lump sum contract, there are many ways in which this contract sum can be adjusted.

Clause 4.4 explains that where the contract sum is adjusted then the adjusted amount shall be taken into account in the next interim certificate. In other words, appropriate payment is made promptly even if it is only a partial payment with further payments taking place as the work in question is completed.

A final adjustment is made under clause 4.5, whereby the contractor provides the A/CA or the QS (if so instructed) with all the documents necessary to calculate the final account. These would include final bills with agreed measurements, finalised claims for loss and expense and so forth. This contractor submission should be not later than six months after the issue of the practical completion certificate. Within three months of receipt of the contractor's documents, the A/CA or QS makes a final ascertainment of the amounts due to the contractor and sends the contractor details.

Clause 4.6 VAT

Unlike 'domestic' employers, most employers in the construction industry can reclaim VAT. The contract sum does not include VAT and, where adjustments are made, such as with variations, VAT is ignored. However, when making payments to the contractor, the employer must include the value of any necessary VAT payments. The employer later reclaims the VAT payments that were made.

Clause 4.7 Construction Industry Scheme (CIS)

This scheme was brought in by the government to try and reduce the high level of tax evasion by sub-contractors.

Clause 4.8 Advance payment

This is an option that must be included in the contract particulars (where wanted) and allows the employer to make an advanced payment to the contractor. Reimbursement by the contractor follows the terms set out in the contract particulars. Advance payment helps the contractor's cash flow and reduces the potential borrowing requirement. Advance payment increases employer risk (if the contractor goes into liquidation) but it should result in lower tender prices.

Clauses 4.9 to 4.15 Interim and final certificates

Regular payments are made to the contractor and the amount due to the contractor is stated on an interim certificate issued by the A/CA together with the basis of how the amount has been calculated. Interim certificates are issued at the dates set out in the contract particulars up to the date of practical completion or one month thereafter. They are then issued at two-monthly intervals during the rectification period. These dates in the contract particulars begin with a determined first date, and are then on the same date in each month. The first date shall not be more than one month after the date of possession.

In order to determine the amount due on the certificate, interim valuations are made by the quantity surveyor as clause 4.11. It is usual for the contractor to actually make an application for payment with all supporting details, and clause 4.12 allows this, provided the contractor's application is not later than seven days before the date of issue of the interim certificate. If the QS disagrees with the amount shown on the contractor's application, a statement must be submitted in similar detail to the contractor's application that identifies the disagreement.

The final date for payment by the employer against an interim certificate is 14 days after the date of the interim certificate. Clause 4.13 goes on to describe in some detail what happens if the employer withholds any money from the payment (not including retention). This detail is provided to comply with HGCRA.

Clause 4.14 gives the contractor the right to suspend the performance of the obligations under the contract where the employer fails to pay in full. There are notice provisions set out in the clause.

A final certificate under clause 4.15 is issued by the A/CA stating the adjusted contract sum, usually within two months of the end of the rectification period. This states the final contract sum and the amount already paid to the contractor. This balance is usually owed to the contractor, but rarely the contractor may owe the employer money. The balance shall be paid within 28 days of the final certificate. However there are some further detailed provisions in the clause where there are still any disagreements as to the final balance due.

Clauses 4.16 and 4.17 Gross valuation

As we have seen above interim valuations are made and certified on interim certificates in order for the contractor to be paid on a monthly basis. Clause 4.16 is a detailed clause showing how the gross valuation is determined. In simple terms it includes the total values of:

- Work properly executed (in other words without defects) by the contractor including work executed against variations which have been valued (subject to retention).
- Site materials that are properly protected against weather (subject to retention).
- **Listed items** (if any) as clause 4.17 (subject to retention).

Plus the following which are not subject to retention:

- Adjustments for a variety of items including fees, certain tests and so on.
- Amounts paid against loss and expense claims to clause 4.23.
- Some payments due to the contractor following on from insurance repairs or the removal of debris.
- Any payments under the fluctuations clauses (which account for price inflation, where included in the contract).

There are then a number of items that may be deducted such as:

- For setting out errors that are not to be rectified.
- Where defective work in the rectification period is allowed to remain.
- Costs incurred by the employer in engaging other contractors to do work that the contractor failed to execute.
- Some matters relating to the fluctuation clauses (where included).

Clause 4.17 deals with listed items. These are materials, goods and/or prefabricated items which are listed by the employer and annexed to the contract bills. These items are paid for by the employer *before* delivery to site.

They must be the property of the contractor and must be properly marked and insured.

Clauses 4.18 to 4.20 Retention

Most contracts provide for an item called retention. This is a sum of money made up of deductions from regular payments. The usual retention rate is between three and five per cent and three per cent is the default percentage in JCT SBC05 unless amended in the contract particulars. The idea of retention is to give the employer some protection against contractor default, particularly against reme-dying defects. It is not the employer's money and should be placed in a separate account for safe keeping. It is eventually repaid to the contractor; half at practical completion and half when all defects are rectified, with the certificate of making good.

Clause 4.21 Fluctuations

During times of high inflation (perhaps over ten per cent), or for contracts lasting a long time (perhaps over two years), it is advisable to include a fluctuations clause. This compensates the contractor for changes in prices of materials, labour and so on. Unfortunately, calculating such additions to monthly payments is time-consuming and tedious. Consequently few contracts have used fluctuations provisions in recent years. JCT SBC05 goes further and allows for fluctuations to be calculated in different ways by means of fluctuation options A to C. If required the fluctuation option must be included in the contract particulars.

Clauses 4.23 to 4.25 Loss and expense

Clause 4.23 is a very important clause. The contractor makes an application to the A/CA if, during the execution of the contract, the contractor incurs or is likely to incur direct loss and/or expense for which the contractor would not be reimbursed under any other clauses owing to a deferment of possession or because progress is materially affected (or is likely to be affected) by a relevant matter. The employer may defer possession as described in clause 2.5 (where this option is included in the contract particulars), but has to 'pay for it' under this clause. Relevant matters are listed in clause 4.24 and were discussed at the begin-ning of this chapter.

Claims for loss and expense relate to deferment of possession or, more usually, relevant matters.

The contractor should make the application as soon as it becomes apparent that regular progress has been affected, and should provide such information as will allow the A/CA to form an opinion, and upon request the contractor pro-vides details of the loss and/or expense necessary to determine what payment (if any) is due.

The A/CA must consider the application and ascertain what (if any) loss and expense have been incurred. It is usual to expect contractors to refer to the clause under which they are claiming. Claims for loss and expense are often complex and the first thing the A/CA must do is to decide if there is a proper entitlement under the contract. Examples of entitlement would be delays arising from the discovery of archaeological remains, variations and so forth. Once entitlement is proved, the quantum or amount of payment can be determined by the A/CA.

Unfortunately many events which cause the contractor to lose money do not easily fit into the listed relevant matters. Similarly many events may be the contractor's 'fault', such as poor planning, the use of inappropriate construction techniques, labour problems, sub-contractor delays, delayed delivery of materials and so on. If these are not the employer's 'fault' or a risk listed as the employer's under the contract, then the contractor is not entitled to payment and the A/CA should politely refuse.

An even more difficult situation often arises and that is when the regular progress of the works is affected for a mixture of reasons – some relevant matters and some the contractor's fault. Separating the two is usually no easy matter, and many meetings and discussions may be required to reach agreement.

If the A/CA agrees that loss and expense is justified, then the relevant amount is added to the contract sum as clause 4.25.

Section 5 Variations

Clause 5.1 Definition of variations

NEC3 ECC does not actually define variations (called changes in NEC) but ICE 7 and JCT SBC05 do define them. JCT SBC05 divides variations into two categories. The first category in clause 5.1.1 contains the usual additions, modifications to quality and so on which are found in most forms of contract. The second category in clause 5.1.2 may give the contractor more difficulties and is harder to determine the effects of. Hence the contractor is given a 'right of reasonable objection'. This second category of variations includes such matters as limitations on working space or working hours.

Variations under clause 5.1.1

These are the alteration or modification of the design, quality or quantity of the works including:

- The addition, omission or substitution of any work.
- The alteration of the kind or standard of any of the materials or goods to be used in the works.
- The removal from the site of any work executed or materials or goods brought thereon by the contractor for the purposes of the works, other than work, materials or goods which are not in accordance with this contract.

Variations under clause 5.1.2

Variations under this part of the clause give the contractor the right to notify a reasonable objection under clause 3.10.1. Such variations are where the employer imposes obligations or restrictions (or changes existing ones) in regard to:

- Access to the site or use of any specific parts of the site
- Limitations of working space
- Limitations of working hours
- The execution or completion of the work in any specific order.

Clauses 5.2 to 5.5 The valuation of variations and provisional sum work

In traditional contracts such as JCT SBC05, it is usual to try to use the BoQ to value variations where this is fair. Clause 5.2 says that the *employer* and the contractor shall agree the value of variations and provisional sums. If they do not agree then the usual method is for the quantity surveyor to value them using the rules in clauses 5.6 to 5.10 below. This does not apply to **variation quotations** as clause 5.3.1 (again discussed below). It is also possible for the employer and contractor to use another method, rather than the QS, such as an external consultant or specialist, if they agree to do so.

Variation quotations as clause 5.3

These were previously called schedule two quotations (S2Q) because the rules for their use are set out in schedule two. The A/CA may want a definite price for a variation, before ordering the variation. Hence where the A/CA instructs, the contractor prepares such a quotation following the provisions of schedule two. However the contractor must be given sufficient information to prepare the S2Q. The contractor can disagree with the application of schedule two, but must notify such disagreement within seven days (unless a different period is agreed). Schedule two basically sets out the breakdown of the quotation.

The employer may decide to proceed with the work in a S2Q and, if so, the A/CA gives a 'confirmed acceptance' to the contractor stating the adjustment to the contract sum and any adjustment to the time for completion.

If a S2Q is not accepted, the contractor is paid a reasonable amount for the cost of preparing it.

Clauses 5.4 and 5.5 Variations: measurement and valuation

Clause 5.4 gives the contractor the right to be present at the measurement of work for the purpose of a valuation of a variation. This might include site measurements, for example of excavation work carried out or formwork fixed and erected.

The valuation of variations is given effect under clause 5.5 by means of an addition to or deduction from the contract sum and hence will be included in the next interim certificate as clause 4.4. In other words, once the value of a variation is agreed, appropriate payment is made promptly.

The valuation rules in clause 5.6

Where a valuation of a variation relates to additional or substituted work, or work where an approximate quantity (items identified as such in the contract bills) is included in the contract bills (and subject to clause 5.8 for CDP work), the rules are based on three decisions. These are whether the additional or substituted work is:

- Of similar character to work set out in the contract bills.
- Is executed under similar conditions.
- Does not significantly change the quantity of work set out in the contract bills.

The result of this decision is as follows:

- Where the answer to all three decisions is yes, then the rates are used *unchanged* from the contract bills.
- Where the work is of similar character to but is *not* executed under similar conditions and/or significantly changes the quantities, then the rates from the contract bills are used as a *basis* for agreement.
- Where the work is *not* of similar character, then the work is valued at fair rates and prices.

There are additional rules concerning approximate quantities (where the only change is in quantity), thus:

- Where the approximate quantity *is* a reasonably accurate forecast of the quantity of work required, the rate or price for the approximate quantity shall be used.
- Where the approximate quantity *is not* a reasonably accurate forecast of the quantity of work required, the rate or price for that approximate quantity shall be used as a *basis* for agreement.

Daywork as clause 5.7

Where work cannot be valued as clause 5.6 or 5.8, then it can be valued as 'daywork' using the lists of daywork published by the Royal Institution of Chartered Surveyors. These are extensive lists of labour and plant items. The contractor inserts a percentage addition to these rates in the contract bills. Using these lists tends to be expensive for the employer, and since payment is generally on an hourly basis there is no real incentive for contractor efficiency. Hence they are usually used for minor or incidental work of relatively short duration. It is usual for the contractor to complete record sheets and have them signed by the A/CA or clerk of works.

Valuation of CDP work as clause 5.8

Where work in the contractor's design portion (where used) is varied, then the cost of associated design work should be included. Additionally, valuation of variations should reflect the value of similar work in the CDP where possible.

Additional provisions in clause 5.10

This clause makes it clear that the valuation of a variation does not include any effect on the progress of the work or for any other direct loss and/or expense for which the contractor would be paid under other provisions of the contract. In other words, it does not include the additional overhead costs of additional time on-site that may result from a variation, such as extended supervision, welfare facilities, site offices and so on. Presumably the contractor would recover these costs using clause 4.23 (loss and expense) as appropriate.

Sections 6 to 8

These sections deal in detail with more complex issues relating to insurance, assignment, third-party rights and termination. These are areas that are usually dealt with by experienced construction professionals, lawyers or experts and are outside the scope of this book.

Section 9 Settlement of disputes

Clause 9.1 suggests that if a dispute arises under the contract, then the parties should 'give serious consideration to mediation.' Mediation is described in Chapter 28.

Adjudication

Statutory adjudication is available as a right. The Housing Grants, Construction and Regeneration Act 1996 (HGCRA) brought with it the right to statutory adjudication for all 'construction contracts' falling within its definition in clause 104, which is basically all construction contracts and also associated design contracts. In addition to this statutory right, the employer nominates adjudication or arbitration as the dispute procedure in the contract particulars.

Clause 9.2 deals with adjudication. It also has specific provisions where a dispute arises under clause 3.18.4 which deals with 'opening up' to search for defects; in other words, work not in accordance with the contract. Schedule four sets out a code of practice to be followed in such circumstances. This is always a difficult area in contracts and JCT SBC05 suggests in clause 9.2.2 that the adjudicator has appropriate expertise and experience relevant to the issue or instruction in dispute. If this is not the case, the adjudicator is expected to appoint an independent expert with such experience to advise and report on the issue.

Arbitration

Clause 9.3 deals with arbitration and goes into some detail on procedures to be followed. In particular it requires the use of the JCT 2005 edition of the Construction Industry Model Arbitration Rules.

Chapter summary

- This chapter covers the clauses of JCT SBC05 of most interest to construction professionals during university study or the first few years of professional practice.

28

Disputes

Aim

This chapter aims to explain the background to disputes and methods of dispute resolution.

Learning outcomes

On completion of this chapter you will be able to:

>> Describe a dispute.
>> Explain how disputes arise and how they can be avoided.
>> Describe the main methods of dispute resolution.
>> Explain some of the issues connected with disputes from the contractor's viewpoint.

Disputes

What is a dispute?

Defining a dispute in the context of construction contracts is not straightforward. In the case of *Amec Civil Engineering Ltd* v. *Secretary of State for Transport* (2005) the judge took over 500 words to define it. Basically a dispute arises when a claim of some kind is not accepted. Disputes can arise in design contracts but they are rare.

Disputes usually arise from construction contracts.

It is important to remember that the construction contract is between the employer and the contractor, and the dispute is between them. The contract administrator administers the contract, and it will usually be a contract administrator's decision or a rejection of a contractor claim, or compensation event in NEC3 ECC (decided by the project manager), that gives rise to the dispute.

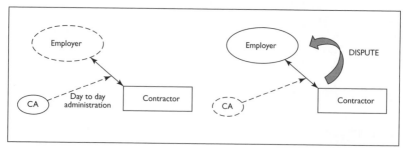

Figure 28.1 Disputes.

After signing the contract, the employer takes a back seat and may not be involved again apart from making payments, until take-over. This is shown on the left-hand side of Figure 28.1. Figure 28.1 shows the relationships with which we are now familiar, with a slight modification to emphasise the main contract relationship (contractor and employer) and the intervention of the contract administrator. Day-to-day relationships and administration are between the contract administrator (called the project manager in NEC3 ECC) and the contractor. At some stage of the construction work the contractor may make a claim for financial loss and expense for example (compensation event in NEC3 ECC), which the contract administrator does not accept. There will then no doubt be many meetings and much correspondence.

If contractors consider they have a valid claim which is still rejected by the contract administrator then the contractor's recourse is potentially to a dispute procedure. However this dispute is under the construction contract, and hence it is *with the employer*. The contract administrator now takes a back seat, although will no doubt be called upon to inform and advise the employer. This is shown on the right-hand side of Figure 28.1.

How do disputes arise?

There are five main causes of disputes:

- There may be no contract terms properly applicable to the event in question.
- The contract terms may be unclear, or ambiguous.
- The contractor may be dissatisfied with the *interpretation* of the contract terms by the contract administrator.
- The contractor may be dissatisfied with the 'quantum' (the amount of money to be received) if an event is agreed for payment.
- The contractor may be dissatisfied with the amount of additional time received to complete the contract.

From the contract administrator's point of view, the contract administrator may consider that:

- The contract terms are clear.
- There is no justified claim.
- The contractor has not made a plausible and reasoned case.

- The contractor has not linked 'cause to effect' to prove the case.
- The quantum awarded is correct.
- The additional time granted (if any) is correct.

We therefore have an impasse, and contractors will have to answer a difficult decision. Do they invoke the disputes procedure or not? This situation is discussed below.

The contractor's position in a dispute

It is possible for the employer to enter into a dispute with the contractor, but this is rare. Contractors who are contemplating entering into a dispute procedure will consider their position very carefully, and will often decide *not* to proceed. These are some of the reasons:

- Disputes involve staff time and costs, which diverts them from productive activity.
- Disputes usually involve lawyers and legal advice, which can be very expensive.
- Disputes are likely to lead to deteriorating relationships with the contract administrator and any site supervisory staff working for the employer, making the current project more difficult to complete efficiently.
- It is difficult not to become adversarial in a dispute, which runs counter to the cooperative behaviour required in NEC3 ECC for example.
- As with all legal procedures, there is no guaranteed outcome. Hence the contractor may have a good chance of a positive outcome, but it can never be certain. Time and money may be spent, relationships may deteriorate, and the contractor may 'lose' the argument.
- Most importantly, a dispute is likely to upset the employer.

Contractors stay in business by getting on tender lists and winning tenders. A dispute with an employer does not help either of these. In fact a contractor who is known for commercial aggressiveness and a fondness for disputes is likely to find it very difficult to secure future projects and stay in business at all.

Because disputes usually have a legal basis and can prove very costly to resolve, employers will often require regular reports on any disputes at their meetings of the board of directors. These directors will be responsible for sales, marketing, finance, HR, operations and other functions, not engineering projects. We can imagine the pressure this places on an engineering director, reporting on disputes, and the very high level of visibility at the most senior level that the contractor receives. A spotlight such as this is the last thing a contractor wants, however valid the case might be.

How to minimise the likelihood of disputes

Disputes are very serious matters and are never entered into lightly. They are usually dealt with at a very senior level in the contractor and employer organisations. Hence in many ways construction professionals will have little direct contact with dispute processes until later in their careers. In the meantime, potential disputes tend to be 'referred up'. However, in addition to avoiding disputes, less-experienced professionals should always report a potential dispute situation to their manager. In difficult or serious cases, legal advice may also be a good idea.

> Less-experienced construction professionals should concentrate on avoiding disputes in the first place.

The crucial role of the construction professional is to minimise the likelihood of disputes in the first place. This can be done at the early stages of the investment process, during design and during contract administration and any site supervision of the contractor (such as a resident engineer, the supervisor in NEC3 ECC or the clerk of works in JCT contracts). Many disputes arise from poor contract preparation, misunderstandings by the contractor and mistakes. The following steps will assist in reducing the likelihood of disputes:

- Thorough site investigation and ground investigation made available to tenderers.
- No conflicts between the works information and site information (these are the terms used in NEC3 ECC but the principle applies to all contracts).
- An established form of contract with minimum amendments.
  - Any amendments must be carefully drafted.
- Enough time for the contractor to submit a well-considered tender.
- Risks understood and risk allocation clear.
- Clearly defined contract relationships.
- Competent people in all organisations.
- A clear and unambiguous specification.
- Clear and comprehensive drawings.
- Transparent and fair pricing and payment mechanisms.
- Correctly assessed payments made on time by the employer.
- Fair application of contract procedures.
- Potential defects assessed properly.
- Defects corrected promptly.
- Prompt, fair and impartial assessment of claims (compensation events) by the contract administrator.
- All contract certificates given promptly.
- Completion assessed fairly.

Dispute resolution

Contract provisions

The Housing Grants, Construction and Regeneration Act 1996 (HGCRA) brought with it the right to statutory adjudication for all 'construction contracts' falling within its definition in clause 104, which is basically all construction contracts and also associated design contracts. HGCRA applies whether or not there are adjudication clauses in the construction contract itself. As a result, construction contracts have been modified to comply with the act. An example of this is option W2 of NEC3 ECC and the NEC3 sub-contract. Most contracts name either adjudication or arbitration as the dispute resolution procedure but, regardless of this, there is always the right to statutory adjudication where HGCRA applies.

Adjudication

Here an adjudicator is appointed and the matters are referred to him or her within very short timescales following detailed rules and a set procedure. One party must serve a written notice on the others in the dispute and send a copy to the adjudicator (appointed by the relevant professional institution). The adjudicator may ask for documents from the parties, appoint expert advisors, make site visits and meet with the parties. Within a fixed period (28 days from referral) the adjudicator must reach a decision and the parties are jointly responsible for meeting the adjudicator's fees and expenses. Adjudication is therefore relatively fast.

The adjudicator's decision can be enforced in court if payment is not made or if other provisions are not met. The important point about adjudication is that the decision itself can be challenged in court and the case examined afresh, so in this sense it is not final. However, the majority of decisions are not challenged in this way, and so adjudication has become a fast and relatively cheap way of resolving disputes.

Arbitration

Until more recent times, arbitration was the normal way of resolving disputes in construction contracts. However, for it to be available, the contract must contain suitable arbitration clauses. A number of arbitration procedures are published by the major bodies such as the Joint Contract Tribunal, and the Arbitration Act 1996 will also apply.

Arbitration is a more formal method than adjudication and has much longer timescales. It is therefore generally more expensive. An arbitrator's decision is final, and can only be challenged in court on its *legal* basis or if there is misconduct on the part of the arbitrator. This is an important difference between arbitration and adjudication; arbitration is usually legally binding. Because expert witnesses can be called in arbitration and both sides generally have legal representatives it really approaches the costs and time of going to court; which is why many parties in dispute now do just that, rather than using arbitration.

Litigation

This means taking a case to court. Legal procedures must be followed and the process takes a considerable time and can be very expensive. Costs of over £50,000 are not unusual and some cases have been seen to cost millions of pounds. Unlike arbitration and adjudication (where the parties agree to the method), court action is started by one party, and the other party must comply with the court's directions.

Before a court hearing takes place there are a number of obligatory legal steps which often result in a settlement between the parties. Both sides to the dispute have to present a statement of case and there follows 'disclosure', when relevant documents are made available to the other side. The process of disclosure often makes one or both parties consider carefully whether to proceed. It can be very sobering to see letters and emails going back potentially for months or years. Many may undermine your case or show your company in a bad light. There is a saying that only about two per cent of cases which are commenced ever actually get to court. They are settled by negotiation on the way.

There are two other features of litigation. The first is that the decision is reported and hence is public. Many companies do not wish their problems and mistakes to be made public. The other feature is the possibility of an appeal to a higher court with further attendant costs and delays. This produces yet more uncertainty.

Alternative methods of dispute resolution (ADR)

Conciliation and mediation

These are private and informal processes. There is no precise meaning to either word and the processes may often be similar because they are agreed between the parties in dispute. The parties in dispute appoint a conciliator/mediator, who is usually someone with appropriate construction knowledge and usually training in mediation/conciliation from an interpersonal perspective.

Conciliation and mediation are informal processes and allow the parties to explore ways of settling the dispute with the assistance of an independent external facilitator (the conciliator/mediator).

The conciliator/mediator seeks to establish common ground between the parties.

The conciliator/mediator works separately with both sides of the dispute, but should not reveal the results of such discussions to the other party without prior permission. After making progress the conciliator/mediator may suggest the parties meet, but only with their mutual agreement. The conciliator/mediator may make a recommendation of a solution to the problem. This will often be some form of settlement between the parties' expectations. The parties do not have to accept the recommendation but, if they do, they can take further steps outside the process to make it legally binding. The process is usually fairly quick and not as costly as most other methods of dispute resolution. Importantly, with the conciliator/mediator's assistance both sides can begin to see the other side's point of view. In this way, conciliation/mediation is more likely to preserve business relationships than any of the other methods of dispute resolution.

JCT SBC05 actually gives mediation further weight because it says in clause 9.1 that 'if a dispute or difference arises under this contract which cannot be resolved by direct negotiations, each party shall give serious consideration to any request by the other to refer the matter to mediation.'

Executive tribunal

This method of ADR has a number of different formats. It is usually constituted by an executive team from the parties (usually the employer and the contractor) together with a neutral advisor. Meetings are held, documents examined and witnesses may be called. The executive teams then seek to reach an agreement. Again this agreement may be put into a suitable form to ensure that it is legally binding. In many ways the executive tribunal is similar to negotiation, only with more participants.

Negotiation

This method of dispute resolution is used frequently, but is little publicised. It often takes place before a formal dispute process has been set in train but, of course, it can be used at any time. The contract administrator should already have ruled on the contractor's claims and claimed costs, under the terms of the contract, and further movement looks unlikely. The contractor may then request a meeting with the employer.

The employer is likely to have had meetings and reports from the contract administrator and frequently the employer's own staff or advisors. The employer is unlikely to want a dispute, which produces cost, delay, diversion from normal business and possible adverse publicity, any more than the contractor will want these things. Hence both sides have an incentive to negotiate and to try and settle the matter as quickly and amicably as possible.

Suppose the contractor has claimed a million pounds under various heads of claim. The contract administrator has rejected some and certified payments against some others. The employer has paid say £600,000 and the contractor's costs are a million pounds, most of which are attributable to 'employer risks' and changes (variations), but some to the contractor's own problems with suppliers, sub-contractors and its own labour force. The contractor will do a risk assessment against the claims, probably rating them from likely to unlikely to succeed. The result may be a realistic maximum figure of £800,000 that the contractor considers justifiable and which should eventually be paid. This means the contractor has £200,000 to 'go for', and few contractors can leave that much money behind them.

The employer's advisors may suggest that the contract administrator has been a little harsh in its judgments, and in a 'fair world' the contractor ought to be paid against another head of claim and some of the quantum (assessment of the amount to be paid) is a little low. The advisors suggest a settlement figure of £700,000. We now see that the difference between the two sides is not £400,000, but £100,000. A meeting to negotiate may well resolve the matter, perhaps at £760,000, and a dispute procedure and its cost are unnecessary. The other point to be made here is that the more the employer pays the less incentive the contractor will have to go to a dispute procedure. In this example, the contractor is £40,000 below the expected figure, but is that worth entering a dispute procedure for? Probably not.

Chapter summary

- Disputes usually arise between the contractor and the employer when the contract administrator rejects a claim or does not certify as much payment as the contractor considers is due.
- Formal dispute procedures are arbitration and adjudication. Adjudication is now the most used.
- The other formal mechanism for settling a dispute is recourse to the courts.
- Contractors are generally wary of invoking a formal dispute procedure because of the adverse effect it may have on their relationship with the employer.
- There are a number of types of alternative dispute resolution (ADR) and they are generally the quickest and cheapest methods, if the parties have the will to settle in this way.
- Well-prepared, comprehensive and fair contracts which are properly administered minimise the likelihood of disputes.

Index

(**Bold** page numbers indicate *major* references)

acceleration in NEC3, 416–17
acceptance in contract law, 290–2
acceptance of contractor's work or
  communication, 391–2
accepting defects in NEC3, *see* defects in
  NEC3
access, giving access to site, 59–60, 405
access date in NEC3, 403, **404**, 405,
  414, 447
accident statistics, 118–19
ACOP to CDM, 120
activities on programmes, 269–70, 369
activity schedules
  generally, 232–3
  NEC3, 369, 431, 435
additional employer risks in NEC3, 471–2
adjudication, 496, 501
adjudicator in NEC3, 378
ADR, 502
advance payment, 373, 490
alliances, 184–7
ambiguities in NEC3, 396–7
amount due to contractor (payment) in
  NEC3, 430
antiquities, *see* archaeology
arbitration, 501
  see also dispute procedures
archaeology, 105, 448, 467, 489
architect/contract administrator, 479,
  486, 487
articles in JCT, 477
articles of agreement, 385, 477
asbestos, 105
assessing the amount due in NEC3, 429
assessment interval in NEC3, 404, 430
assessment of compensation events in
  NEC3
  by project manager, 462
  the prices, 459–60
asset life, 14
assumption, correcting a PM, in NEC3,
  452, 457
avoiding compensation events, 443, **444**

badgers, *see* protected species
balancing cost, risk and complexity, 13

bats, *see* protected species
battle of the forms, 293, 294
bill of quantities
  as a pricing/payment mechanism,
    228–30
  CESMM3, 250–2
  contractual status of, 257
  described, **249–50**
  NEC3, 370, 385, 431, 436
birds, *see* protected species
BoQ, *see* bill of quantities
breach of contract
  by employer in NEC3, 452
  remedies for, 307–9, 313
buildability, 137
builders, *see* contractors
business needs and opportunities, 8

capacity in law, 301
capital programmes, 24
  cyclical nature of, 25
cases and precedent in law, 286
cash flow, how contractors adjust it, 85
causation, *see* negligence
CDM 2007, **120–2**
  and clients, 122, 126
  definition of design, 166
  generally, Chapter 8
CDM coordinator, 123
  method statements and risk assessment,
    see specific entries
ceiling price, 238
certainty in contract formation, 289
CESMM3, 250–2
change, 22
changing a decision in NEC3, 448
checking programmes, 280, 281
claims by the contractor, 394, 492,
  493
clause numbering in NEC3, 367
clerk of works in JCT, 479
clients under CDM, 122, 126
collateral warranties, 156
commencement, 261
commercial settlement, 324
communications in NEC3, 390